T0140105

Information Fusion and Data Science

Series editor
Henry Leung, University of Calgary, Calgary, Alberta, Canada

This book series provides a forum to systematically summarize recent developments, discoveries and progress on multi-sensor, multi-source/multi-level data and information fusion along with its connection to data-enabled science. Emphasis is also placed on fundamental theories, algorithms and real-world applications of massive data as well as information processing, analysis, fusion and knowledge generation. The aim of this book series is to provide the most up-to-date research results and tutorial materials on current topics in this growing field as well as to stimulate further research interest by transmitting the knowledge to the next generation of scientists and engineers in the corresponding fields. The target audiences are graduate students, academic scientists as well as researchers in industry and government, related to computational sciences and engineering, complex systems and artificial intelligence. Formats suitable for the series are contributed volumes, monographs and lecture notes.

More information about this series at http://www.springer.com/series/15462

Zhongliang Jing • Han Pan • Yuankai Li
Peng Dong

Non-Cooperative Target Tracking, Fusion and Control

Algorithms and Advances

Zhongliang Jing
School of Aeronautics and Astronautics
Shanghai Jiao Tong University
Shanghai, China

Han Pan
School of Aeronautics and Astronautics
Shanghai Jiao Tong University
Shanghai, China

Yuankai Li
University of Electronic Science
and Technology of China
Chengdu, Sichuan, China

Peng Dong
School of Aeronautics and Astronautics
Shanghai Jiao Tong University
Shanghai, Shanghai, China

ISSN 2510-1528 ISSN 2510-1536 (electronic)
Information Fusion and Data Science
ISBN 978-3-030-08081-5 ISBN 978-3-319-90716-1 (eBook)
https://doi.org/10.1007/978-3-319-90716-1

Softcover re-print of the Hardcover 1st edition 2018

Printed on acid-free paper

This Springer imprint is published by the registered company Springer International Publishing AG part of Springer Nature.
The registered company address is: Gewerbestrasse 11, 6330 Cham, Switzerland

Preface

Non-cooperative target (NCT) refers to the objects with unknown state or attributes in air, marine, space, etc. The study on NCT has experienced explosive growth in the last few decades. Specially, the research on non-cooperative satellite is receiving an always increasing attention from the space science as well as industrial world due to its both theoretical and practical implications. With the advances in imaging, computing, and filtering theory, the study on non-cooperative target involves a field at the intersection of adaptive filtering, target tracking, information fusion, image processing, etc. The combined efforts in both the signal-processing, dynamics and control communities have led to technical breakthroughs in a wide variety of topics. Space missions like Orbital Express launched in 2007 open up a new era for on-orbit servicing (OOS), which is defined by rendezvous, proximity operations and station keeping, capture, docking, fluid transfer, and orbit replaceable unit (ORU) transfer. However, the on-orbit servicing on non-cooperative satellite is still an open research area facing many technical challenges.

There is a significant gap between what we can learn from textbooks and what the current state of the art is in the field of NCT. This monograph is intended to put together the latest research on the methods of tracking, fusion, and control for non-cooperative target with specific attention to applications, thus filling a gap in the available literature on non-cooperative target. It should also be useful to advanced undergraduates in aerospace science, or electrical engineering. It can be used as a complement to an introductory monograph that provides up-to-date applications in engineering. The particular themes and emphases in this monograph have grown out of the author's experience at Shanghai Jiao Tong University. The coverage is somewhat less comprehensive. We apologize for any important gaps in the contents. We were always aware of the fact that this has only given us a glimpse of the situation and highlighted merely a handful of representative research concepts. Thus, we have considered a monograph featured by (1) focusing on algorithms and its applications for non-cooperative target and (2) promoting further research in some topics.

The monograph is divided into four parts. Part I contains background and fundamentals that are basic to much of the subject. The presentation in this part covers and discusses the main elements of underlying nonlinear filtering theory. Part II is devoted to some topic of much importance in target tracking and fusion, including sensor registration, multi-target tracking, robust tracking, and a unified framework via joint target detection, tracking, and classification. In Part III, some real-world problems in visual tracking and fusion, such as image deblurring, visual tracking, image fusion, and focus measures, are investigated, showing that these have been of important research areas. Part IV presents the advances in spacecraft control for tracking. In this part, the concept of control-tracking integration is discussed, and a typical control-based tracking scheme and algorithms are formed with application to a general problem of active satellite tracking.

Shanghai, China Zhongliang Jing
Shanghai, China Han Pan
Chengdu, China Yuankai Li
Shanghai, China Peng Dong

Acknowledgments

This monograph has developed a set of algorithms on non-cooperative target tracking, fusion and control by Prof. Zhongliang Jing and his PhD students and postdoctoral fellows over the past 20 years. We would like to express special gratitude to them: Hongjian Zhang, Wei Huang, Bo Yang, Yongqing Qi, Xiao Yun, Rongli Liu, Minzhe Li, and Shiqiang Hu, who also made a contribution to this monograph. Special thanks go to Prof. Henry Leung for his devotion to our monograph, and his encouragement and suggestions have been invaluable to this monograph. More specially, we thank Mr. Bing Yi for his great help in typing monograph and in preparation of some of the figures. We also owe thanks to a number of other colleagues who reviewed parts of the manuscript, such as Lingfeng Qiao, Xuanguang Ren, Miaomiao Hu, and Wuji Liu.

This monograph has grown out of more than a decade of planning, discussing, and experimenting. A few of the figures, used in this monograph, have appeared in some of our publications. We are thankful to IEEE and Elsevier for their copyright policy in allowing the authors to reuse these plots.

The research work in this monograph is jointly supported by the National Natural Science Foundation of China (Nos. 60375008, 60775022, 61175028, 61673262, 61603249), the key project of Science and Technology Commission of Shanghai Municipality (No. 16JC1401100), and other institutions.

Contents

Part III Visual Tracking and Fusion

About the Authors

Prof. Zhongliang Jing received his BS, MS, and PhD degrees from Northwestern Polytechnical University, Xi'an, China, in 1983, 1988, and 1994, respectively, all in electronics and information technology.

From 1983 to 1985, he was with the China Leihua Electronics Technology Institute, Aviation Industries of China. From 1988 to 2000, he was with the Department of Automatic Control, Northwestern Polytechnical University. In 2000, as the Cheung Kong professor he joined the faculty of the School of Electronic Information and Electrical Engineering, Shanghai Jiao Tong University, where he was Associate Dean from 2001 to 2004. From 2005 to present, he has been the Executive Dean of the School of Aeronautics and Astronautics, the Institute of Aerospace Science and Technology at Shanghai Jiao Tong University. From 2013 to present, he has been Director of the Engineering Research Center of Aerospace Science and Technology, Ministry of Education of the People's Republic of China. From 2014 to present, he has been Director of the Shanghai Collegiate Key Laboratory of Aerospace Science and Technology. From 1997 to 1998, he was a research associate and senior visiting scholar of the University of California at Berkeley. In 1999, he was a visiting professor of Southern Illinois University at Edwardsville, USA.

His research interests include adaptive filtering, target tracking, information fusion, avionics integration, and aerospace control. He has published over 200 journal papers and 7 monographs. He is currently Editor-in-Chief of the *Aerospace Systems*, Springer. He has been the editorial board member of various journals such as the *International Journal of Space Science and Engineering*, the *Information and Control*, the *Systems Engineering and Electronics*, and the *Manned Spaceflight*. He also served as the editorial board member for various journals such as the *Science China Information Sciences*, and the *Chinese Optics Letters*.

Han Pan received his BS and MS degree from Xidian University, Xi'an, China, in 2006 and 2009. He received his PhD degree from Shanghai Jiao Tong University, China, in 2014. From 2014 to 2016, he was a postdoctoral fellow at the School of Aeronautics and Astronautics, Shanghai Jiao Tong University.

Currently, he is an Associate Professor at Shanghai Jiao Tong University, Shanghai, China. He has authored more than 15 journal papers and published an invited book chapter in Springer. He actively serves as a reviewer for numerous academic journals. His research interests include information fusion, optimization algorithms, and space robotics.

Yuankai Li received his BS degree from Hefei University of Technology, Hefei, China, in 2002; MS degree from Chongqing University of Posts and Telecommunications, Chongqing, China, in 2006; and PhD degree from Shanghai Jiao Tong University, Shanghai, China, in 2011, all in control science and engineering.

During 2011, he was a Research Engineer in Shanghai Institute of Satellite Engineering, China. From 2012 to 2013, he was a Postdoctoral Fellow at the Department of Aerospace Engineering, Ryerson University, Toronto, Canada.

Currently, he is an Associate Professor in the School of Aeronautics and Astronautics, University of Electronic Science and Technology of China.

His research interests include nonlinear estimation and control, autonomous target tracking and GNC systems, particularly in aerospace vehicles and planetary rovers and their applications to defense technology, on-orbit service, and asteroid exploration.

Peng Dong received his BS and PhD degrees from the School of Astronautics, Northwestern Polytechnical University, Xi'an, China, in 2008 and 2014, respectively. From 2013 to 2015, he is a postdoctoral fellow at the School of Aeronautics and Astronautics, Shanghai Jiao Tong University.

Currently, he is a research assistant at the School of Aeronautics and Astronautics, Shanghai Jiao Tong University. His research interests include nonlinear filtering, target tracking, and information fusion. He has published over 15 journal papers.

Part I
Background and Fundamentals

Chapter 1
Introduction

1.1 What is Non-cooperative Target?

Non-cooperative target (NCT) refers to the objects with unknown state or attributes in air, space, marine, etc. For example, plenty of space missions involved in operation with non-cooperative target, i.e., servicing a malfunctioning satellite, refueling a powerless spacecraft, collecting and removing space debris. For ground-to-air or air-to-air activities, radar surveillance systems are developed to detect birds, weather and unmanned aircraft systems (UAS) and hot balloons, which are labeled as non-cooperative objects [14]. Much of these objects can be traced to a few fundamental properties, such as unknown geometric appearance, uncooperative and non-communicative.

The research on non-cooperative target is relatively short, spanning roughly 20 years. At the end of the 1990s, some methods for automatic target recognition were developed [6, 10]. With the development of radar surveillance systems [2], the automatic detection, positioning and recognition for non-cooperative target has been one of its important functions. There is no doubt that the exploitation of redundant and complementary information, and the available context were essential for monitoring the state of non-cooperative target [7, 9]. One of the most famous accidents by NCT is the ditching of US Airways flight 1549 on the Hudson River about 8.5 miles from LaGuardia Airport (LGA), New York City, New York. Careful accident analysis [5] by United States National Transportation Safety Board (NTSB) indicated that "the ingestion of large birds into each engine, which resulted in an almost total loss of thrust in both engines" is the probable cause of the accident. It turns out that this is the most general requirement on detecting and preventing ambiguity and confusion from non-cooperative target.

Another important research area is on-orbit servicing [3], refers to the maintenance of space systems in orbit such as docking, berthing, refueling, repairing, upgrading, transporting, rescuing, and orbital debris removal. In 2007, the Orbital

Z. Jing et al., *Non-Cooperative Target Tracking, Fusion and Control*,
Information Fusion and Data Science, https://doi.org/10.1007/978-3-319-90716-1_1

Express mission [11] sponsored by the Defense Advanced Research Projects Agency (DARPA) opened up a new era in space-on-orbit servicing. To reduce size, complexity and ultimately cost of satellite systems, DARPA's Phoenix Program [4] was created for unique and economically valuable orbital real estate that could be recycled for other uses. This is an extension of the Orbital Express project in several aspects. Yet, despite their favorable mission objectives, the operations on non-cooperative target may cause serious difficulties in theory and applications [1, 13]. Inter-satellite tracking remains a difficult and challenging problem in operations and applications.

There has been significant progress toward developing methods within some of the individual disciplines. Autonomous rendezvous and capturing [12] (AR&C) is a key enabling technology in the context of capturing a non-cooperative satellite. However, these operations are very risky and difficult to carry out in space, since the target's motion is not well known in advance. Some state estimation and tracking methods began to merge and receive increased attention. The advantages of these methods are being demonstrated in the growing literature, but they themselves need to be validated in a more effective way. It is a challenging problem due to several features of the natural dynamics in space. Thus, it is important to design and construct ground simulation for autonomous rendezvous and proximity operation of spacecraft. Another constraint is that space vehicles have an overwhelmingly large percentage of their total energy in their translational motion.

1.2 Challenges by Non-cooperative Target

The technical challenges of non-cooperative target can be outlined as follows:

1. Sensors and actuators technologies.
2. Dynamics with flight, contact, combined with a whole body.
3. Guidance, navigation, and control.
4. Simulation and validation.

It should be noted that the research on non-cooperative target is young and some areas need to be better defined. Some of new problems will be encountered and solved while continuing research. Some of new research directions have been emerged, such as soft robot for on-orbit servicing [8]. The selected topics in this monograph are aimed to simulate scientific discussion.

1.3 Plan of the Monograph

The emphasis of this monograph is almost exclusively on the methods for tracking, fusion, and control of non-cooperative target. Because NCT typically involves the estimation of certain quantities based on direct or indirect measurements, the estimation process often contains uncertainty by measurement error, processing errors by contact, or corruption. This monograph integrates mathematical and statistical theory with applications and practical methods, including topics like random finite set, adaptive filters, optimization, and neural networks.

Part I presents a brief introduction to non-cooperative target and some tools that are needed for the development of target tracking, visual tracking, fusion and control in subsequent chapters. Chapter 1 provides the definition of non-cooperative target. Here, we give two practical examples, both which are quite important in general. Furthermore, the technical challenges of non-cooperative target are discussed. Chapter 2 gives fundamental nonlinear filter algorithms that pertain to tracking, fusion, and control. Experts and scientists in the field will benefit from the overview of relevant tracking and control methods.

Part II investigates some specialized topics on target tracking and fusion. Chapter 3 describes a sensor registration method via dynamic bias estimation with Gaussian mean shift algorithm. The sufficient condition for convergence of this method is provided. Chapter 4 provides an adaptive cubature information filter for target tracking and multi-sensor fusion. This filter can jointly estimate the dynamic state and time varying measurement noise for Gaussian nonlinear state space models. Chapter 5 develops a multi-target tracking method with Gaussian mixture cardinalized probability hypothesis density (CPHD) filter and gating technique. Chapter 6 investigates the problem of the localization of multiple emitters from passive angle measurements, which was solved with sequential PHD filter. We evaluated the performance in different sensor settings. Chapter 7 treats a general framework of joint target detection, tracking, and classification, which employs both finite set statistics and generalized Bayesian risk. Chapter 8 presents a novel method of redundant filtering for active satellite tracking, and performs complete tracking error evaluation. Chapter 9 develops a conservativeness optimization method named optimal-switched H_∞ robust filtering for space target tracking.

Part III addresses some problems that might be encountered in visual tracking and fusion. Chapter 10 provides an efficient algorithm for constrained image deblurring with sparse proximal Newton splitting method. This chapter serves to introduce image processing method for removing blurring artifacts from images. Chapter 11 is devoted to a general framework of joint decision and estimation for simultaneous visual object recognition and tracking. Chapter 12 provides an incremental visual object tracking method with ℓ_1 norm approximation. The resulting subspace update is performed on Grassmann manifold. Chapter 13 develops a quincunx-sampled discrete wavelet frame. We applied the proposed method on image fusion and discussed its characteristics. Chapter 14 investigates a multi-focus image fusion method using pulse coupled neural network. Chapter 15 presents an efficient scheme to assess focus measures for multi-focus image fusion.

Part IV investigates spacecraft control for tracking. The control-tracking integration is considered, and some control-based methods for active satellite tracking are presented. Chapter 16 provides a dynamic optimal sliding-mode control method for active satellite tracking, built on a relative motion model using osculating reference orbit. Chapter 17 addresses a general control-based tracking problem of control-based spacecraft maneuver-aided tracking. A six-degree-of-freedom optimal dynamic inversion control (ODIC) law is developed to achieve the tracking scheme.

References

1. Benninghoff H, Boge T (2015) Rendezvous involving a non-cooperative, tumbling target-estimation of moments of inertia and center of mass of an unknown target. In: 25th international symposium on space flight dynamics, vol 25
2. Blacknell D, Griffiths H (2013) Radar automatic target recognition and non-cooperative target recognition. The Institution of Engineering and Technology
3. Flores Abad A, Ma O, Pham K, Ulrich S (2014) A review of space robotics technologies for on-orbit servicing. Prog Aerosp Sci 68:1–26
4. Henshaw CG (2014) The darpa phoenix spacecraft servicing program: overview and plans for risk reduction. In: International symposium on artificial intelligence, robotics and automation in space
5. Hersman D, Hart C, Sumwalt R (2010) Loss of thrust in both engines after encountering a flock of birds and subsequent ditching on the hudson river. Technical Report, Accident Report NTSB/AAR-10/03, National Transportation Safety Board, Washington DC
6. Jacobs SP, O'Sullivan JA (1997) High resolution radar models for joint tracking and recognition. In: IEEE national conference radar. IEEE, New York, pp 99–104
7. Jing Z, Pan H, Qin Y (2013) Current progress of information fusion in China. Chin Sci Bull 58(36):533–4540
8. Jing Z, Qiao L, Pan H, Yang Y, Chen W (2017) An overview of the configuration and manipulation of soft robotics for on-orbit servicing. Sci China Inf Sci 60(5):050201
9. Khaleghi B, Khamis A, Karray FO, Razavi SN (2013) Multisensor data fusion: a review of the state-of-the-art. Inf Fusion 14(1):28–44
10. Li HJ, Yang SH (1993) Using range profiles as feature vectors to identify aerospace objects. IEEE Trans Antennas Propag 41(3):261–268
11. Ogilvie A, Allport J, Hannah M, Lymer J (2008) Autonomous satellite servicing using the orbital express demonstration manipulator system. In: Proceedings of the 9th international symposium on artificial intelligence, robotics and automation in space, pp 25–29
12. Timmons KK, Ringelberg JC (2008) Approach and capture for autonomous rendezvous and docking. In: IEEE conference on aerospace. IEEE, New York, pp 1–6
13. Yoshida K, Dimitrov D, Nakanishi H (2006) On the capture of tumbling satellite by a space robot. In: IEEE/RSJ international conference on intelligent robots and systems. IEEE, New York, pp 4127–4132
14. Yuan X (2014) Distributing non-cooperative object information in next generation radar surveillance systems. Master's thesis, University of Waterloo

Chapter 2
Nonlinear Filter

2.1 Introduction

The aim of filters is to get accurate estimates of the useful signal from the signal with noises. Based on the measurement of the observable signal of the system and using some statistical optimal method, the theory of filtering can be treated as the theory and method of estimating the state of the system according to certain optimization criteria.

When the system model is linear, the state noise and measurement noise conform to the Gaussian distribution, the mean and variance of the distribution can be propagated and updated by the classical Kalman filter. In this case, the Kalman filter algorithm is the optimal state estimation algorithm. In practice, the state equation and the measurement equation are generally nonlinear, and the optimal analytical solution cannot be obtained by the classical Kalman filter. There are some suboptimal solutions to nonlinear filtering. Among them, the extended Kalman filter is the most popular suboptimal filtering algorithm. However, since extended Kalman filter simply linearizes all nonlinear functions to first-order Taylor series and needs to calculate the Jacobian matrix of the model, the extended Kalman filter is very difficult to achieve and can easily lead to filter instability when the model is a complex nonlinear equation. In view of this situation, references [1, 2, 9, 16] proposed a class of deterministic sampling algorithms called sigma points Kalman filter (SPKF), which does not need to linearize the nonlinear system and to compute the Jacobin matrix. The algorithm is simple and the amount of calculation is moderate. For the linear model, the two have the same filtering precision. However, the estimate accuracy of SPKF is higher than that of the extended Kalman filter (EKF) for the nonlinear model. Particle filter [5] is a filter based on random sampling. This method approximates the posterior probability density distribution by randomly picking up a set of particles (with the corresponding weights) by Monte Carlo method and accomplishes the optimization by sequential prediction and state updating estimation of state variables. This method is not limited by nonlinear

Z. Jing et al., *Non-Cooperative Target Tracking, Fusion and Control*,
Information Fusion and Data Science, https://doi.org/10.1007/978-3-319-90716-1_2

and non-Gaussian problems, however it is computationally intensive. The above algorithms are all based on the optimal single-target Bayesian filter. In addition, the H_∞ filter [6] can be used when there is uncertainty in the system model.

In addition to the point estimates mentioned above, the multi-target estimation problem is also an important issue for filtering. The multi-target estimation method based on the random finite set (RFS) [15] has attracted the attention of many researchers. This method models multi-target state and multi-target measurement as RFS, and adopts the optimal multi-target Bayesian (MTB) filter to propagate the posterior density of multi-target states. The computational complexity of the optimal MTB filter based on RFS is high due to the treatment of multi-target densities and multiple integration. The probability hypothesis density (PHD) filter [13] propagating only the first order moment (intensity) of the RFS of the states was presented to approximate the multiple target densities and multiple integration. A more general method called cardinalized PHD (CPHD) filter [14], which jointly estimates the cardinality distribution and the intensity, is also proposed to deal with the problem. The multi-target multi-Bernoulli (MeMBer) filter [15] and the cardinality-balanced MeMBer (CBMeMBer) filter [22] were also proposed to approximate the MTB filter. However, these methods are not designed to estimate the trajectories of targets. Recently, the tractable and mathematically principled approaches based on the labeled RFS, which is called the δ generalized labeled multi-Bernoulli (δ-GLMB) filter [23], are proposed to distinguish the individual targets.

This chapter will give brief introduction to parts of the above methods including the SPKF filter, the PF, the H_∞ robust filtering, the PHD filter, and the CPHD filter.

2.2 Nonlinear Filter for Point Estimation

2.2.1 Sigma Point Kalman Filter

We consider the nonlinear state space system form given by

$$\begin{aligned} x_k &= f(x_{k-1}) + w_{k-1} \\ z_k &= h(x_k) + r_k \end{aligned} \tag{2.1}$$

where x_k is the n-dimension state vector, $f(\cdot)$ is the nonlinear state function, w_k is the process noise with zero mean and covariance Q_k, z_k is the d-dimension measurement vector, $h(\cdot)$ is the nonlinear measurement function, r_k is the measurement noise with zero mean and covariance R_k.

The aim of Bayesian filtering is to estimate the posterior distribution $p(x_k|Z_k)$ of states x_k, where Z_k are defined as $Z_k = \{z_1, \ldots, z_k\}$. The recursion solutions for this problem can be summarized as follows:

1. Initialization: The prior distribution is given by $p(x_0)$.

2. Prediction: The Chapman-Kolmogorov equation provides the connection between prior density $p(x_k|Z_{k-1})$ and posterior density, that is

$$\begin{aligned} p(x_k|Z_{k-1}) &= \int p(x_k|x_{k-1}, Z_{k-1})p(x_{k-1}|Z_{k-1})dx_{k-1} \\ &= \int p(x_k|x_{k-1})p(x_{k-1}|Z_{k-1})dx_{k-1} \end{aligned} \tag{2.2}$$

3. Update: When the new measurement comes, the posterior density can be calculated by

$$p(x_k|Z_k) = \frac{p(z_k|x_k)p(x_k|Z_{k-1})}{p(z_k|Z_{k-1})} \tag{2.3}$$

where $p(z_k|x_k, \Lambda_k)$ denotes the likelihood, and $p(z_k|Z_{k-1})$ is the normalization constant computed by

$$p(z_k|Z_{k-1}) = \int p(z_k|x_k)p(x_k|Z_{k-1})dx_k \tag{2.4}$$

Theoretically, the posterior probability density distribution can be obtained recursively, however the integral equation is not tractable for analytic solution. Under certain conditions, such as the state noise and measurement noise are additive Gaussian, and the posterior probability density distribution is Gaussian distribution at each time step, the analytical solution can be obtained by Kalman filtering. In many practical cases, the state model and the measurement model are not necessarily linear and Gaussian. At this time, we need to approximate the solution to obtain the next optimal solution. For nonlinear systems, approximations such as Gaussian approximations (GA) are often used for Gaussian systems and Monte Carlo methods are used for non-Gaussian systems [8].

Since it is easier to approximate the probability density distribution of the nonlinear function than to approximate the nonlinear function itself, the approaches of using determinate sampling methods to approximate the nonlinear distribution to solve the nonlinear problems have drawn wide attention [9]. As with the EKF, the SPKF is also a class of the recursive Bayesian estimation methods, however the SPKF does not have to linearize the nonlinear state equation and observation equation. Instead, it uses a set of defined sampling points to approximate a posteriori probability.

The sigma point Kalman filters, including the unscented Kalman filter (UKF) [9], the central difference Kalman filter (CDKF) [16], the Cubature Kalman filter (CKF) [1], and the Gauss–Hermite Kalman filter (GHKF) [2], share a similar structure. The SPKF recursion at time k contains two major steps: prediction (time update) and measurement update. Then the common process of the SPKF can be summarized as follows.

1. Prediction: Compute the predicted $\hat{x}_{k|k-1}$ and $P_{k|k-1}$. P_{k-1} can be factorized by

$$P_{k-1} = S_{k-1}S_{k-1}^T \tag{2.5}$$

where $S_{k-1} \in R^{n\times n}$ is a lower triangular matrix called the square root of P_{k-1}. In the time update step, a set of sigma points are generated by

$$\chi_{k-1}^j = \hat{x}_{k-1} + S_{k-1}\xi_j \quad j = 1, \ldots, N_s \tag{2.6}$$

where N_s is the number of sigma points and ξ_j is the vector points of the same dimension as the state vector. The predicted sigma points are

$$\chi_{k|k-1}^{j,*} = f(\chi_{k-1}^j) \tag{2.7}$$

so the predicted state is

$$\hat{x}_{k|k-1} = \sum_{j=1}^{N_s} \omega_j \chi_{k|k-1}^{j,*} \tag{2.8}$$

and the associated predicted error covariance is

$$P_{k|k-1} = \sum_{j=1}^{N_s} \omega_j \chi_{k|k-1}^{j,*} (\chi_{k|k-1}^j)^T - \hat{x}_{k|k-1}\hat{x}_{k|k-1}^T + Q_{k-1} \tag{2.9}$$

where ω_j denotes the weight of the sigma point.

2. Measurement update: $P_{k|k-1}$ can be factorized by

$$P_{k|k-1} = S_{k|k-1}S_{k|k-1}^T \tag{2.10}$$

Then a set of sigma points are generated by

$$\chi_{k|k-1}^j = \hat{x}_{k|k-1} + S_{k|k-1}\xi_j \quad j = 1, \ldots, N_s \tag{2.11}$$

The propagated sigma points are given by

$$Z_{k|k-1}^j = h(\chi_{k-1}^j), \tag{2.12}$$

The predicted measurement and the innovation covariance matrix are

$$\hat{z}_{k|k-1} = \sum_{j=0}^{2n} \omega_j^m Z_{k|k-1}^j \tag{2.13}$$

$$P_{zz,k|k-1} = \sum_{j=1}^{N_s} \omega_j Z^j_{k|k-1} (Z^j_{k|k-1})^T - \hat{z}_{k|k-1} \hat{z}^T_{k|k-1} + R_k \tag{2.14}$$

The cross covariance can be obtained by

$$P_{xz,k|k-1} = \sum_{j=1}^{N_s} \omega_j (\hat{x}_{k|k-1} - \chi^j_{k|k-1})(\hat{z}_{k|k-1} - Z^j_{k|k-1})^T \tag{2.15}$$

The Kalman gain can be obtained by

$$K_k = P_{xz,k|k-1} P^{-1}_{zz,k|k-1} \tag{2.16}$$

The updated state and the corresponding error covariance is given by

$$\hat{x}_k = \hat{x}_{k|k-1} + K_k (z_k - \hat{z}_{k|k-1}) \tag{2.17}$$

$$P_k = P_{k|k-1} + K_k P_{zz,k|k-1} K_k^T \tag{2.18}$$

It should be noted that different choices of ξ_j and ω_j can lead to different class of the SPKF. The values for ξ_j and ω_j for each of the SPKF filters can be seen in [8].

2.2.2 *Particle Filter*

In the particle filter (PF), a set of particles (with corresponding weights) is randomly selected to approximate the posterior probability density distribution by Monte Carlo method, then the state variables are estimated by the sequential prediction and update. The particle filter can be applied to nonlinear and non-Gaussian problems, however, its computational complexity is relatively large [8].

Consider the first-order Markov process x_k, whose estimated moments are determined by the conditional posterior density $p(x_k|Z_k)$, with an initial density $p(x_0)$. Given an arbitrary posterior PDF, the general representation of the nonlinear function of moments is

$$I(x_k) = \int g(x_k) p(x_k|Z_k) dx_k \tag{2.19}$$

where $g(x_k)$ is a function that can represent any moment generating function.

The conditional posterior density $p(x_k|Z_k)$ is usually unknown or difficult to sample. If we have an easily sampled importance density $q(x_k|Z_k)$ with the same support as $p(x_k|Z_k)$ and its maximum is almost the same place as $p(x_k|Z_k)$, then $I(x_k|Z_{k-1})$ can be obtained by

$$I(x_k) = \int g(x_k)\frac{p(x_k|Z_k)}{q(x_k|Z_k)}q(x_k|Z_k)dx_k$$
$$= \int g(x_k)\omega(x_k)q(x_k|Z_k)dx_k \tag{2.20}$$

where $\omega(x_k)$ is the weight function defined by

$$\omega(x_k) \triangleq \frac{p(x_k|Z_k)}{q(x_k|Z_k)} \tag{2.21}$$

According to the Bayes' rule, $\omega(x_k)$ becomes

$$\omega(x_k) = \frac{p(z_k|x_k)p(x_k|Z_{k-1})}{p(z_k|Z_{k-1})}\frac{1}{q(x_k|Z_k)} \tag{2.22}$$

Since $p(z_k|Z_{k-1})$ is a normalization term, we can obtain

$$\omega(x_k) \propto \frac{p(z_k|x_k)p(x_k|Z_{k-1})}{q(x_k|Z_k)} \tag{2.23}$$

where $\propto$ means a proportionality.

If we draw a set of samples $\{x_k^{(i)}, i = 1, \ldots, N_s\}$ from the importance density $q(x_k|Z_k)$, we can approximate $q(x_k|Z_k)$ via the set of Dirac delta function

$$q(x_k|Z_k) = \frac{1}{N_s}\sum_{i=1}^{N_s}\delta(x_k - x_k^{(i)}) \tag{2.24}$$

and (2.19) can be obtained by

$$I(x_k) = \int g(x_k)\omega(x_k)\frac{1}{N_s}\sum_{i=1}^{N_s}\delta(x_k - x_k^{(i)})dx_k$$
$$= \frac{1}{N_s}\sum_{i=1}^{N_s}\omega_k^{(i)}g(x_k^{(i)}) \tag{2.25}$$

where $\omega_k^{(i)} = \omega(x_k^{(i)})$ defined by (2.21). If $\omega_k^{(i)}$ is given by (2.23), a set of unnormalized weights are given by

$$\tilde{\omega}_k^{(i)} = \frac{p(z_k|x_k^{(i)})p(x_k^{(i)}|Z_{k-1})}{q(x_k^{(i)}|Z_k)} \tag{2.26}$$

The weights can be normalized by

$$\omega_k^{(i)} = \frac{1/N_s \tilde{\omega}_k^{(i)}}{1/N_s \sum_{i=1}^{N_s} \tilde{\omega}_k^{(i)}} \tag{2.27}$$

Then (2.25) can be rewritten as

$$I(x_k) = \sum_{i=1}^{N_s} \omega_k^{(i)} g(x_k^{(i)}) \tag{2.28}$$

Recursive estimation methods based on the sequential estimation of the weights are called sequential importance sampling (SIS) particle filters. Using the Chapman-Kolmogorov theorem, $p(x_k|Z_{k-1})$ and $q(x_k|Z_k)$ in (2.23) can be expanded as

$$\omega(x_k) \propto \frac{p(z_k|x_k) \int p(x_k|x_{k-1}) p(x_{k-1}|Z_{k-1}) dx_{k-1}}{\int q(x_k|x_{k-1}, Z_k) q(x_{k-1}|Z_{k-1}) dx_{k-1}} \tag{2.29}$$

For a Markov process, we have

$$q(x_k|x_{k-1}, Z_k) = q(x_k|x_{k-1}, z_k) \tag{2.30}$$

If we draw a set of samples $\{x_{k-1}^{(i)}, i = 1, \ldots, N_s\}$ from $q(x_{k-1}|Z_{k-1})$, then we can obtain

$$\tilde{\omega}_k(x_k) = \sum_{i=1}^{N_s} \frac{p(z_k|x_k) p(x_k|x_k^{(i)}) p(x_{k-1}^{(i)}|Z_{k-1})}{q(x_k|x_{k-1}^{(i)}, z_k) q(x_{k-1}^{(i)}|Z_{k-1})} \tag{2.31}$$

according to (2.29). From the definition of $\omega_k(x_k)$ given in (2.21), it follows that

$$\tilde{\omega}_k(x_k) = \sum_{i=1}^{N_s} \omega_{k-1}^{(i)} \frac{p(z_k|x_k) p(x_k|x_k^{(i)})}{q(x_k|x_{k-1}^{(i)}, z_k)} \tag{2.32}$$

Then the posterior distribution can be given by

$$p(x_k|Z_k) = \sum_{i=1}^{N_s} \omega_{k-1}^{(i)} \frac{p(x_k|x_k^{(i)})}{q(x_k|x_{k-1}^{(i)}, z_k)} p(z_k|x_k) q(x_k|Z_k) \tag{2.33}$$

Generating samples from such that

$$x_k^{(i)} \sim q(x_k|x_{k-1}^{(i)}, z_k) \tag{2.34}$$

$$q(x_k|Z_k) = \frac{1}{N_s} \sum_{i=1}^{N_s} \delta(x_k - x_k^{(i)}) \tag{2.35}$$

results in

$$p(x_k|Z_k) = \sum_{i=1}^{N_s} \frac{\tilde{\omega}_{k-1}^{(i)}}{N_s} \frac{p(x_k|x_k^{(i)})}{q(x_k|x_{k-1}^{(i)}, z_k)} p(z_k|x_k^{(i)})\delta(x_k - x_k^{(i)}) \tag{2.36}$$

The weights $\frac{\tilde{\omega}_{k-1}^{(i)}}{N_s}$ can be normalized by

$$\omega_k^{(i)} = \frac{1/N_s \tilde{\omega}_{k-1}^{(i)}}{1/N_s \sum_{i=1}^{N_s} \tilde{\omega}_{k-1}^{(i)}} \tag{2.37}$$

Then we have

$$p(x_k|Z_k) = \omega_k^{(i)} \sum_{i=1}^{N_s} \delta(x_k - x_k^{(i)}) \tag{2.38}$$

with the SIS recursive weight update equation

$$\omega_k^{(i)} = \omega_{k-1}^{(i)} \frac{p(x_k|x_k^{(i)})}{q(x_k|x_{k-1}^{(i)}, z_k)} p(z_k|x_k^{(i)}) \tag{2.39}$$

The first two moments of x_k with respect to $p(x_k|Z_k)$ are given by

$$\hat{x}_k = \sum_{i=1}^{N_s} \omega_k^{(i)} x_k^{(i)} \tag{2.40}$$

$$P_k = \sum_{i=1}^{N_s} \omega_k^{(i)} (x_k^{(i)} - \hat{x}_k)(x_k^{(i)} - \hat{x}_k)^T \tag{2.41}$$

There are several significant practical problems in the above particle filter procedure. In the propagation of the particle filter, the variance of the weight of SIS increases with time. After a few iterations, with the exception of a few particles, the weight of the rest of the particles is negligible. This problem is often referred to as the particle degeneracy problem. One way to reduce the problem of particle degeneracy is to introduce a resampling step [6] after the particle's weights have been updated. However, this leads to a new problem known as sample impoverishment. The repeated selection of the large weighted particles makes the sampling result contains many repeated points, which makes it lose the diversity of the particles. Specifically, since resampling duplicates high-weighted particles too much, which results in decreasing particle diversity, after several recursions,

Table 2.1 The SIS particle filter

1. Initialize filter: $x_0^{(i)} \sim q(x_0), \quad i = 1, \ldots, N_s, \omega_0^{(i)} = 1/N_s$.
2. Sequential importance sampling
1) Draw new samples: $x_k^{(i)} \sim q(x_k \vert x_{k-1}^{(i)}, z_k)$
2) Generate unnormalized importance weights: $\tilde{\omega}_k(x_k) = \sum_{i=1}^{N_s} \omega_{k-1}^{(i)} \frac{p(z_k \vert x_k) p(x_k \vert x_k^{(i)})}{q(x_k \vert x_{k-1}^{(i)}, z_k)}$
3) Normalize the weights: $\omega_k^{(i)} = \frac{\tilde{\omega}_{k-1}^{(i)}}{\sum_{i=1}^{N_s} \tilde{\omega}_{k-1}^{(i)}}$
3. Resample the particles
4. Regularize the resampled particles
5. Calculate moments: $\hat{x}_k = \sum_{i=1}^{N_s} \omega_k^{(i)} x_k^{(i)}, \quad P_k = \sum_{i=1}^{N_s} \omega_k^{(i)} (x_k^{(i)} - \hat{x}_k)(x_k^{(i)} - \hat{x}_k)^T$

other valid particles will be depleted by the resampling step until the last one is left. Eventually, all of the particles will collapse to the same value. This problem can be solved by regularization [17]. The procedure for a typical SIS particle filter is presented in Table 2.1.

2.2.3 H_∞ Robust Filter

Robust filter is an effective point estimation approach for maneuvering target tracking to acquire accurate and precise real-time target trajectory [4], a typical way of which is to use a H_∞ robust filter. The filter is essentially a state estimator via output measurement, which may guarantee the H_∞ norm of transfer function between estimation error and stochastic uncertainties (noises and external disturbances) to be minimized [18, 19]. Such H_∞ criterion makes the filter acquire the best estimates of the worst-possible case. Without requirement of the exact statistical properties of random uncertainties, the H_∞ robust filter only needs the energy of noises and outer disturbances bounded. Due to simplicity of assumption, the H_∞ filter has already been applied in many other areas [11, 20, 25] to cope with the estimation problems with various types of uncertainties.

2.2.3.1 H_∞ Filtering Algorithm

Write system model for target tracking as

$$X_{k+1} = F_k X_k + G_k w_k + D \tag{2.42}$$

$$Z_k = H_k X_k + v_k \tag{2.43}$$

where F, G, and H are matrices independent with the system states. Z denotes the measurement vector. w and v are process and measurement noises assumed white with covariance matrix W and V, respectively.

The objective is to find a filter that estimates the state X_k from the measurement Z_k such that the estimation error $\hat{X}_{k|k}$ minimizes a H_∞ norm criterion, specifically, a H_∞ error transfer function

$$T_\infty = \left\| \frac{\hat{X}_{k|k} - \bar{X}_k}{X_k^Z - \bar{X}_k} \right\|_\infty \tag{2.44}$$

where X^Z is the pseudo real states built from measurements. In this chapter, we define that a variable with sharp hat represents its estimate, with bar for true value and tilde for estimation error. Clearly, the deviation from X^Z to the true value of states $\bar{X}_k$ is yielded by system input uncertainties, so minimizing (2.44) can be specified by satisfying

$$\max \frac{\|\tilde{X}_{k|k}\|_2^2}{\|w_k\|_2^2 + \|v_k\|_2^2 + \|D\|_2^2} \le \gamma^2 \tag{2.45}$$

for some minimum $\gamma \in R$. The inequality aims to minimizing the maximum energy transfer efficiency from overall input uncertainties to the estimation error, describing a worst-case performance criterion.

A typical solution for state estimation to (2.42) and (2.43) constrained by (2.45) is the H_∞ filter (H∞F) based on game theory [3, 7]. Derived by using the H_∞ criterion to build the H_∞ filtering gain, the filter resembles the structure of the standard Kalman filter (KF) and can be performed by the steps as follows.

Step 1: *State Prediction.*
One-step prediction state and prediction covariance are

$$\hat{X}_{k+1|k} = F_k \hat{X}_{k|k} \tag{2.46}$$

$$P_{k+1|k} = F_k P_{k|k} F_k^T + G_k W_k G_k^T \tag{2.47}$$

Step 2: *Gain Adjustment.*
Calculate H_∞ prediction covariance by

$$\Sigma_{k+1|k} = (P_{k+1|k}^{-1} - \gamma^{-2} I)^{-1} \tag{2.48}$$

in which γ satisfies (2.45). I represents the unit matrix.

Step 3: *Measurement Innovation.*
Measurement residue and covariance are

$$\hat{Z}_{k+1} = Z_{k+1} - H_{k+1} \hat{X}_{k+1|k} \tag{2.49}$$

$$P_{k+1}^Z = H_{k+1} \Sigma_{k+1|k} H_{k+1}^T + V_{k+1} \tag{2.50}$$

Step 4: *Estimate Update*.
H_∞ filtering gain, state estimate and covariance are

$$K_{k+1} = \Sigma_{k+1|k} H_{k+1}^T (P_{k+1}^Z)^{-1} \tag{2.51}$$

$$\hat{X}_{k+1|k+1} = \hat{X}_{k+1|k} + K_{k+1} \tilde{Z}_{k+1} \tag{2.52}$$

$$P_{k+1|k+1} = (\Sigma_{k+1|k}^{-1} + H_{k+1}^T V_{k+1}^{-1} H_{k+1})^{-1} \tag{2.53}$$

From the criterion (2.45), it is clear that γ evaluates the filtering robustness. Decreasing γ will enhance the robustness, but implied from (2.53), it increases the mean square error of the state estimates simultaneously, that is, the filtering optimality is weakened. If $\gamma \to \infty$, $\Sigma_{k+1|k}$ reverts to $P_{k+1|k}$ and the H∞F degenerates to the traditional KF. Hence, choosing γ is to balance the filtering performance between robustness and optimality. However, constant γ may lead to conservativeness of the estimation results, and also a proper value is inconvenient to find. That motivates appearance of many other types of H_∞ filters.

2.2.3.2 Conservativeness and Optimization

A problem of H_∞ filter is that the estimation results yielded by minimizing the H_∞ norm criterion are conservative, that is, too much filtering optimality is sacrificed to obtain the robustness. The reason is that the bounds of uncertainties defined in the H_∞ criterion are given with a priori values, but it is unrealistic in practice and the uncertainty bounds are difficult to be determined accurately.

In order to improve the optimality of the H_∞ filters, a switched H_∞ filter (SH∞F) is studied recently [10, 12, 24], which will be discussed with application to active satellite tracking in Chap. 8, Part II of the book.

This filtering approach constructs the filter switched between H_∞ robust and optimal filtering mode by judging an inequality condition related with real measurement covariance, so that the filter robustness becomes finite-horizon and the optimality is enhanced from the time span point of view. Allowing for resetting the Kalman gain, the filter can improve the excessive cost of filtering optimality and augment the tracking accuracy of maneuvering target with fast transient variations.

In SH∞F, the switching mechanism actually replaces the uncertainty bounds with a switching inequality. However, it introduces an auxiliary switching parameter to determine the switching threshold [10, 12]. The value of the parameter varies with system uncertainties and is usually given as a priori. Although that is simpler than to find the uncertainty bounds, mismatch of the given values will remarkably reduce estimation precision of the switched filters. For this reason, it would be desirable if the switching parameter can be obtained on-line optimally, so that the SH∞F may be independent with any system conditions that need a priori value.

To achieve that objective, an optimal-switched H_∞ filter (SH∞F) is developed and we discuss it with application to maneuvering spacecraft tracking in Chap. 9.

The filter is built upon the H_∞ filter (H∞F) provided by Theodor et al. [21] and the switched structure given by Li et al. [10, 12] embedded an extra mechanism to on-line calculate optimal value of the switching parameter at each iteration. The extra mechanism is established by optimizing a quadratic optimality-robustness cost function (ORCF), which is defined with the weighted sum of the measurement estimation square error and the measurement prediction covariance. The two parts represent robustness and optimality, reflecting the estimation accuracy and precision of the filter, respectively, and their proportional relation is clarified by a designed non-dimensional weight factor (WF). To implement the OSH∞F, WF is the only parameter that needs given to state quantitatively that how much filtering optimality or robustness will be considered in the estimation results. That prevents the decrement of filtering performance from any switching parameter mismatch, and the results can be regarded as optimal in the sense of WF. Independent with any information of system uncertainties to set parameter, the OSH∞F is a generalized version of the standard H∞F. It has been verified that the OSH∞F has better tracking performances than the usual H∞F and SH∞F, and that the filtering results depend on the WF only.

2.3 Random Finite Set Based Filter

2.3.1 Probability Hypothesis Density Filter

The multi-target state and multi-target measurement based on the finite set statistics (FISST) are random finite sets

$$X_k = \left\{x_{k,1}, x_{k,2}, \ldots, x_{k,N_k}\right\} \subset F(\mathrm{X})$$

$$Z_k = \left\{z_{k,1}, z_{k,2}, \ldots, z_{k,M_k}\right\} \subset F(\mathrm{Z})$$

$$\mathrm{Z}_{1:k} = \bigcup_{i=1}^{k} Z_i$$

where $F(\mathrm{X})$ and $F(\mathrm{Z})$ are the respective collections of all finite subset of X and Z. N_k and M_k are, respectively, the number of targets and measurement at time k.

Uncertainty in a multi-target system based on FISST is characterized by modeling the multi-target state X_k and multi-target measurement Z_k as Random Finite Sets (RFS). In a similar vein to the single-target dynamical model, the randomness in the multi-target evolution and observation are captured in the multi-target Markov transition density $f_{k|k-1}(\cdot|\cdot)$ and multi-target sensor likelihood function $h_k(\cdot|\cdot)$, respectively. Denote the multi-target posterior density function as $p(X_k|\mathrm{Z}_{1:k})$. Then, the optimal multi-target Bayes filter propagates the multi-target posterior in time via the recursion:

$$\begin{aligned} p(X_k|Z_{1:k-1}) &= \int f_{k|k-1}(X_k|X)p(X|Z_{1:k-1})\mu(dX) \\ p(X_k|Z_{1:k}) &= \frac{h_k(Z_k|X_k)p(X_k|Z_{1:k-1})}{\int h_k(Z_k|X)p(X|Z_{1:k-1})\mu(dX)} \end{aligned} \tag{2.54}$$

where $\mu(T) = \sum_{i=0}^{\infty} \frac{\lambda^i(\chi^{-1}(T)\cap \mathrm{X}^i)}{i!}$ is an appropriate reference measure on $F(\mathrm{X})$. λ^i is the i-th product Lebesque measure, and $\chi : \uplus_{i=0}^{\infty}\mathrm{X}^i \mapsto F(\mathrm{X})$ is a mapping of vector to sets defined by $\chi\left([x_1, x_2, \ldots, x_i]^{\mathrm{T}}\right) = \{x_1, x_2, \ldots, x_i\}$. This reference measure is often used in point process theory.

The Bayes recursion involves multiple set integrals on the space of finite sets, which are computationally intractable. The PHD filter is more tractable than the optimal multi-target filtering. It is a recursion propagating the first order statistical moment of the posterior multiple targets state. For an RFS X on X with probability distribution P, its first order moment is a nonnegative function υ on X, called the intensity or PHD. For each region $S \subset \mathrm{X}$, the PHD has the property that:

$$\int |X \cap S| P(DX) = \int_S \upsilon(x) Dx. \tag{2.55}$$

Since the integral domain of the intensity function is the single-target state space, its propagation requires much less computational cost than the multi-target posterior.

The recursion of PHD filter is as follows: given the posterior intensity D_{k-1} at time step $k-1$, the predicted intensity function $D_{k|k-1}$ and updated intensity function D_k are formulated as

$$D_{k|k-1}(x) = \int p_S(x') f_{k|k-1}(x|x') D_{k-1}(x')dx' + \gamma_{k|k-1}(x) \tag{2.56}$$

$$\begin{aligned} D_k(x) =&[1 - p_D(x)]D_{k|k-1}(x) \\ &+ \sum_{z\in Z_k} \frac{p_D(x)g_k(z|x)D_{k|k-1}(x)}{\kappa_k(z) + \int p_D(x)g_k(z|x)D_{k|k-1}(x)dx} \end{aligned} \tag{2.57}$$

where $f_{k|k-1}(x|x')$ is the single-target transition density, $p_S(x')$is the probability of target survival, $\gamma_{k|k-1}(x)$is the intensity of target birth, $p_D(x)$is the probability of target detection, Z_k is the multi-target measurement set, $g_k(z|x)$ is the single-target measurement likelihood, and $\kappa_k(z)$ is the intensity of clutter.

2.3.2 Cardinalized Probability Hypothesis Density Filter

The CPHD filter can jointly propagate the cardinality (the number of targets) and the first-order moment of the multi-target posterior which is known as the PHD. Suppose D_{k-1} is the multi-target intensity and p_{k-1} is the cardinality distribution at time $k-1$, then the predicted intensity $D_{k|k-1}$ and cardinality distribution $p_{k|k-1}$ are formulated by

$$p_{k|k-1}(n) = \sum_{j=0}^{n} p_{\Gamma,k}(n-j)\Pi_{k|k-1}[D_{k-1}, p_{k-1}](j) \tag{2.58}$$

$$D_{k|k-1}(x) = \int p_{S,k}(\varsigma) f_{k|k-1}(x|\varsigma) D_{k-1}(\varsigma) d\varsigma + \gamma_k(x) \tag{2.59}$$

where $\Pi_{k|k-1}[D, p](j) = \sum_{l=j}^{\infty} C_j^l \frac{<p_{s,k},D>^j <1-p_{s,k},D>^{l-j}}{<1,D>^l} p(l)$, $<\cdot,\cdot>$ denotes the inner product, $f_{k|k-1}(\cdot|\varsigma)$ is the single-target transition function, ς is the previous state of single-target, $p_{S,k}(\varsigma)$ is the survival probability of single target state ς at time k, $\gamma_k(\cdot)$ and $p_{\Gamma,k}(\cdot)$ are the intensity function and cardinality distribution of new born target, respectively, $C_i^l = \frac{l!}{i!(l-i)!}$ is the binomial coefficient.

When the new measurements come, the updated cardinality distribution p_k and intensity D_k are given by

$$p_k(n) = \frac{\Upsilon_k^0[D_{k|k-1}, Z_k](n) \cdot p_{k|k-1}(n)}{< \Upsilon_k^0[D_{k|k-1}, Z_k], p_{k|k-1} >} \tag{2.60}$$

$$\begin{aligned} D_k(x) = {} & \left[1 - p_{D,k}(x)\right] \frac{< \Upsilon_k^1[D_{k|k-1}, Z_k], p_{k|k-1} >}{< \Upsilon_k^0[D_{k|k-1}, Z_k], p_{k|k-1} >} D_{k|k-1}(x) \\ & + \sum_{z=Z_k} \frac{< \Upsilon_k^1[D_{k|k-1}, Z_k \backslash \{z\}], p_{k|k-1} >}{< \Upsilon_k^0[D_{k|k-1}, Z_k], p_{k|k-1} >} \psi_{k,z}(x) D_{k|k-1}(x) \end{aligned} \tag{2.61}$$

where

$$\begin{aligned} \Upsilon_k^u[D, Z](n) = {} & \sum_{j=0}^{\min(|Z|,n)} (|Z|-j)! p_{K,k}(|Z|-j) P_{j+u}^n \cdot \\ & \frac{< 1 - p_{D,k}, D>^{n-(j+u)}}{< 1, D>^n} \sigma_j \left(\Xi_k(D, Z)\right) \end{aligned} \tag{2.62}$$

$$\psi_{k,z}(x) = \frac{< 1, \kappa_k >}{\kappa_k(z)} g_k(z|x) p_{D,k}(x) \tag{2.63}$$

$$\kappa_k(z) = \lambda_c \cdot V \cdot u(z) \tag{2.64}$$

$$\Xi_k(D, Z) = \left\{< D, \psi_{k,z} >: z \in Z\right\} \tag{2.65}$$

$$\sigma_j(Z) = \sum_{S \subseteq Z, |S|=j} \left(\prod_{\varsigma \in S} \varsigma\right) \tag{2.66}$$

and $g_k(\cdot|x)$ is the likelihood function of the single target, $p_{D,k}(x)$ is the detection probability of state x, $\kappa_k(\cdot)$ is the clutter distribution, V is the volume of surveillance region, λ_c is the average number of clutter returns per unit hyper volume, $u(\cdot)$ is a point process, $p_{K,k}(\cdot)$ is the cardinality distribution of the clutter.

References

1. Arasaratnam I, Haykin S (2009) Cubature Kalman filters. IEEE Trans Autom Control 54(6):1254–1269
2. Arasaratnam I, Haykin S, Elliott RJ (2007) Discrete-time nonlinear filtering algorithms using gausshermite quadrature. Proc IEEE 95(5):953–977
3. Banavar RN (1992) A game theoretic approach to linear dynamic estimation. Doctoral Dissertation, University of Texas at Austin
4. Bishop AN, Pathirana PN, Savkin AV (2007) Radar target tracking via robust linear filtering. IEEE Signal Process Lett 14(12):1028–1031
5. Cappe O, Godsill SJ, Moulines E (2007) An overview of existing methods and recent advances in sequential Monte Carlo. Proc IEEE 95(5):899–924
6. Dan S (2006) Optimal state estimation: Kalman, H_∞, and nonlinear approaches. Wiley-Interscience, New York
7. Einicke GA, White LB (1999) Robust extended Kalman filtering. IEEE Trans Signal Process 47(9):2596–2599
8. Haug AJ (2012) Bayesian estimation and tracking: a practical guide. Springer, New York
9. Julier SJ, Uhlmann JK (2004) Unscented filtering and nonlinear estimation. Proc IEEE 92(3):401–422
10. Li YK, Jing ZL, Hu SQ (2010) Redundant adaptive robust tracking of active satellite and error evaluation. IET Control Theory Appl 4(11):2539–2553
11. Li W, Gong D, Liu M, Chen J, Duan D (2013) Adaptive robust Kalman filter for relative navigation using global position system. IET Radar Sonar Navig 7(5):471–479
12. Li Y, Jing Z, Liu G (2014) Maneuver-aided active satellite tracking using six-DOF optimal dynamic inversion control. IEEE Trans Aerosp Electron Syst 50(1):704–719
13. Mahler R (2003) Multitarget Bayes filtering via first-order multitarget moments. IEEE Trans Aerosp Electron Syst 39(4):1152–1178
14. Mahler RPS (2007) PHD filters of higher order in target number. IEEE Trans Aerosp Electron Syst 43(4):1523–1543
15. Mahler RPS (2007) Statistical multisource-multitarget information fusion. Artech House, Norwood
16. Norgaard M, Poulsen NK, Ravn O (2000) New developments in state estimation for nonlinear systems. Automatica 36(11):1627–1638
17. Ristic B, Arulampalam S, Gordon N (2003) Beyond the Kalman filter-particle filters for tracking applications. IEEE Trans Aerosp Electron Syst 19(7):37–38

18. Seo J, Yu MJ, Park CG, Lee JG (2006) An extended robust h_∞ filter for nonlinear constrained uncertain systems. IEEE Trans Signal Process 54(11):4471–4475
19. Shen X, Deng L (1997) Game theory approach to discrete h_∞ filter design. IEEE Trans Signal Process 45(4):1092–1095
20. Soken HE, Hajiyev C (2010) Pico satellite attitude estimation via robust unscented Kalman filter in the presence of measurement faults. ISA Trans 49(3):249–256
21. Theodor U, Shaked U, Souza CED (1994) A game theory approach to robust discrete-time H_∞-estimation. IEEE, New York
22. Vo BT, Vo BN, Cantoni A (2009) The cardinality balanced multi-target multi-Bernoulli filter and its implementations. IEEE Trans Signal Process 57(2):409–423
23. Vo BN, Vo BT, Phung D (2014) Labeled random finite sets and the Bayes multi-target tracking filter. IEEE Trans Signal Process 62(24):6554–6567
24. Xiong K, Zhang H, Liu L (2008) Adaptive robust extended Kalman filter for nonlinear stochastic systems. IET Control Theory Appl 2(3):239–250
25. Zhong M, Zhou D, Ding SX (2010) On designing h_∞ fault detection filter for linear discrete time-varying systems. IEEE Trans Autom Control 55(7):1689–1695

Part II
Target Tracking and Fusion

Chapter 3
Dynamic Bias Estimation with Gaussian Mean Shift Registration

3.1 Introduction

Information fusion can increase the reliability of systems and improve detection probability in multi-sensor tracking systems. However, because of the sensor bias, direct data fusion may lead to inaccurate estimation of position of a target. The measurement bias, which includes axis misalignment causing azimuth and elevation bias and range offset bias, is the major source of sensor bias [17, 18, 25]. Accurate sensor registration is an essential step before the fusion step of multi-sensor [25]. Large registration errors may generate ghost targets on the same target and it will degrade performance of the system due to the tracking errors [17, 23]. Figure 3.1 [22] gives the geometry of registration bias. It can be seen from the figure that sensor bias has an immediate effect on the reported target position. Let T_k be the target. $\{r_A, \theta_A\}$ and $\{r_B, \theta_B\}$ are measurements of the range and azimuth for T_k received from sensor A and sensor B, respectively. $\{\Delta r_A, \Delta\theta_A\}$ and $\{\Delta r_B, \Delta\theta_B\}$ are the corresponding sensor biases. $T_{A,k}$ and $T_{B,k}$ denote the biased measurements of T_k from sensor A and sensor B, respectively.

Batched methods [17] can be used when the bias of the sensor is time invariant and deterministic. Other off-line registration algorithms such as the exact maximum likelihood registration algorithm [28], the maximum likelihood registration algorithm [18], and the new least squares registration algorithm [27] are presented to deal with sensor bias.

On-line bias estimate methods have also been studied in the past. A direct method is the augmented state Kalman filter (ASKF) via augmenting states of all the targets and sensor bias into a single vector. However, it is computationally infeasible [15], which makes it unavailable. The decoupled Kalman filter [26], which is called van Doorn and Blom (VDB) technique [15], is proposed to reduce the complexity of the augmented vector for sensor bias estimation. The modified VDB method called Kastella VDB (KVDB) [15] which uses an approximation is presented in [12]. The extended Kalman filter (EKF) method is proposed to estimate the target state,

Z. Jing et al., *Non-Cooperative Target Tracking, Fusion and Control*,
Information Fusion and Data Science, https://doi.org/10.1007/978-3-319-90716-1_3

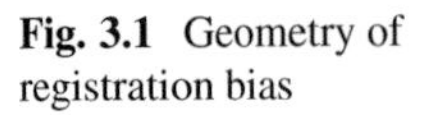
Fig. 3.1 Geometry of registration bias

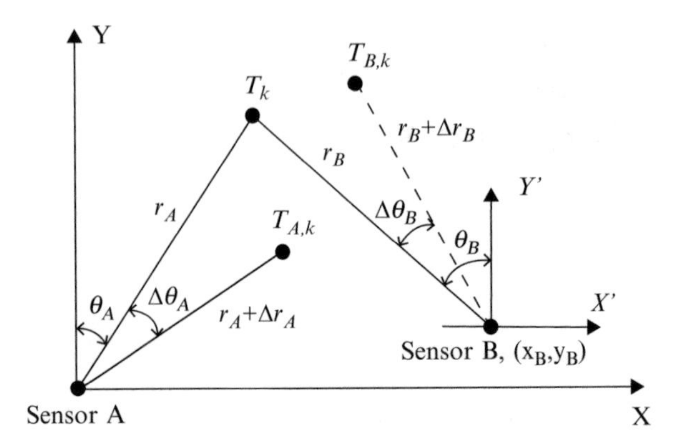

position, and orientation bias, using a nonlinear coordinated turn maneuver model [16]. Under the circumstance of the absence of information from local sensors, another method [25] is presented based on Kalman filter.

The real-time estimation is needed because the bias of sensors may change due to technical maintenance or other causes [14, 17]. The equivalent measurements method is proposed by formulating the sensor registration and track fusion as a Bayesian estimation problem. Using augmented state and measurement equations, the unscented Kalman filter (UKF) method is presented to estimate the sensor bias and target state simultaneously. The previous methods may be useless or inefficient to estimate the dynamic bias when sensor bias is a time varying process and changes slowly and continuously. The "exact" (EX) method [15] is presented to handle the dynamic sensor bias estimation for multiple targets and multiple frames by a minimum mean square error estimator. However, as the number of targets decreased the EX method becomes ineffective. In addition, after transforming the measurements into Cartesian coordinates, the correlation of measurement noise is introduced.

In this chapter, we present an approach based on the mean shift procedure proposed in [11]. Mean shift is a nonparametric clustering method widely used in many areas [5, 6, 20, 24] and an iterative method which shifts the data point u to the average v of data points [5]. Fashing and Tomasi show that the mean shift procedure is a quadratic bound maximization for all kernels [10] and it has been applied to problems such as image segmentation [6], target tracking [7–9, 20], and fusion [4]. The mean shift method is also an expectation-maximization (EM) algorithm when the kernel is Gaussian according to reference [3]. It can be seen from this chapter that when the measurement noise is white Gaussian and mutually independent, the Gaussian mean shift can be derived from the joint likelihood density function. The presentation of this chapter is based on the work in [21, 22].

The purpose of this work is to propose a new sensor bias estimation algorithm which is called the Gaussian mean shift registration (GMSR) algorithm. Gaussian mean shift algorithm and extended Kalman filter are combined to obtain the sensor bias estimation dynamically. The GMSR method is a recursive algorithm. It can be used to process the sensor's dynamic bias estimation from a single-target measurement to achieve a more accurate estimation than the EX method.

This chapter is organized as follows. The Gaussian mean shift is developed and the sufficient condition for convergence of the iterative Gaussian mean shift procedure is given in Sect. 3.2. The Gaussian mean shift algorithm combined with the EKF is used to estimate the dynamic bias in Sect. 3.3. Simulation results are given in Sect. 3.4. Section 3.5 gives the conclusion.

3.2 Gaussian Mean Shift Algorithm

3.2.1 *Gaussian Mean Shift as Maximum Likelihood*

In the surveillance region, we consider $n(n > 2)$ sensors of which the measurements include the range and azimuth for common targets. The observed model [18] with bias is given by

$$z(k) = h\left[X(k)\right] + \boldsymbol{\beta}(k) + w(k), \tag{3.1}$$

where $k = 1, 2, \ldots, N$, N is the time index; $z(k) = \left[z_1(k)^{\mathrm{T}}, \ldots, z_n(k)^{\mathrm{T}}\right]^{\mathrm{T}}$ is the measurement from all n sensors; $h\left[X(k)\right] = \left[h_1\left[X(k)\right]^{\mathrm{T}}, \ldots, h_n\left[X(k)\right]^{\mathrm{T}}\right]^{\mathrm{T}}$ is the known nonlinear measurement function; $\boldsymbol{\beta}(k) = \left[\boldsymbol{\beta}_1(k)^{\mathrm{T}}, \ldots, \boldsymbol{\beta}_n(k)^{\mathrm{T}}\right]^{\mathrm{T}}$ is the dynamic bias which is independent from $x(k)$; $w(k) = \left[w_1(k)^{\mathrm{T}}, \ldots, w_n(k)^{\mathrm{T}}\right]^{\mathrm{T}}$ is zero mean Gaussian random noise; $R = \mathrm{diag}\left[R_1, \ldots, R_n\right]$ is the noise covariance matrix, $R_i = \mathrm{diag}\left[\sigma_{ir}^2, \sigma_{i\theta}^2\right]$, $i = 1, 2, \ldots, n$; and σ_{ir}^2, $\sigma_{i\theta}^2$ are the noise variances of range and azimuth, respectively.

Given $[z(k); k = 1, 2, \ldots, N]$, then the joint likelihood density function [18] is

$$p\left[z(k)|X(k), \boldsymbol{\beta}(k)\right] = \frac{1}{(2\pi)^n\sqrt{|R|}} \exp\left\{-\frac{1}{2}\left[z(k) - \bar{z}(k)\right]^{\mathrm{T}} R^{-1}\left[z(k) - \bar{z}(k)\right]\right\}, \tag{3.2}$$

where $\bar{z}(k) = [\bar{z}_1(k)^{\mathrm{T}}, \ldots, \bar{z}_n(k)^{\mathrm{T}}]^{\mathrm{T}}$, $\bar{z}_i(k) = h_i[X(k)] + \boldsymbol{\beta}_i(k)$.

Given a set $\{z(j); i = 1, \ldots, m, m \leq N\}$ of m data points in the d-dimensional real Euclidean space $\mathbb{E}^d$, the multivariate kernel density estimator [6] with Gaussian kernel $K(\cdot)$ is given by

$$f_K(\boldsymbol{\xi}) = \frac{1}{m}\sum_{j=1}^{m} K\left[\boldsymbol{\xi}, z(j), R\right], \tag{3.3}$$

where

$$K\left[\boldsymbol{\xi}, z(j), R\right] = \frac{1}{(2\pi)^n\sqrt{|R|}} \exp\left\{-\frac{1}{2}\left[\boldsymbol{\xi} - z(j)\right]^{\mathrm{T}} R^{-1}\left[\boldsymbol{\xi} - z(j)\right]\right\}. \tag{3.4}$$

In order to find a model for $f_K(\boldsymbol{\xi})$, the stationary points should be sought by the gradient of the density estimator

$$\frac{\partial f_K(\boldsymbol{\xi})}{\partial \boldsymbol{\xi}} = \frac{1}{m}\sum_{j=1}^{m} K\left[\boldsymbol{\xi}, z(j), R\right] R^{-1}\left[z(j) - \boldsymbol{\xi}\right] = f_K(\boldsymbol{\xi}) R^{-1} m(\boldsymbol{\xi}), \tag{3.5}$$

where

$$\frac{\sum_{j=1}^{m} K\left[\boldsymbol{\xi}, z(j), R\right] z(j)}{\sum_{j=1}^{m} K\left[\boldsymbol{\xi}, z(j), R\right]} - \boldsymbol{\xi}. \tag{3.6}$$

The repeated movement of data points to the sample means is called the mean shift algorithm [5, 11]. Equation (3.6) means that the mean shift vector $m(\boldsymbol{\xi})$ points toward the direction of gradient ascent and it is proportional to the normalized density gradient estimate [6, 24]. Let $y_l = \boldsymbol{\xi}$, and Eq. (3.6) becomes the iterative equation

$$y_{l+1} = y_l + m(y_l), \tag{3.7}$$

where

$$y_{l+1} = \frac{\sum_{j=1}^{m} K\left[y_l, z(j), R\right] z(y_l)}{\sum_{j=1}^{m} K\left[y_l, z(j), R\right]}, \quad l = 1, 2, \ldots. \tag{3.8}$$

and y_{l+1} is the weighted mean at y_l which is computed by the Gaussian kernel K, y_l is the center of the initial position for the kernel [6]. After the iteration, the convergence on a stationary point where the gradient of density estimate is zero is guaranteed [20].

3.2.2 Convergence Analysis

Under the assumption of Gaussian noise, the Gaussian mean shift algorithm can be established by joint likelihood function converges in the sensor registration. The definitions and theorems of convex and concave functions can be found in some documents. They are useful in the following theorems, here is introduced in order to understand the lemma.

Definition 3.1 Let $\phi : \Phi \mapsto \mathbb{E}$, where Φ is a nonempty open convex set in $\mathbb{E}^d$ [2]. The function ϕ is said to be convex on Φ if

$$\phi\left[\lambda x_1 + (1-\lambda)x_2\right] \leq \lambda\phi(x_1) + (1-\lambda)\phi(x_2), \tag{3.9}$$

for each $x_1,\ x_2 \in \Phi$ and for each $\lambda \in (0,\ 1)$. The function $\phi : \Phi \mapsto \mathbb{E}$ is called concave on Φ if $-\phi$ is convex on Φ.

Lemma 3.1 *Let Φ be a nonempty open convex set in $\mathbb{E}^d$, and let $\phi : \Phi \mapsto \mathbb{E}$ be twice differentiable on Φ [2]. Then ϕ is convex if and only if the Hessian matrix is positive semidefinite at each point in Φ.*

The lemma can be used to check convexity or concavity of a twice differentiable function [2, 19]. Thus, it is sufficient to check whether the Hessian matrix is positive semidefinite or negative semidefinite.

Theorem 3.1 *If the Kernel $K(\boldsymbol{\theta}, \boldsymbol{\mu}, \Sigma)$ is Gaussian and $\boldsymbol{\theta} = [\theta_1, \ldots, \theta_d]^T$ is mutually independent, i.e., the correlative coefficient $\rho(\theta_p, \theta_q) = 0$ for $p \neq q$, $\boldsymbol{\mu}$ is the mean of $\boldsymbol{\theta}$, $\Sigma = diag[\Sigma_1, \ldots, \Sigma_d]$ is the covariance, $\Sigma_t > 0$, $t,\ p,\ q = 1, 2, \ldots, d$. The sequences $\{y_l\}$ and $\{f_K(y_l)\}$ converge and $\{f_K(y_l)\}$ is monotonically increasing, $l = 1, 2, \ldots$.*

The proof can be seen in Appendix. The Gaussian mean shift algorithm, developed by Theorem 3.1, extends the "Theorem 1," given by Comaniciu and Meer [6], from a strict convex kernel to a segmented convex concave kernel. The multivariate kernel density estimator $\{f_K(y_l)\}$ is discussed in the whole domain where it is concave in $\mathbb{E}_1^d$ (see Appendix) and convex in $\mathbb{E}_2^d$, while in "Theorem 1" it is convex [see Eqs. (3.28)–(3.30) in Appendix] [6].

3.3 Bias Estimation Based on Gaussian Mean Shift

We consider a system consisting of a static sensor with zero mean Gaussian noise and a sufficient condition for convergence theorems in the two-dimensional region. The Gaussian mean shift is used to estimate the dynamic deviation. The dynamic equation for the target is given by a discretized continuous white noise acceleration (DCWNA) model

$$X(k+1) = FX(k) + v(k), \tag{3.10}$$

where $X = [x, \dot{x}, y, \dot{y}]^{\mathrm{T}}$, $F = \mathrm{diag}\left[F_x, F_y\right]$, $F_x = F_y = [1, T; 0, 1]$, and $v(k)$ is a zero-mean white process noise with covariance. $Q = \mathrm{diag}\left[Q_x, Q_y\right]$, $Q_x = Q_y = [\frac{1}{3}T^3, \frac{1}{2}T^2; \frac{1}{2}T^2, T]\tilde{q}$ with the sampling interval T and the power spectral densities $\tilde{q}$.

The dynamic model for the bias vector [15] is given by

$$\boldsymbol{\beta}(k+1) = F_b(k)\boldsymbol{\beta}(k) + v_b(k), \tag{3.11}$$

where $F_b(k) = \mathrm{diag}\,[F_{b1}(k), \ldots, F_{bn}(k)]$ is the transition matrix, and $v_b(k)$ is zero-mean white noise with variance $Q_b(k) = \mathrm{diag}\,[Q_{b1}, \ldots, Q_{bn}(k)]$.

For $j = 1, 2, \ldots, m,\ m \le N$ by Eq. (3.8), we can obtain the bias estimator

$$\hat{\boldsymbol{\beta}}^*_{l+1} = \frac{\sum_{j=N-m+1}^{N} K\left(\hat{\boldsymbol{\beta}}^*_l, \boldsymbol{\delta}(j), R\right) \times \boldsymbol{\delta}(j)}{\sum_{j=N-m+1}^{N} K\left(\hat{\boldsymbol{\beta}}^*_l, \boldsymbol{\delta}(j), R\right)}, \ l = 1, 2, \ldots, \tag{3.12}$$

where

$$\boldsymbol{\delta}(j) = z(j) - \hat{z}(j), \tag{3.13}$$

$$z(j) = h\left[X(j)\right] + \boldsymbol{\beta}(j) + \boldsymbol{\omega}(j), \tag{3.14}$$

$$\hat{z}(j) = h\left[\hat{X}(j)\right] + \hat{\boldsymbol{\beta}}(j), \tag{3.15}$$

where $\hat{\boldsymbol{\beta}}^*_l$ is the l-th estimate of $\hat{\boldsymbol{\beta}}$, $\boldsymbol{\beta}_l$ denotes the bias of i-th sensor in Eq. (3.1), $\boldsymbol{\delta}(j)$ is a residual measurement that includes a residual bias, $\hat{\boldsymbol{\beta}}^*_{l+1}$ is a bias estimate which is the weighted mean at $\hat{\boldsymbol{\beta}}^*_l$. After the iteration, $\hat{\boldsymbol{\beta}}^*_{l+1}$ converges at the mode of the density distribution $\boldsymbol{\delta}(j)$ (that is $\hat{\boldsymbol{\beta}}^*_{\mathbf{conv}}$).

According to Eqs. (3.4), (3.12), and (3.15), we have

$$\hat{\boldsymbol{\beta}}^*_{l+1} = \frac{\sum_{j=N-m+1}^{N} \exp\left\{-\frac{1}{2}\left[\hat{\boldsymbol{\beta}}^*_l - \boldsymbol{\delta}(j)\right]^{\mathrm{T}} R^{-1}\left[\hat{\boldsymbol{\beta}}^*_l - \boldsymbol{\delta}(j)\right]\right\} \cdot \boldsymbol{\delta}(j)}{\sum_{j=N-m+1}^{N} \exp\left\{-\frac{1}{2}\left[\hat{\boldsymbol{\beta}}^*_l - \boldsymbol{\delta}(j)\right]^{\mathrm{T}} R^{-1}\left[\hat{\boldsymbol{\beta}}^*_l - \boldsymbol{\delta}(j)\right]\right\}}. \tag{3.16}$$

Equation (3.16) may be sensitive to the round-off error, therefore the modified estimator is given by

$$\hat{\boldsymbol{\beta}}^*_{l+1} = \frac{\sum_{j=N-m+1}^{N} \exp\left\{-\frac{1}{2M_1} G_j\right\} \cdot \boldsymbol{\delta}(j)}{\sum_{j=N-m+1}^{N} \exp\left\{-\frac{1}{2M_1} G_j\right\}}, \tag{3.17}$$

where

$$G_j = \left[\hat{\boldsymbol{\beta}}^*_l - \boldsymbol{\delta}(j)\right]^{\mathrm{T}} R^{-1}\left[\hat{\boldsymbol{\beta}}^*_l - \boldsymbol{\delta}(j)\right], \tag{3.18}$$

$$M_1 = \sum_{j=N-m+1}^{N} |\boldsymbol{\delta}(j)| = |\boldsymbol{\delta}(1)| + \cdots + |\boldsymbol{\delta}(m)|. \tag{3.19}$$

The modified natural exponential function does not change the exponential property using the normalization method, and the convergence remains unchanged. By Eq. (3.7), the bias estimation iterative equation can be rewritten as

$$\hat{\boldsymbol{\beta}}^*_{l+1} = \hat{\boldsymbol{\beta}}^*_l + m(\hat{\boldsymbol{\beta}}^*_l). \tag{3.20}$$

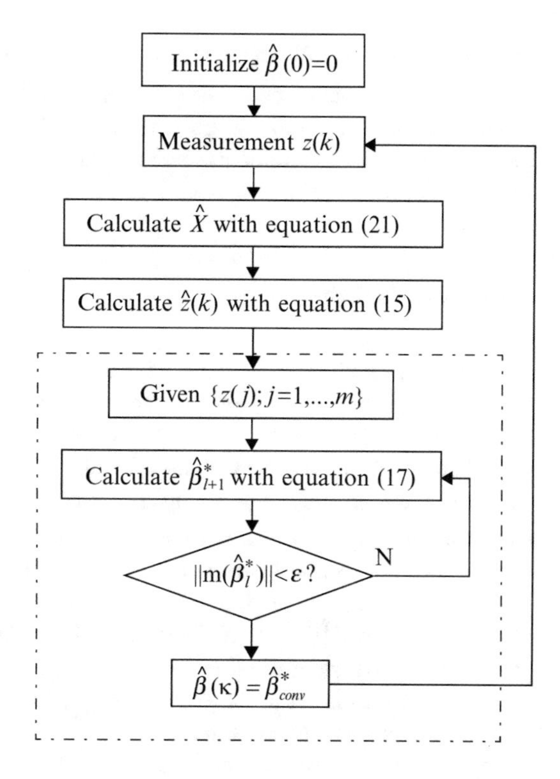

Fig. 3.2 Flowchart of the Gaussian mean shift registration algorithm

Using Eqs. (3.1), (3.10), and (3.11), the target state $\hat{X}$ of Eq. (3.15) is estimated by the EKF [1]

$$\hat{X}(k+1|k+1) = \hat{X}(k+1|k)+W(k+1)\left[z(k+1) - H(k+1)\hat{X}(k+1|k) - \hat{\boldsymbol{\beta}}(k)\right], \tag{3.21}$$

where

$$H(k+1) = a! \tag{3.22}$$

$$P(k+1|k) = F(k)P(k|k)F(k)^{\mathrm{T}} + Q(k), \tag{3.23}$$

$$S(k+1) = R(k+1) + H(k+1)P(k+1|k)H(k+1)^{\mathrm{T}}, \tag{3.24}$$

$$W(k+1) = P(k+1|k)H(k+1)^{\mathrm{T}}S(k+1)^{-1}, \tag{3.25}$$

$$P(k+1|k) = P(k+1|k) - W(k+1)S(k+1)W(k+1)^{\mathrm{T}}. \tag{3.26}$$

The flowchart of the Gaussian mean shift registration algorithm is summarized in Fig. 3.2, where the part represented by a dashed line gives the Gaussian mean shift procedure. The initial estimate of vector $\hat{\boldsymbol{\beta}}(0)$ is set to zero. We use the current estimate $\hat{\boldsymbol{\beta}}(k)$ to compensate sensor measurements $z(k)$ for each $k = 1, 2, \ldots, N$. The EKF is utilized to estimate the target state. Once $\hat{z}(k)$ is computed, the Gaussian mean shift algorithm is implemented to estimate a new vector $\hat{\boldsymbol{\beta}}^*_{l+1}$ by Eq. (3.17) until the bias estimation converges. The standard threshold constraint vector $m(\hat{\boldsymbol{\beta}}^*_l)$ is stopped within a certain range. The lower the threshold of ε, the more accurate the estimation is. In fact, the average number of average displacement iterations is very small, about 4 [8].

3.4 Simulations

We consider a simulation case of two sensors. The sensors are located at $(0, 0)$ and $(0, 50)$ in kilometers and the measurements of them are time-synchronized. The initial state and error covariance are given by $X_0 = [30{,}000\,\text{m}, 40\,\text{m/s}, 30{,}000\,\text{m}, 40\,\text{m/s}]^{\text{T}}$ and $P(0|0) = \text{diag}\left[(1000\,\text{m})^2, (50\,\text{m/s})^2, (1000\,\text{m})^2, (50\,\text{m/s})^2\right]$, respectively. The initial positions of sensors and targets are shown in Fig. 3.3. The bias is set to $b_1 = b_2 = [50\,\text{m}, 8\,\text{mrad}]^{\text{T}}$. The standard deviation of the measurement noise variances of the range and azimuth measurements are set to $\sigma_r = 2\,\text{m}$ and $\sigma_{i\theta}\theta = 0.5\,\text{mrad}$. $F_{b1}(k) = F_{b2}(k) = 0.99 \times I_{2\times2}$, $Q_{b1} = Q_{b2} = \text{diag}\left[(2\,\text{m})^2, (0.3\,\text{mrad})^2\right]$. The sampling interval is $T = 1$ s, and the power spectral densities are $\tilde{q} = 6\,\text{m}^2/\text{s}^3$. Suppose that the initial covariance matrix of the sensor bias is set to $\Sigma_b(0|0) = \text{diag}\left[(200\,\text{m})^2, (300\,\text{mrad})^2, (200\,\text{m})^2\right]$ and there are six targets signed as in Fig. 3.3.

Figure 3.4 gives the actual trajectories of the real lines, and the dotted lines represent the estimated trajectories and the unregistered trajectories. The estimation of range and azimuth deviation are shown in Figs. 3.5 and 3.6, respectively. The

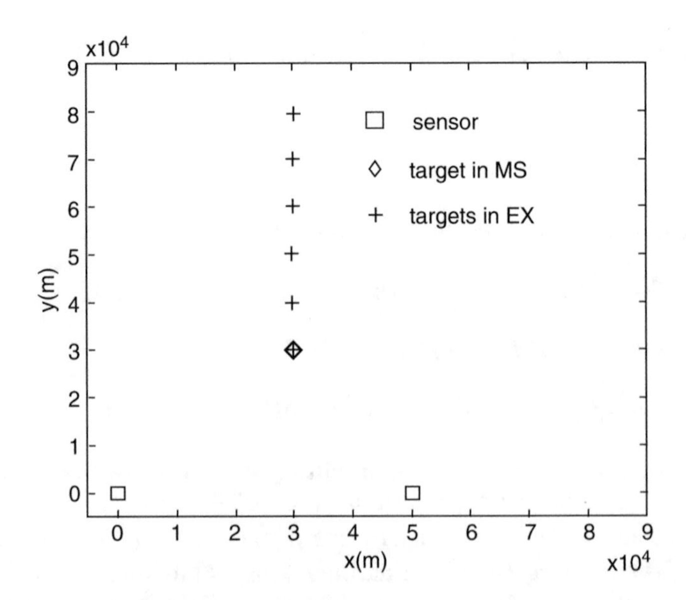

Fig. 3.3 Initial position of sensors and targets

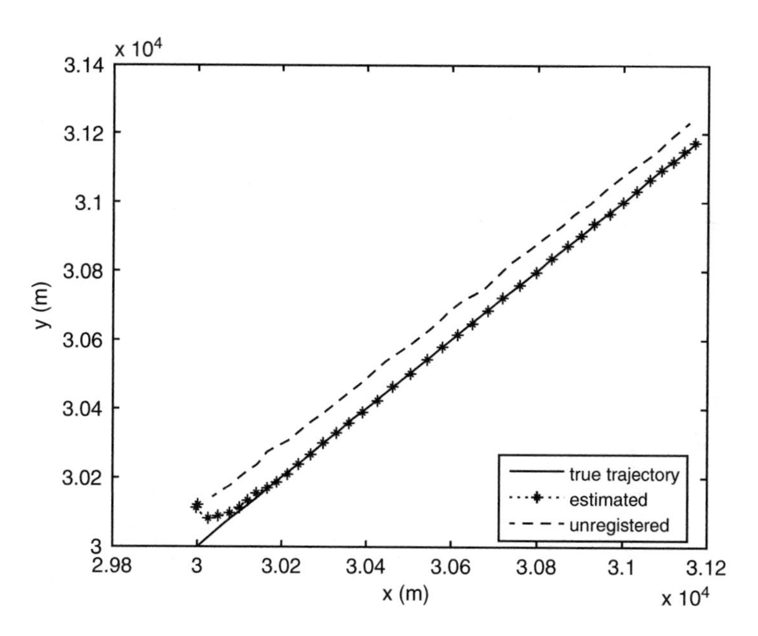

Fig. 3.4 Target trajectory

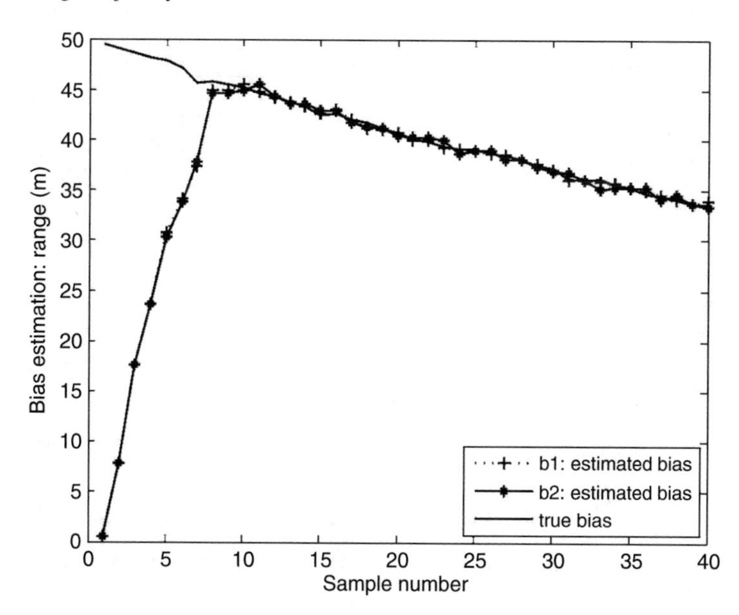

Fig. 3.5 Range bias estimation versus true value

results of the 100 Monte Carlo operation shown in Figs. 3.5 and 3.6 show that the GMSR algorithm can accurately estimate the dynamic deviations of $k > 10$.

Figures 3.7 and 3.8 show that the RMS error of the GMSR method is close to the square root of the Cramer-Rao lower bound (CRLB) of the KVDB [15] method.

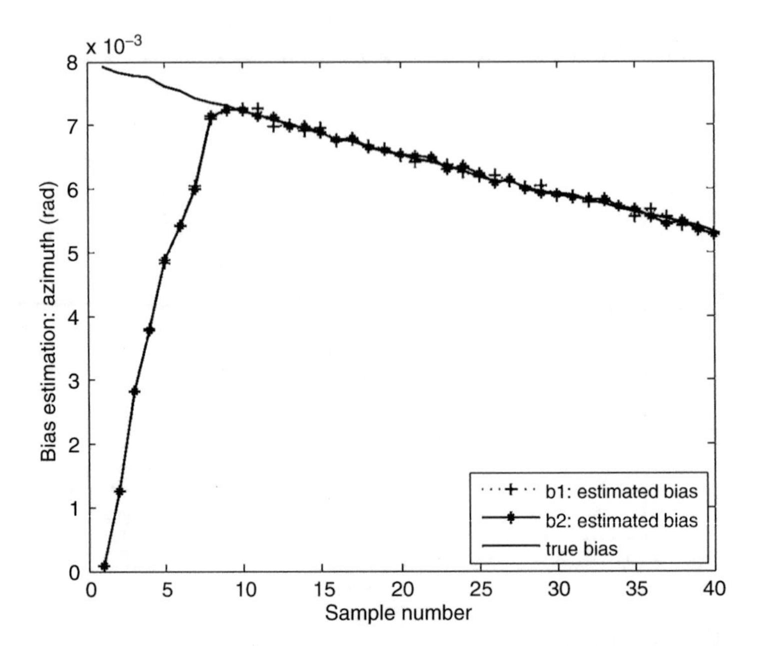

Fig. 3.6 Azimuth bias estimation versus true value

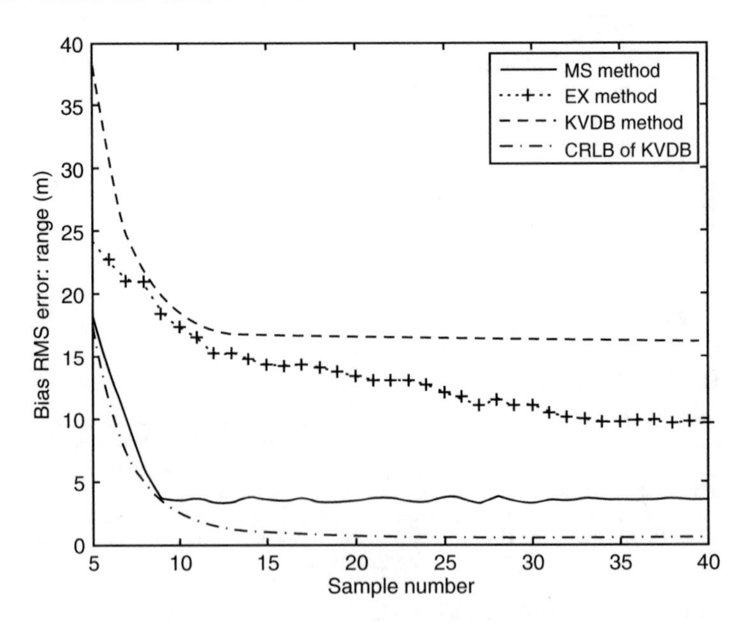

Fig. 3.7 Range bias RMS error and square root of CRLB of KVDB

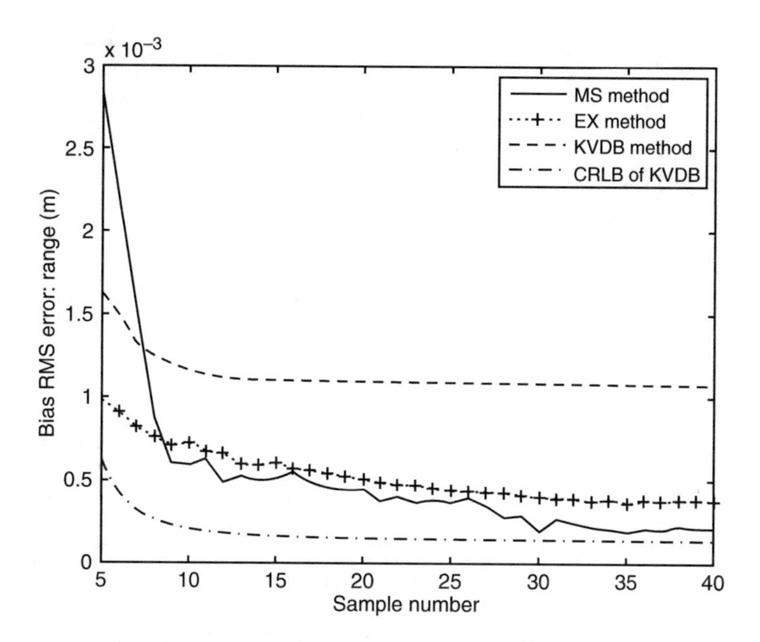

Fig. 3.8 Azimuth bias RMS error and square root of CRLB of KVDB

This means that the GMSR algorithm can estimate the dynamic deviation more effectively than the EX method or the KVDB method. The reason for using the CRLB of the KVDB method is that the deviation RMS error of the GMSR method exceeds the CRLB of the EX method. Figure 3.7 shows that the estimation of the range deviation of the GMSR has a significant improvement compared with the estimation of the deviation of the EX method. In Fig. 3.8, the mean square root error of the azimuth deviation of the GMSR method for $k > 10$ is also lower than that of the EX method. For the estimation of the deviation of the $k > 35$, GMSR and EX methods seems to be stable. For longer sampling intervals ($T = 10\,\mathrm{s}$), the RMS error of distance and azimuth deviation is shown in Figs. 3.9 and 3.10. The results show that EX method can produce better results, but still worse than GMSR method.

The calculation between the GMSR method and the EX method is very difficult to produce an exact expression of the number of operations. The method of calculating GMSR concentration in iterative computation of exponential function [13], and the inverse and matrix multiplication is the core of EX method. In the previous example, the average iteration number of the Gaussian mean is 3.6. Table 3.1 shows that the calculation of GMSR method is about 58 of the EX method ($T = 1\,\mathrm{s}$, $N = 40$), and it is tested on a personal computer running on 2.4 GHz Pentium 4 of Windows 2000.

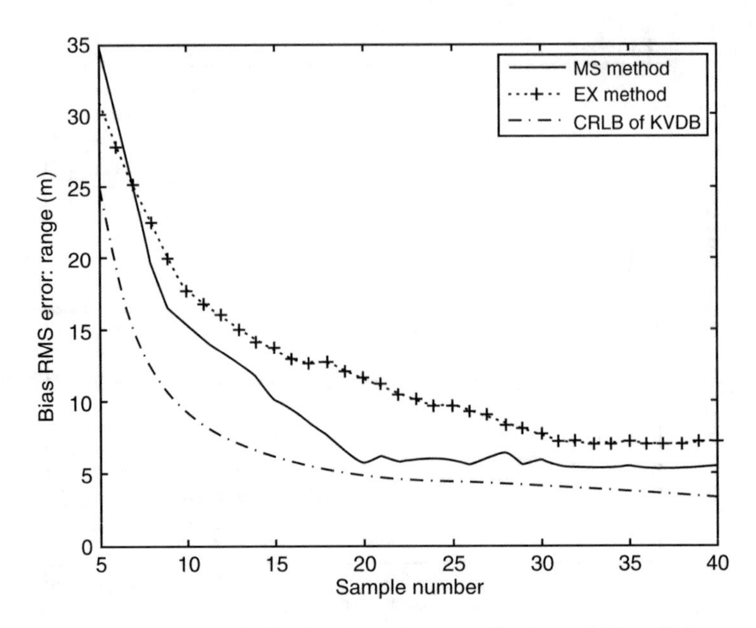

Fig. 3.9 Comparison of range bias RMS error for long sampling interval ($T = 10\,\mathrm{s}$)

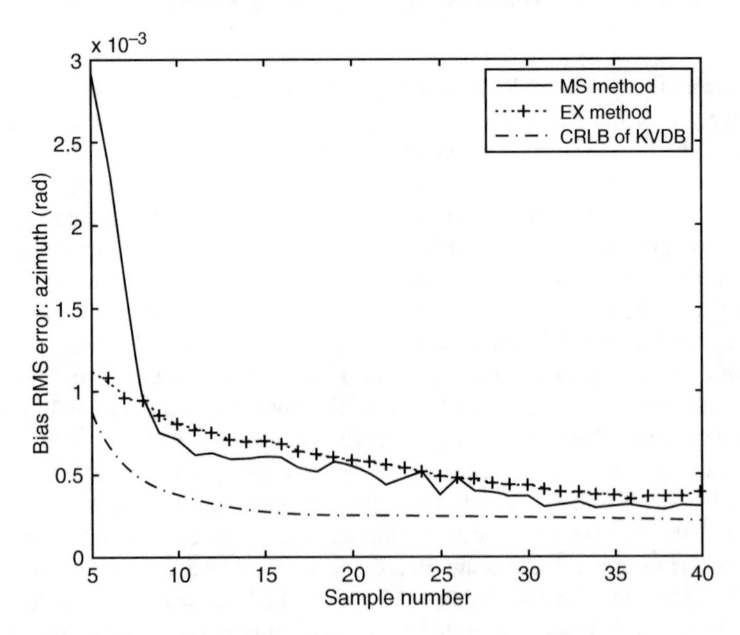

Fig. 3.10 Comparison of azimuth bias RMS error for long sampling interval ($T = 10\,\mathrm{s}$)

Table 3.1 Comparison between the GMS method and the EX method

Monte Carlo runs	GMS (s)	EX (s)	GMS/EX (s)
100	5.290625	9.081704	0.5826
100	5.275442	9.068125	0.5818
100	5.271382	9.053701	0.5822
100	5.285998	9.098367	0.5810
100	5.277204	9.087044	0.5807

3.5 Conclusion

In this chapter, Gaussian mean shift algorithm for Gaussian noise with zero mean is studied, and sufficient conditions for the convergence of Gaussian kernel are given. As an application of Gaussian mean shift algorithm, the dynamic deviation of Gaussian mean shift estimation is proposed. The algorithm recursively estimates the dynamic deviation based on the trajectory of the target. Monte Carlo simulations show that the root mean square error of the GMSR algorithm reduces the bias. Besides, computational time significantly increases compared to the previous approach. The new calculation method is valid because the root-mean square error of the GMSR method is closer to the CRLB than other methods.

Acknowledgements We are thankful to the international society for optics and photonics (SPIE) for the copyright policy in allowing the authors to reuse the article, i.e., Yongqing Qi, Zhongliang Jing, Shiqiang Hu and Haitao Zhao; "New method for dynamic bias estimation: gaussian mean shift registration," Optical Engineering, vol. 47, no. 2, pp. 026401-026401-8, 2008.

Appendix

The sequence $f_K(y_l)$ is bounded because m is finite. The Gaussian kernel

$$K(\boldsymbol{\theta}, \boldsymbol{\mu}, \Sigma) = C_1 \exp\left\{-\frac{1}{2}(\boldsymbol{\theta} - \boldsymbol{mu})^{\mathrm{T}} \Sigma^{-1} (\boldsymbol{\theta} - \boldsymbol{mu})\right\}, \tag{3.27}$$

where C_1 is a constant coefficient, and Σ is the $d \times d$ bandwidth matrix. The gradient and the Hessian matrix of the kernel are given by

$$G_{\nabla} = \nabla K(\boldsymbol{\theta}, \boldsymbol{\mu}, \Sigma) = K(\boldsymbol{\theta}, \boldsymbol{\mu}, \Sigma)\, \Sigma^{-1}(\boldsymbol{\mu} - \boldsymbol{\theta}), \tag{3.28}$$

$$H_{\nabla^2} = \nabla^2 K(\boldsymbol{\theta}, \boldsymbol{\mu}, \Sigma) = K(\boldsymbol{\theta}, \boldsymbol{\mu}, \Sigma)\, \Sigma^{-1}\left[(\boldsymbol{\mu} - \boldsymbol{\theta})(\boldsymbol{\mu} - \boldsymbol{\theta})^{\mathrm{T}} - \Sigma\right]\Sigma^{-1}. \tag{3.29}$$

Let $\boldsymbol{\gamma} = \boldsymbol{\mu} - \boldsymbol{\theta}$, $\boldsymbol{\gamma} = [\gamma_1, \ldots, \gamma_d]^{\mathrm{T}}$, so $\rho(\gamma_p, \gamma_q) = 0$. The third term in the right-hand side of Eq. (3.29) is defined as

$$\Theta = (\boldsymbol{\mu} - \boldsymbol{\theta})(\boldsymbol{\mu} - \boldsymbol{\theta})^{\mathrm{T}} - \Sigma = \mathrm{diag}\left[\gamma_1^2 - \Sigma_1, \ldots, \gamma_d^2 - \Sigma_d\right]. \tag{3.30}$$

The space $\mathbb{E}^d$ can be decomposed into two domains

$$\mathbb{E}_1^d = \left\{\left(\gamma_t^2 - \Sigma_t\right) \le 0\right\}, \tag{3.31}$$

$$\mathbb{E}_2^d = \left\{\left(\gamma_t^2 - \Sigma_t\right) > 0\right\}. \tag{3.32}$$

When $\left(\gamma_t^2 - \Sigma_t\right) \in \mathbb{E}_1^d$, Θ is a negative definite matrix, and the Hessian matrix is also a negative definite matrix. By the lemma, the Gaussian kernel $K\ (\boldsymbol{\theta}, \boldsymbol{\mu}, \Sigma)$ is concave on $\mathbb{E}_1^d$. The sum of concave functions is concave, too [9].

As $l \ge 1$, according to first-order Taylor series expansion formula around y_l, we have

$$f_K(y_{l+1}) = f_K(y_l) + \nabla f_K(y^*)^{\mathrm{T}}(y_{l+1} - y_l), \tag{3.33}$$

where $y^* = \lambda y_l + (1 - \lambda)y_{l+1},\ \lambda \in (0, 1)$.

$$\lim_{\lambda \to 1}\left[\nabla f_K(y^*)^{\mathrm{T}}(y_{l+1} - y_l)\right] = \nabla f_K(y_l)^{\mathrm{T}}\frac{R\nabla f_K(y_l)}{f_K(y_l)} = \frac{R\nabla f_K(y_l)^2}{f_K(y_l)} > 0. \tag{3.34}$$

Assuming $f_K(y_{l+1}) \le f_K(y_l)$, by Eq. (3.33),

$$\nabla f_K(y^*)^{\mathrm{T}}(y_{l+1} - y_l) = f_K(y_{l+1}) - f_K(y_l) \le 0 \tag{3.35}$$

It is paradoxical with the hypothesis. Thus, $f_K(y_l)$ is monotonically increasing, i.e.,

$$f_K(y_{l+1}) > f_K(y_l) \tag{3.36}$$

Assuming $f_K(y_l)$ converges to the point $f_K(y_c)$, $\nabla f_K(y_c) = 0$. By Eq. (3.33),

$$\lim_{f_K(y_l) \to f_K(y_c)}(y_{l+1} - y_l) = \frac{R\nabla f_K(y_c)}{f_K(y_c)} = 0 \tag{3.37}$$

So, $\{y_l\}$ is convergent.

When $(\gamma_t^2 - \Sigma_t) \in \mathbb{E}_2^d$, the Hessian matrix is a positive definite matrix and $f_K(y_l)$ is a convex function. The proof is similar to that of "Theorem 1" [6].

Without losing generality, the density of the sequence F points to the direction of the gradient and converges to the peak [6] when the convex set is located at the starting point. However, we have the Hessian $H_{\nabla^2} < 0$ when the gradient is $G_\nabla = 0$. The extreme point lies in the concave set. At the end of the procedure, y_l always falls into the concave set until it converges at the extreme point y_c.

References

1. Bar-Shalom Y, Li XR, Kirubarajan T (2004) Estimation with applications to tracking and navigation. Wiley, New York
2. Bazaraa MS, Shetty CM (1979) Nonlinear programming: theory and algorithms. Wiley, New York
3. Carreira-Perpinan MA (2007) Gaussian mean-shift is an em algorithm. IEEE Trans Pattern Anal Mach Intell 29(5):767–776
4. Chen H, Meer P (2005) Robust fusion of uncertain information. IEEE Trans Syst Man Cybern Part B Cybern 35(3):578–586
5. Cheng Y (1995) Mean shift, mode seeking, and clustering. IEEE Trans Pattern Anal Mach Intell 17(8):790–799
6. Comaniciu D, Meer P (2002) Mean shift: a robust approach toward feature space analysis. IEEE Trans Pattern Anal Mach Intell 24(5):603–619
7. Comaniciu D, Ramesh V, Meer P (2000) Real-time tracking of non-rigid objects using mean shift. In: Proceedings of IEEE conference on computer vision and pattern recognition (CVPR), vol 2. IEEE, New York, pp 142–149
8. Comaniciu D, Ramesh V, Meer P (2003) Kernel-based object tracking. IEEE Trans Pattern Anal Mach Intell 25(5):564–577
9. Elgammal A, Duraiswami R, Davis LS (2003) Efficient kernel density estimation using the fast gauss transform with applications to color modeling and tracking. IEEE Trans Pattern Anal Mach Intell 25(11):1499–1504
10. Fashing M, Tomasi C (2005) Mean shift is a bound optimization. IEEE Trans Pattern Anal Mach Intell 27(3):471–474
11. Fukunaga K, Hostetler L (1975) The estimation of the gradient of a density function, with applications in pattern recognition. IEEE Trans Inf Theory 21(1):32–40
12. Kastella K, Yeary B, Zadra T, Brouillard R, Frangione E (2000) Bias modeling and estimation for GMTI applications. In: Proceedings of the third international conference on information fusion, vol 1. IEEE, New York, pp 1–7
13. Lainiotis DG, Katsikas SK, Likothanasis S (1988) Adaptive deconvolution of seismic signals-performance, computational analysis, parallelism. IEEE Trans Acoust Speech Signal Process 36(11):1715–1734
14. Li W, Leung H (2004) Simultaneous registration and fusion of multiple dissimilar sensors for cooperative driving. IEEE Trans Intell Transp Syst 5(2):84–98
15. Lin X, Bar-Shalom Y, Kirubarajan T (2004) Exact multisensor dynamic bias estimation with local tracks. IEEE Trans Aerosp Electron Syst 40(2):576–590
16. Nabaa N, Bishop RH (1999) Solution to a multisensor tracking problem with sensor registration errors. IEEE Trans Aerosp Electron Syst 35(1):354–363
17. Okello NN, Challa S (2004) Joint sensor registration and track-to-track fusion for distributed trackers. IEEE Trans Aerosp Electron Syst 40(3):808–823
18. Okello N, Ristic B (2003) Maximum likelihood registration for multiple dissimilar sensors. IEEE Trans Aerosp Electron Syst 39(3):1074–1083
19. Peressini AL, Sullivan FE, Uhl JJJ (1993) The mathematics of nonlinear programming. Springer, Berlin
20. Polat E, Ozden M (2006) A nonparametric adaptive tracking algorithm based on multiple feature distributions. IEEE Trans Multimedia 8(6):1156–1163
21. Qi Y (2008) Research on multi-platform multi-sensor registration algorithm. Ph.D. thesis, Shanghai Jiao Tong University
22. Qi Y, Jing Z, Hu S, Zhao H (2008) New method for dynamic bias estimation: Gaussian mean shift registration. Opt Eng 47(2):026401
23. Ristic B, Okello N (2003) Sensor registration in ECEF coordinates using the MLR algorithm. In: Proceedings of the 6th international conference on information fusion, pp 135–142

24. Subbarao R, Meer P (2006) Nonlinear mean shift for clustering over analytic manifolds. In: Proceedings of the IEEE computer society conference on computer vision and pattern recognition (CVPR). IEEE Computer Society, New York, pp 1168–1175
25. Taghavi E, Tharmarasa R, Kirubarajan T, Bar-Shalom Y, Mcdonald M (2016) A practical bias estimation algorithm for multisensor-multitarget tracking. IEEE Trans Aerosp Electron Syst 52(1):2–19
26. van Doorn BA, Blom HA (1993) Systematic error estimation in multisensor fusion systems. In: Optical engineering and photonics in aerospace sensing. International Society for Optics and Photonics, Bellingham, pp 450–461
27. Zheng ZW, Zhu YS (2004) New least squares registration algorithm for data fusion. IEEE Trans Aerosp Electron Syst 40(4):1410–1416
28. Zhou Y, Leung H, Yip, PC (1997) An exact maximum likelihood registration algorithm for data fusion. IEEE Trans Signal Process 45(6):1560–1573

Chapter 4
Target Tracking and Multi-Sensor Fusion with Adaptive Cubature Information Filter

4.1 Introduction

The hidden states of nonlinear dynamic system can be estimated through a series of noisy measurements by Bayesian filter. When the noise is Gaussian, Bayesian filter can be approximated by the extended Kalman filter (EKF). The information form of EKF called extended information filter (EIF) is proposed in [20]. The EIF propagates the information matrix which is the inverse of the error matrix rather than the error matrix itself. Compared to the EKF, the EIF is more effective when the dimension of measurement vector is larger than the state vector and it is effective for multi-sensor fusion [16, 20]. It can be initialized with limited prior knowledge. However, both EKF and EIF can only achieve first-order precision in approximating the nonlinear function with the use of state and measurement Jacobians matrix. A class of filters called sigma points filters (SPF) have been proposed to overcome this drawback. They include the quadrature Kalman filter [5, 13], the unscented Kalman filter [14], the central difference Kalman filter [21], and the cubature Kalman filter [4]. These filters use a set of deterministic sample points named sigma points to capture the mean and covariance instead of approximating the nonlinear function. SPF can get higher approximation accuracy than EKF. The relevant information forms of SPF such as the quadrature information filter (QIF), the unscented information filter (UIF) [27], the central difference information filter (CDIF) [17], and the cubature information filter (CIF) [23] have also been considered. The number of sigma points used is as follows: the QIF used 3^n sigma points, the UIF used $2n+1$ sigma points, the CDIF used $2n+1$ sigma points, and CIF used $2n$ points, where n denotes the dimension of state. Hence, CIF is considered more effective.

CIF suffers from non-positive-definite covariance due to the finite word-length effect that may result in filters divergence. The square root cubature information filter (SRCIF) [2, 8] is proposed to deal with this issue. The square root information filters propagate the square root of covariance instead of the full covariance.

Z. Jing et al., *Non-Cooperative Target Tracking, Fusion and Control*, Information Fusion and Data Science, https://doi.org/10.1007/978-3-319-90716-1_4

SRCIF has the following benefits [3, 12, 18, 28]: (1) improved numerical accuracy; (2) doubled order precision; (3) preservation of symmetry; (4) availability of square roots.

Both CIF and SRCIF assume that the prior knowledge of measurement noise statistics is completely known, however, this information may not be available in practice. Recently, reference [22] proposed a variational Bayesian approximation adaptive Kalman filter (VB-AKF), which can approximate the joint posterior distribution of the state and the noise variances by a factorized free form distribution. The VB-AKF can deal with non-stationary noise with lower computation load. In [22], the noise variance is assumed to be a diagonal matrix and the priori distribution of noise is modelled as the product of inverse Gamma. The VB method is further extended to nonlinear Gaussian filters along with full noise covariance matrix estimation, where the priori distribution of noise variance is modelled by the inverse Wishart distribution [25]. VB method has also been used in other estimation problems [1, 9, 19, 24, 29, 30].

In this chapter, we introduce an adaptive cubature information filter (VB-ACIF) based on variational Bayesian approximation, which is proposed in [10] early. The square root form (VB-ASCIF) of it is also derived. In information nonlinear filters, we focus on the information matrix. We choose Wishart distribution rather than inverse Wishart distribution to approximate the distribution of the inverse of measurement noise variance, as the Wishart distribution is the conjugate prior distribution of the unknown information matrix of a Gaussian model. Through this distribution together with statistical linearization technique, the VB-ACIF that propagates information state vector and information matrix is derived. In order to achieve better numerical characteristics, the square root form of VB-ACIF is obtained by QR decomposition, Cholesky factor update, and efficient least squares. The simulation results show that the proposed approach outperforms the conventional CIF and SRCIF with unknown measurement noise statistics.

The rest of this chapter is organized as follows. Section 4.2 gives a brief description of the standard CIF. Section 4.3 presents the CIF based on variational Bayesian approximation and the square root form of it. Numerical results and analysis are given in Sect. 4.4 and the conclusion is given in Sect. 4.5.

4.2 Bayesian Filter for Unknown Measurement Noise Statistics

We consider the nonlinear state space system form given by

$$x_k = f(x_{k-1}, w_{k-1}) \tag{4.1a}$$

$$z_k = h(x_k, r_k) \tag{4.1b}$$

where x_k is the n-dimension state vector, $f(\cdot)$ is the nonlinear state function, $w_k \sim N(0, Q_k)$ is the Gaussian process noise with zero mean and covariance Q_k, z_k is the d-dimension measurement vector, $h(\cdot)$ is the nonlinear measurement function, $r_k \sim N(0, R_k)$ is the Gaussian measurement noise with zero mean and covariance R_k, the measurement information matrix related to covariance R_k is defined as the inverse of it, that is $\Lambda_k = R_k^{-1}$. The construction of suitable dynamic model for information matrix Λ_k will be discussed later, and the distribution of Λ_k is governed by $\Lambda_k \sim p(\Lambda_k|\Lambda_{k-1})$.

The aim of Bayesian filtering with unknown measurement noise is to estimate the joint posterior distribution $p(x_k, \Lambda_k|Z_k)$ of states x_k and information matrix Λ_k, where Z_k are defined as $Z_k = \{z_1, \ldots, z_k\}$. The recursion solutions for this problem can be summarized as follows:

(1) Initialization: The prior distribution is given by $p(x_0, \Lambda_0)$.
(2) Prediction: The Chapman–Kolmogorov equation provides the connection between prior density $p(x_k, \Lambda_k|Z_{k-1})$ and posterior density, that is

$$
\begin{aligned}
p(x_k, \Lambda_k|Z_{k-1}) = \int & p(x_k, \Lambda_k|x_{k-1}, \Lambda_{k-1}, Z_{k-1}) \\
& \times p(x_{k-1}, \Lambda_{k-1}|Z_{k-1}) dx_{k-1} d\Lambda_{k-1}.
\end{aligned}
\tag{4.2}
$$

Suppose that the state x_k and information matrix Λ_k are independent from each other. Giving the first-order Markov dynamic model (4.1), the above equation can be rewritten as

$$
\begin{aligned}
p(x_k, \Lambda_k|Z_{k-1}) = \int & p(x_k|x_{k-1}) p(\Lambda_k|\Lambda_{k-1}) \\
& \times p(x_{k-1}, \Lambda_{k-1}|Z_{k-1}) dx_{k-1} d\Lambda_{k-1}.
\end{aligned}
\tag{4.3}
$$

(3) Update: When the new measurement comes, the posterior density can be calculated by

$$
p(x_k, \Lambda_k|Z_k) = \frac{p(z_k|x_k, \Lambda_k) p(x_k, \Lambda_k|Z_{k-1})}{p(z_k|Z_{k-1})}
\tag{4.4}
$$

where $p(z_k|x_k, \Lambda_k)$ denotes the likelihood, and $p(z_k|Z_{k-1})$ is the normalization constant computed by

$$
p(z_k|Z_{k-1}) = \int p(z_k|x_k, \Lambda_k) p(x_k, \Lambda_k|Z_{k-1}) dx_{k-1} d\Lambda_{k-1}.
\tag{4.5}
$$

Since it is hard to obtain the analytical solutions for nonlinear systems, approximations such as VB approximation are often used. In the following subsection, a simple introduction to variational approximation will be given.

4.3 Tracking and Fusion with Adaptive Cubature Information Filter Using Variational Approximation

4.3.1 Variational Approximation

Assume that x_k and Λ_k are independent, so the joint posterior distribution can be approximated with the VB approximation by

$$p(x_k, \Lambda_k|Z_k) \approx q(x_k)q(\Lambda_k) \tag{4.6}$$

where $q(x_k)$ and $q(\Lambda_k)$ is unknown approximating density. The log marginal probability of $p(Z_k)$ can be decomposed as

$$\begin{aligned} \ln p(Z_k) = {} & L(q(x_k)q(\Lambda_k)) \\ & + \mathrm{KL}[q(x_k)q(\Lambda_k) \parallel p(x_k, \Lambda_k|Z_k)] \end{aligned} \tag{4.7}$$

where $L(q(x_k)q(\Lambda_k))$ is a lower bound on the log likelihood function $\ln p(Z_k)$ defined by

$$L(q(x_k)q(\Lambda_k)) = \int q(x_k)q(\Lambda_k) \ln\left(\frac{p(x_k, \Lambda_k, Z_k)}{q(x_k)q(\Lambda_k)}\right) dx_k d\Lambda_k, \tag{4.8}$$

and the "KL" means Kullback-Leibler divergence

$$\begin{aligned} & \mathrm{KL}[q(x_k)q(\Lambda_k) \parallel p(x_k, \Lambda_k|Z_k)] \\ & = -\int q(x_k)q(\Lambda_k) \ln\left(\frac{p(x_k, \Lambda_k|Z_k)}{q(x_k)q(\Lambda_k)}\right) dx_k d\Lambda_k. \end{aligned} \tag{4.9}$$

It should be noticed that maximizing lower bound (4.8) is equivalent to minimizing the KL divergence [7], that is

$$\hat{q}(x_k)\hat{q}(\Lambda_k) = \underset{q(x_k)q(\Lambda_k)}{\operatorname{argmin}} \ \mathrm{KL}[q(x_k)q(\Lambda_k) \parallel p(x_k, \Lambda_k|Z_k)]. \tag{4.10}$$

The difference is that maximizing $L(q(x_k)q(\Lambda_k))$ involves the operations on complete-data log-likelihood $p(x_k, \Lambda_k, Z_k)$ rather than the true posterior. Maximization will be obtained by the following equations which is called VB-marginals [26]

$$\ln \hat{q}(x_k) = \underset{\hat{q}(\Lambda_k)}{E} [\ln p(x_k, \Lambda_k, Z_k)] + \mathrm{const}, \tag{4.11a}$$

$$\ln \hat{q}(\Lambda_k) = \underset{\hat{q}(x_k)}{E} [\ln p(x_k, \Lambda_k, Z_k)] + \mathrm{const}. \tag{4.11b}$$

A common way to solve these equations is by iteration such as iterative VB (IVB) algorithm [6]. The iteration can be done by alternating between them until settling at a fixed point. The proof of its convergence is given in [6].

4.3.2 Conventional CIF with Known Measurement Noise

In CIF, the parameters of interest are the information state vector y_k and the associated information matrix Y_k. That is

$$Y_k \triangleq P_k^{-1}, \hat{y}_k \triangleq P_k^{-1}\hat{x}_k = Y_k\hat{x}_k \tag{4.12}$$

where $\hat{x}_k$ is the estimation of the state and P_k is the state error covariance.

Assume that the posterior density function $p(x_{k-1}|z_{1:k-1}) = N(x_{k-1}|\hat{x}_{k-1}, P_{k-1})$ is known at time $k-1$, then the CIF recursion at time k contains two major steps: time update and measurement update.

(1) Time update: factorize

$$P_{k-1} = S_{k-1}S_{k-1}^T \tag{4.13}$$

where $S_{k-1} \in R^{n \times n}$ is a lower triangular matrix called the square root of P_{k-1}.

In time update step, a set of cubature points are generated by

$$\chi_{i,k-1} = S_{k-1}\xi_i + \hat{x}_{k-1} \tag{4.14}$$

where $i = 1, \ldots, m$, $m = 2n$, $\xi_i = \sqrt{n/2} \times [1]_i$, and $[1]_i$ denotes the i-th element of set $[1]$, e.g., if $[1] \in R^2$, then it represents the following set

$$\left\{ \begin{pmatrix} 1 \\ 0 \end{pmatrix}, \begin{pmatrix} 0 \\ 1 \end{pmatrix}, \begin{pmatrix} -1 \\ 0 \end{pmatrix}, \begin{pmatrix} 0 \\ -1 \end{pmatrix} \right\}. \tag{4.15}$$

The predicted cubature points are

$$\chi_{i,k|k-1} = f(\chi_{i,k-1}), \tag{4.16}$$

so the predicted state and predicted error covariance are given by

$$\hat{x}_{k|k-1} = \frac{1}{m}\sum_{i=1}^{m} \chi_{i,k|k-1}, \tag{4.17}$$

$$\begin{aligned} P_{k|k-1} = &\frac{1}{m}\sum_{i=1}^{m} \chi_{i,k|k-1}\chi_{i,k|k-1}^T \\ &- \hat{x}_{k|k-1}\hat{x}_{k|k-1}^T + Q_{k-1}. \end{aligned} \tag{4.18}$$

Then the predicted information state vector and its associated information matrix can be computed by

$$\hat{y}_{k|k-1} = Y_{k|k-1}\frac{1}{m}\sum_{i=1}^{m}\chi_{i,k|k-1}, \tag{4.19}$$

$$\begin{aligned} Y_{k|k-1} &= P_{k|k-1}^{-1} \\ &= \left[\frac{1}{m}\sum_{i=1}^{m}\chi_{i,k|k-1}\chi_{i,k|k-1}^{T} - \hat{x}_{k|k-1}\hat{x}_{k|k-1}^{T} + Q_{k-1}\right]^{-1}. \end{aligned} \tag{4.20}$$

(2) Measurement update: factorize

$$P_{k|k-1} = S_{k|k-1}S_{k|k-1}^{T}. \tag{4.21}$$

A set of cubature points for measurement update are generated by

$$\chi_{i,k|k-1}^{*} = S_{k|k-1}\xi_i + \hat{x}_{k|k-1}. \tag{4.22}$$

The propagated cubature points are given by

$$Z_{i,k|k-1} = h\left(\chi_{i,k|k-1}^{*}\right) \tag{4.23}$$

and predicted measurement is

$$\hat{z}_{k|k-1} = \frac{1}{m}\sum_{i=1}^{m}Z_{i,k|k-1}, \tag{4.24}$$

and the cross covariance matrix is

$$P_{xz,k|k-1} = \frac{1}{m}\sum_{i=1}^{m}\chi_{i,k|k-1}^{*}Z_{i,k|k-1}^{T} - \hat{x}_{k|k-1}\hat{z}_{k|k-1}^{T}. \tag{4.25}$$

By utilizing the statistical linear error propagation methodology, we define pseudo-measurement matrix $\bar{H}_k$ as

$$\bar{H}_k \triangleq P_{k|k-1}^{-1}P_{xz,k|k-1}, \tag{4.26}$$

then the information state contribution i_k and its associated information matrix I_k are derived by

$$i_k = \bar{H}_k^{T}\Lambda_k(\tilde{z}_k + \bar{H}_k\hat{x}_{k|k-1}) \tag{4.27a}$$

$$I_k = \bar{H}_k^{T}\Lambda_k\bar{H}_k \tag{4.27b}$$

where $\tilde{z}_k = z_k - \hat{z}_{k|k-1}$ is innovation. So the updated information state vector and its associated information matrix are

$$\hat{y}_k = \hat{y}_{k|k-1} + i_k, \tag{4.28a}$$

$$Y_k = Y_{k|k-1} + I_k. \tag{4.28b}$$

The state vector and covariance matrix can be recovered by

$$\hat{x}_k = Y_k^{-1} \hat{y}_k, \tag{4.29a}$$

$$P_k = Y_k^{-1}. \tag{4.29b}$$

4.3.3 Variational Approximation for Cubature Information Filter

CIF assumes that the measurement noise is prior known, which is usually unknown in practice. The VB-ACIF algorithm uses VB approximation in the framework of CIF for unknown measurement noise. The state prediction is the same as CIF prediction, and the propagation of measurement information matrix is added. However, in the updated step, the form of VB approximation is given and iterative method is used to solve the coupled joint posterior distribution. The detailed derivations of VB-ACIF algorithm are as follows.

Suppose that the joint posterior distribution of x_k and Λ_k conditional on Z_{k-1} is approximated at time $k-1$ with the following form

$$\begin{aligned} p(x_{k-1}, \Lambda_{k-1}|Z_{k-1}) = N(x_{k-1}|\hat{x}_{k-1}, P_{k-1}) \\ \times \mathrm{W}(\Lambda_{k-1}|\nu_{k-1}, V_{k-1}) \end{aligned} \tag{4.30}$$

where $\mathrm{W}(\Lambda|\nu, V)$ denotes that Λ obeys Wishart distribution governed by the degrees of freedom ν and symmetric, positive defined $d \times d$ scale matrix V. The form of Gaussian and Wishart distribution is given in Appendix.

Wishart distribution is a conjugate prior for the unknown information matrix of a Gaussian with known mean [11]. It is a nature choice that the conjugacy guarantees that the posterior is of the same functional form as the prior. So if we choose the prior approximation distribution as in (4.30) and confirm that the predictive distribution is Gaussian-Wishart, the form of posterior distributions will remain the same.

The state dynamic model (4.1) ensures that the predictive distribution of state is Gaussian. That is,

$$p(x_k|Z_{k-1}) = N(x_k|\hat{x}_{k|k-1}, P_{k|k-1}) \tag{4.31}$$

where $\hat{x}_{k|k-1}$ and $P_{k|k-1}$ can be obtained by (4.17) and (4.18), respectively.

The dynamic model of Λ is chosen to make sure the predictive distribution of measurement information matrix is Wishart. For the dynamic of measurement information matrix, we simply assume a transfer rule for the sufficient statistics of Wishart distribution from the back step to the current step. Similar to [22], the expected measurement information matrix is chosen to be constant, a dynamic model for distribution of measurement information matrix can be given as

$$\nu_{k|k-1} = \rho \nu_{k-1} \tag{4.32a}$$

$$V_{k|k-1} = B V_{k-1} B^T \tag{4.32b}$$

where ρ is a discount parameter satisfying $0 < \rho \leq 1$. The parameter $\rho = 1$ refers to stationary variances with no decay of information and the lower values increase their assumed time-fluctuations with larger decay information. In order to keep the measurement information matrix unchanged, a reasonable choice for B is $B = I/\sqrt{\rho}$, where I is the identity matrix. Given the dynamic model (4.32), the distribution of $p(\Lambda_k|\Lambda_{k-1})$ is

$$p(\Lambda_k|\Lambda_{k-1}) = \mathrm{W}(\Lambda_k|\rho\nu_{k-1}, BV_{k-1}B^T), \tag{4.33}$$

so the expectation is $E(\Lambda_k|\Lambda_{k-1}) = E(\Lambda_{k-1}) = \nu_{k-1}V_{k-1}$. Note the dispersion is increased through discounting the degrees of freedom parameter.

We can see that the predictive steps for state and measurement information matrix are separable and independent. The joint predictive distribution becomes

$$\begin{aligned} p(x_k, \Lambda_k|Z_{k-1}) &= p(x_k|Z_{k-1})p(\Lambda_k|Z_{k-1}) \\ &= N(x_k|\hat{x}_{k|k-1}, P_{k|k-1})\mathrm{W}(\Lambda_k|\nu_{k|k-1}, V_{k|k-1}). \end{aligned} \tag{4.34}$$

Notice that the state and measurement information matrix are coupled in the likelihood $p(z_k|x_k, \Lambda_k)$ and the joint posterior distribution $p(x_k, \Lambda_k|Z_k)$ does not have tractable forms. The VB approximation will be used to obtain the joint posterior distribution. The free-form of VB approximation for joint posterior distribution is given by (4.6), where $q(x_k)$ is a Gaussian distribution and $q(\Lambda_k)$ is a Wishart distribution. The joint distribution of x_k, Λ_k, and Z_k is

$$p(x_k, \Lambda_k, Z_k) = p(x_k, \Lambda_k|Z_k)p(Z_k). \tag{4.35}$$

By substituting (4.4) into (4.35), we have

$$\begin{aligned} p(x_k, \Lambda_k, Z_k) &= \frac{p(z_k|x_k, \Lambda_k)p(x_k, \Lambda_k|Z_{k-1})}{p(z_k|Z_{k-1})}p(Z_k) \\ &= p(z_k|x_k, \Lambda_k)p(x_k, \Lambda_k|Z_{k-1})p(Z_{k-1}). \end{aligned} \tag{4.36}$$

According to (4.34) and (4.36), the VB-marginals (4.11a) and (4.11b) can be rewritten as

$$\ln \hat{q}(x_k) = \underset{\hat{q}(\Lambda_k)}{E}[\ln(p(z_k|x_k, \Lambda_k)p(x_k|Z_{k-1}))]+\text{const} \tag{4.37a}$$

$$\ln \hat{q}(\Lambda_k) = \underset{\hat{q}(x_k)}{E}[\ln(p(z_k|x_k, \Lambda_k)p(\Lambda_k|Z_{k-1}))]+\text{const} \tag{4.37b}$$

where the likelihood $p(z_k|x_k, \Lambda_k)$ is given by

$$p(z_k|x_k, \Lambda_k) = N(z_k|h(x_k), \Lambda_k^{-1}). \tag{4.38}$$

The solutions to VB-marginals cannot be solved directly, but can be obtained iteratively by solving only one of the parameters and fixing the other to its last estimated value.

Fixing the $\hat{q}(\Lambda_k)$, we have

$$\hat{i}_k = \bar{H}_k^T \hat{\Lambda}_k(\tilde{z}_k + \bar{H}_k\hat{x}_{k|k-1}) \tag{4.39}$$

$$\hat{I}_k = \bar{H}_k^T \hat{\Lambda}_k \bar{H}_k \tag{4.40}$$

where

$$\hat{\Lambda}_k = \underset{\hat{q}(\Lambda_k)}{E}(\Lambda_k) = \nu_k V_k \tag{4.41}$$

and $\hat{y}_{k|k-1}$, $Y_{k|k-1}$ can be obtained by (4.19) and (4.20).

Fixing the $\hat{q}(x_k)$, we can obtain

$$\nu_k = \nu_{k|k-1} + 1 \tag{4.42}$$

$$V_k^{-1} = V_{k|k-1}^{-1} + P_{zz,k} + (z_k - \hat{z}_k)(z_k - \hat{z}_k)^T \tag{4.43}$$

and $\hat{z}_k$ is given as

$$\hat{z}_k = \frac{1}{m}\sum_{i=1}^{m} Z_{i,k} \tag{4.44}$$

where

$$Z_{i,k} = f(\chi_{i,k}) \tag{4.45}$$

and $\chi_{i,k}$ is the cubature points

$$\chi_{i,k} = S_k\xi_i + \hat{x}_k \tag{4.46}$$

where S_k is the square root of P_k satisfying $P_k = S_k S_k^T$. $P_{zz,k}$ is given by

$$P_{zz,k} = Z_k^* Z_k^{*T} \tag{4.47}$$

where

$$Z_k^* = \frac{1}{\sqrt{m}} [Z_{1,k} - \hat{z}_k, \ldots, Z_{m,k} - \hat{z}_k]. \tag{4.48}$$

The above procedure is one cycle of the VB-ACIF, which is summarized in Algorithm 1 (see Table 4.1). This method can be applied recursively when new measurements are coming.

4.3.4 *Variational Approximation for Square-Root Cubature Information Filter*

This subsection presents a derivation of square root version for VB-ACIF (VB-SRACIF). The matrix in VB-ACIF such as the information matrix Y_k of state and the sufficient statistics V_k for measurement information matrix should be kept symmetric nonnegative definite in the recursion process. However, the algorithm is always implemented in the finite-precision processor in practice, which will result in round-off errors. The round-off errors may destroy those properties, and the alternative method is propagating only half of the elements of the Y_k and V_k, such as square root factors of them. The use of square root can improve numerical accuracy, achieve doubled order precision, and guarantee symmetric nonnegative definite [15], which will improve the stability of VB-ACIF. Another benefit of square root method for VB-SRACIF is that the square root factor is available explicitly. So the covariance matrix factorization for VB-ACIF in the time and measurement updates are not needed anymore.

To deduce the VB-SRACIF, the following three techniques are used: QR decomposition, Cholesky factor update, and efficient least squares, which we use qr, cholupdate, and "\" to indicate them, respectively.

QR Decomposition The square root of covariance matrix is calculated by Cholesky decomposition on P. The QR decomposition can be defined as follows: the matrix $D^{m\times n}$ with linearly independent columns can be uniquely factored as $D = QR$ in which the columns of $Q^{m\times n}$ are orthonormal basis, and $R^{n\times n}$ is an upper triangular matrix with positive diagonal entries. The operator $qr\{\cdot\}$ returns the upper triangular part R of QR decomposition. If $P = AA^T$ $(A \in R^{n\times m})$ is known, the square root of P can be obtained through QR decomposition. When $A^T = QR$, we have

$$P = AA^T = (QR)^T (QR) = R^T R = SS^T. \tag{4.49}$$

Table 4.1 VB-ACIF algorithm

(1) Time update: Evaluate the predicted time update parameters through (4.13)–(4.20) and (4.32a)–(4.32b).

(2) Measurement update: Calculate the update parameters by (4.26)–(4.25), VB approximation: Set $\hat{y}_k^{(0)} = \hat{y}_{k|k-1}$, $Y_k^{(0)} = Y_{k|k-1}$, $\nu_k = \nu_{k|k-1} + 1$, $V_k^{(0)} = V_{k|k-1}$, iterate the following N steps $j = 1, \ldots, N$

$$\hat{\Lambda}_k^{(j+1)} = \nu_k V_k^{(j)}$$

$$\hat{y}_k^{(j+1)} = \hat{y}_{k|k-1} + \bar{H}_k^T \hat{\Lambda}_k^{(j+1)} (\tilde{z}_k + \bar{H}_k \hat{x}_{k|k-1})$$

$$Y_k^{(j+1)} = Y_{k|k-1} + \bar{H}_k^T \hat{\Lambda}_k^{(j+1)} \bar{H}_k$$

$$P_k^{(j+1)} = (Y_k^{(j+1)})^{-1}, \quad \hat{x}_k^{(j+1)} = P_k^{(j+1)} \hat{y}_k^{(j+1)}$$

$$\chi_{i,k}^{(j+1)} = S_k^{(j+1)} \xi_i + \hat{x}_k^{(j+1)}$$

$$Z_{i,k}^{(j+1)} = f(\chi_{i,k}^{(j+1)}), \quad \hat{z}_k^{(j+1)} = \frac{1}{m} \sum_{i=1}^{m} Z_{i,k}^{(j+1)}$$

$$Z_k^{*,(j+1)} = \frac{1}{\sqrt{m}} [Z_{1,k}^{(j+1)} - \hat{z}_k^{(j+1)}, \ldots, Z_{m,k}^{(j+1)} - \hat{z}_k^{(j+1)}]$$

$$P_{zz,k}^{(j+1)} = Z_k^{*,(j+1)} (Z_k^{*,(j+1)})^T$$

$$(V_k^{(j+1)})^{-1} = V_{k|k-1}^{-1} + P_{zz,k}^{(j+1)} + (z_k - \hat{z}_k^{(j+1)})(z_k - \hat{z}_k^{(j+1)})^T$$

And set $\hat{y}_k = \hat{y}_k^{(N)}$, $Y_k = Y_k^{(N)}$, $V_k = V_k^{(N)}$, $\hat{x}_k = \hat{x}_k^{(N)}$, $P_k = P_k^{(N)}$.

Then the square root of P is $S{=}R^T$ and can be calculated by $S = qr\{A\}^T$. The computational complexity is $O(mn^2)$ for QR decomposition.

Cholesky Factor Update The operator $cholupdate\{u, \pm 1\}$ means the Cholesky factor update for $P{=}SS^T$ with update vector u, that is $chol(SS^T \pm uu^T)$. If u is a matrix, then can update the columns of $cholupdate\{\cdot\}$ in turn. If $u \in R^{n \times m}$, the computational complexity is $O(ML^2)$.

Efficient Least Squares The solution of least squares problem $Ax = b$ is equal to the solution of the equation $(AA^T)x = A^T b$. This can be solved efficiently using QR decomposition by pivoting, which can be implemented using MATLABs leftdivide operator "\",

$$Q_k = S_{Q_k} S_{Q_k}^T, \hat{\Lambda}_k = S_{\hat{\Lambda},k} S_{\hat{\Lambda},k}^T,$$

$$Y_k = S_{Y,k} S_{Y,k}^T, Y_{k|k-1} = S_{Y,k|k-1} S_{Y,k|k-1}^T.$$

We define some notations as follows:

$$P_k = S_k S_k^T, P_{k|k-1} = S_{k|k-1} S_{k|k-1}^T,$$

$$V_k = S_{V,k} S_{V,k}^T, V_{k|k-1} = S_{V,k|k-1} S_{V,k|k-1}^T,$$

$$V_k^{-1} = S_{IV,k} S_{IV,k}^T, V_{k|k-1}^{-1} = S_{IV,k|k-1} S_{IV,k|k-1}^T.$$

The VB-SRACIF algorithms can be summarized in Algorithm 2 (see Tables 4.2 and 4.3).

The following equation can be obtained from (4.20)

$$S_{Y,k|k-1} S_{Y,k|k-1}^T = (S_{k|k-1} S_{k|k-1}^T)^{-1} = S_{k|k-1}^{-T} S_{k|k-1}^{-1}, \tag{4.50}$$

so we have

$$S_{Y,k|k-1} = S_{k|k-1}^{-T}. \tag{4.51}$$

By using efficient least squares, it yields

$$S_{Y,k|k-1} = S_{k|k-1}^T \backslash I \tag{4.52}$$

without matrix inverse. In a similar way, we can obtain

$$S_k^{(j+1)} = (S_{Y,k}^{(j+1)})^T \backslash I, \tag{4.53}$$

$$S_{IV,k|k-1} = S_{V,k|k-1}^{-T} \backslash I, \tag{4.54}$$

$$S_{V,k}^{(j+1)} = (S_{IV,k}^{(j+1)})^T \backslash I. \tag{4.55}$$

From (4.17) and (4.19), we can obtain

$$\hat{y}_{k|k-1} = S_{Y,k|k-1} S_{Y,k|k-1}^T \hat{x}_{k|k-1}. \tag{4.56}$$

Substituting the square root factors on both sides of (4.32b), we get

$$S_{V,k|k-1} S_{V,k|k-1}^T = B S_{V,k-1} S_{V,k-1}^T B^T = B S_{V,k-1} (B S_{V,k-1})^T. \tag{4.57}$$

Table 4.2 VB-ASCIF algorithm

(1) Time update: Evaluate propagated cubature points and predicted state according to (4.14) and (4.17)

$$X_{k|k-1}=\frac{1}{\sqrt{m}}[\chi_{1,k|k-1}-\hat{x}_{k|k-1},\ldots,\chi_{m,k|k-1}-\hat{x}_{k|k-1}]$$

$$S_{k|k-1}=qr\{[\,X_{k|k-1}\;\;S_{Q_{k-1}}\,]^T\}^T$$

$$S_{Y,k|k-1}=S_{k|k-1}^T\backslash I$$

$$\hat{y}_{k|k-1}=S_{Y,k|k-1}S_{Y,k|k-1}^T\hat{x}_{k|k-1}$$

$$\nu_{k|k-1}=\rho\nu_{k-1},\quad S_{V,k|k-1}=BS_{V,k-1}$$

(2) Measurement update:

$$\chi_{i,k|k-1}^{*}=S_{k|k-1}\xi+\hat{x}_{k|k-1}$$

$$Z_{i,k|k-1}=h(\chi_{i,k|k-1}^{*}),\quad \hat{z}_{k|k-1}=\frac{1}{m}\sum_{i=1}^{m}Z_{i,k|k-1}$$

$$X_{k|k-1}^{*}=\frac{1}{\sqrt{m}}[\chi_{1,k|k-1}^{*}-\hat{x}_{k|k-1},\ldots,\chi_{m,k|k-1}^{*}-\hat{x}_{k|k-1}]$$

$$Z_{k|k-1}^{*}=\frac{1}{\sqrt{m}}[Z_{1,k|k-1}-\hat{z}_{k|k-1},\ldots,Z_{m,k|k-1}-\hat{z}_{k|k-1}]$$

$$P_{xz,k|k-1}=X_{k|k-1}^{*}Z_{k|k-1}^{*T}$$

$$\bar{H}_k=S_{Y,k|k-1}S_{Y,k|k-1}^T P_{xz,k|k-1}$$

VB approximation, see Table 4.3.

The measurement update equation is

$$\begin{aligned}\hat{y}_k^{(j+1)}&=\hat{y}_{k|k-1}+\bar{H}_k^T\hat{\Lambda}_k^{(j+1)}(\tilde{z}_k+\bar{H}_k\hat{x}_{k|k-1})\\&=\hat{y}_{k|k-1}+\bar{H}_k^T S_{\hat{\Lambda},k}^{(j+1)}\left(S_{\hat{\Lambda},k}^{(j+1)}\right)^T(\tilde{z}_k+\bar{H}_k\hat{x}_{k|k-1}).\end{aligned}\tag{4.58}$$

Table 4.3 VB approximation of VB-ASCIF

Set $\hat{y}_k^{(0)} = \hat{y}_{k|k-1}$, $S_{Y,k}^{(0)} = S_{Y,k|k-1}$, $\nu_k{=}\nu_{k|k-1} + 1$, $S_{V,k}^{(0)} = S_{V,k|k-1}$, $\nu_{k|k-1}{=}\rho\nu_{k-1}$, iterate the following N steps:

$$S_{\hat{\Lambda},k}^{(j+1)} = \sqrt{\nu_k} S_{V,k}^{(j)}$$

$$\hat{y}_k^{(j+1)} = \hat{y}_{k|k-1} + \bar{H}_k^T S_{\hat{\Lambda},k}^{(j+1)} (S_{\hat{\Lambda},k}^{(j+1)})^T (\tilde{z}_k + \bar{H}_k \hat{x}_{k|k-1})$$

$$S_{Y,k}^{(j+1)} = cholupdate\{S_{Y,k|k-1}, \bar{H}_k^T S_{\hat{\Lambda},k}^{(j+1)}, +\}$$

$$S_k^{(j+1)} = (S_{Y,k}^{(j+1)})^T \backslash I$$

$$\hat{x}_k^{(j+1)} = S_k^{(j+1)} (S_k^{(j+1)})^T \hat{y}_k^{(j+1)}$$

$$\chi_{i,k}^{(j+1)} = S_k^{(j+1)} \xi_i + \hat{x}_k^{(j+1)}$$

$$Z_{i,k}^{(j+1)} = f(\chi_{i,k}^{(j+1)}), \quad \hat{z}_k^{(j+1)} = \frac{1}{m} \sum_{i=1}^{m} Z_{i,k}^{(j+1)}$$

$$S_{IV,k|k-1} = S_{V,k|k-1}^T \backslash I$$

$$Z_k^{*,(j+1)} = \frac{1}{\sqrt{m}} [Z_{1,k}^{(j+1)} - \hat{z}_k^{(j+1)}, \ldots, Z_{m,k}^{(j+1)} - \hat{z}_k^{(j+1)}]$$

$$S_{IV,k}^{(j+1)} = qr\{[\, S_{IV,k|k-1} \; Z_k^{*,(j+1)} \; (z_k - \hat{z}_k^{(j+1)}) \,]\}$$

$$S_{V,k}^{(j+1)} = (S_{IV,k}^{(j+1)})^T \backslash I$$

And set $\hat{y}_k = \hat{y}_k^{(N)}$, $S_{Y,k} = S_{Y,k}^{(N)}$, $S_{V,k} = S_{V,k}^{(N)}$, $\hat{x}_k = \hat{x}_k^{(N)}$, $S_k = S_k^{(N)}$.

The square root of updated information matrix can be derived as follows:

$$\begin{aligned} Y_k^{(j+1)} &= Y_{k|k-1} + \bar{H}_k^T \hat{\Lambda}_k^{(j+1)} \bar{H}_k \\ &= S_{Y,k|k-1} S_{Y,k|k-1}^T + \bar{H}_k^T S_{\hat{\Lambda},k}^{(j+1)} \left(S_{\hat{\Lambda},k}^{(j+1)} \right)^T \bar{H}_k \\ &= S_{Y,k|k-1} S_{Y,k|k-1}^T + \bar{H}_k^T S_{\hat{\Lambda},k}^{(j+1)} (\bar{H}_k^T S_{\hat{\Lambda},k}^{(j+1)})^T. \end{aligned} \tag{4.59}$$

Since $S_{Y,k|k-1}$ is the Cholesky factor of the information matrix $Y_{k|k-1}$, the updated Cholesky factor can be calculated by using the Cholesky update.

Equation (4.43) can be rewritten as

$$\begin{aligned}\left(V_k^{(j+1)}\right)^{-1} &= V_{k|k-1}^{-1} + P_{zz,k}^{(j+1)} + \left(z_k - \hat{z}_k^{(j+1)}\right)\left(z_k - \hat{z}_k^{(j+1)}\right)^T \\ &= S_{IV,k|k-1}S_{IV,k|k-1}^T + Z_k^{*,(j+1)}\left(Z_k^{*,(j+1)}\right)^T \\ &\quad + \left(z_k - \mu_k^{(j+1)}\right)\left(z_k - \mu_k^{(j+1)}\right)^T \\ &= \left[S_{IV,k|k-1}\ Z_k^{*,(j+1)}\ \left(z_k - \mu_k^{(j+1)}\right)\right] \\ &\quad \times \left[S_{IV,k|k-1}\ Z_k^{*,(j+1)}\ (z_k - \mu_k^{(j+1)})\right]^T. \end{aligned} \tag{4.60}$$

For $S_{IV,k}$ is the square root factor of the matrix $(V_k^{(j+1)})^{-1}$, so it can be calculated by using QR decomposition. The sufficient statistics $S_{V,k}$ of Wishart distribution can be recovered by (4.55).

4.3.5 *Multi-Sensor Fusion with Adaptive Cubature Information Filter*

The fused estimation for multi-sensor is just the linear combination of local information contributions, that is

$$\hat{y}_k = \hat{y}_{k|k-1} + \sum_{j=1}^{N_S} i_{j,k}, \tag{4.61a}$$

$$Y_k = Y_{k|k-1} + \sum_{j=1}^{N_S} I_{j,k}. \tag{4.61b}$$

Since the state and measurement information matrix are coupled in update step, multi-sensor state estimation method in (4.61a) and (4.61b) is not suitable here. For a distributed sensor network with N_S sensors, the information state and its associated information matrix are already obtained in the iteration step, so an alternative way of handling this problem is to use the linear combinations of local information state $\hat{y}_{j,k}$ and its associated information matrix $Y_{j,k}$

$$\hat{y}_k = \sum_{j=1}^{N_S} \hat{y}_{j,k} \tag{4.62a}$$

$$Y_k = \sum_{j=1}^{N_S} Y_{j,k} \tag{4.62b}$$

Suppose that we have a distributed sensor network with N_S sensors, the square root form of multiple sensor fusion can be formulated as

$$\hat{y}_k = \sum_{j=1}^{N_S} \hat{y}_{j,k}, \tag{4.63}$$

$$S_k = qr\left\{\left[S_{Y,k}^{(1)}, \ldots, S_{Y,k}^{(N_S)}\right]^T\right\}^T. \tag{4.64}$$

4.4 Simulations

4.4.1 Single Sensor with Constant Acceleration Model

In this subsection, we consider a simple scenario with linear constant acceleration (CA) state model and nonlinear measurement model to verify efficiency of our proposed algorithms. The state $x = [p_x, \dot{p}_x, p_y, \dot{p}_y]^T$ consists of position (p_x, p_y) and velocity $(\dot{p}_x, \dot{p}_y)$. The state transition function follows the linear Gaussian dynamics and is given by

$$x_k = Fx_{k-1} + Gw_{k-1} \tag{4.65}$$

where

$$F = \begin{bmatrix} 1 & T \\ 0 & 1 \end{bmatrix} I_2$$

$$G = \begin{bmatrix} T^2/2 & T & 0 & 0 \\ 0 & 0 & T^2/2 & T \end{bmatrix}^T$$

$$w_{k-1} \sim N\left(0, \begin{pmatrix} w_x^2 & 0 \\ 0 & w_y^2 \end{pmatrix}\right)$$

and $w_x^2 = w_y^2 = 0.1$ is the process noise, T is the sampling period, I_2 is the 2×2 identity matrix.

The target trajectory was generated by model (4.65) with the following true initial state

$$x_0 = [2000\,\text{m}, 10\,\text{m/s}, 10{,}000\,\text{m}, -50\,\text{m/s}]^T$$

In the simulations, initial states for filters were chosen randomly from $N(x_0, P_0)$ in each turn, where

$$P_0 = \text{diag}([100^2\,\text{m}^2, 10^2\,\text{m}^2/\text{s}^2, 100^2\,\text{m}^2, 10^2\,\text{m}^2/\text{s}^2])$$

The sensor is placed at $(p_{x,s}, p_{y,s}) = (0, 0)$ m, and the measurement consists of range and bearing, so the observation with noise is formulated by

$$z_k = \begin{bmatrix} \sqrt{(p_{x,k} - p_{x,s})^2 + (p_{y,k} - p_{y,s})^2} \\ \arctan \dfrac{p_{y,k} - p_{y,s}}{p_{x,k} - p_{x,s}} \end{bmatrix} + r_k \tag{4.66}$$

where the measurement noise covariance $R_k = \text{diag}([\sigma_p^2, \sigma_\theta^2])$.

The SCIF, VB-ACIF, and VB-ASCIF methods were assessed in our simulations. In order to verify the comprehensive performance of the three methods, the constant noise, rectangle noise, and sine noise of measurement were used. For performance comparison, we use the root mean-squared error (RMSE) of position and velocity as performance metric. The RMSE of position at time k is defined as

$$\text{RMSE}_p(k) = \sqrt{\frac{1}{N}\sum_{n=1}^{N}((p_{x,k}^n - \hat{p}_{x,k}^n)^2 + (p_{y,k}^n - \hat{p}_{y,k}^n)^2)}$$

where N denotes the number of Monte Carlo runs, $(p_{x,k}^n, p_{y,k}^n)$ and $(\hat{p}_{x,k}^n, \hat{p}_{y,k}^n)$ denote the true and estimated positions at the n_{th} Monte Carlo run. The definition of RMSE in velocity is similar to the RMSE of position. The averaged RMSE (ARMSE) of position is defined as follows:

$$\text{ARMSE}_p = \sqrt{\frac{1}{N}\sum_{n=1}^{N}\frac{1}{K}\sum_{k=1}^{K}((p_{x,k}^n - \hat{p}_{x,k}^n)^2 + (p_{y,k}^n - \hat{p}_{y,k}^n)^2)}$$

where K is the simulation step of a certain Monte Carlo. The definition of ARMSE in velocity is similar to the RMSE of position.

4.4.1.1 Constant Noise

We assume that measurement noise does not vary with time, and the standard deviations were set as $\sigma_p = 10\,\text{m}$ and $\sigma_\theta = 0.2°$. The measurement noise of SCIF is $R_k = \text{diag}([(10\,\text{m})^2, (0.2°)^2])$, which is the true variance of measurement noise. The initial sufficient statistics for VB-ACIF are $\nu_0 = 1$ and $V_0 = 2R_k^{-1}$. The freedom of VB-ASCIF is the same as VB-ACIF, and $S_{V,0} = \text{chol}(V_0)$. The number of fixed point iterations per time step was 2 for both VB-ACIF and VB-ASCIF.

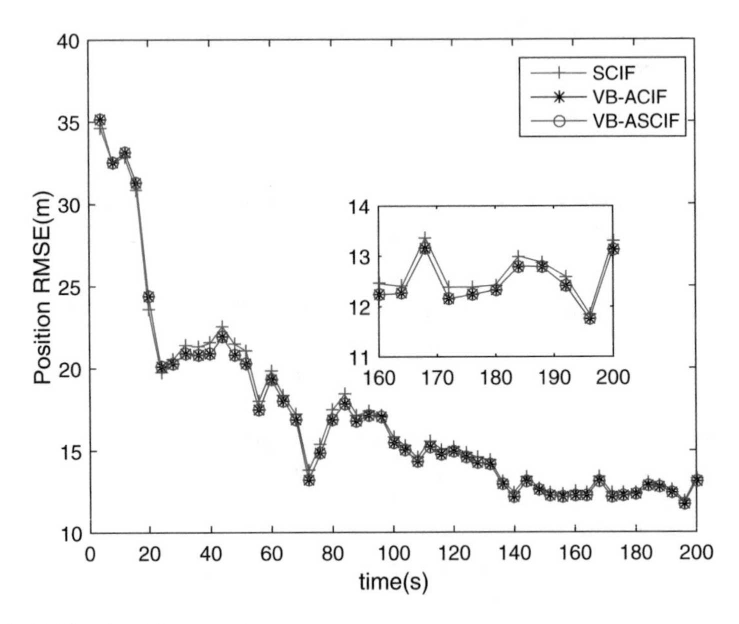

Fig. 4.1 RMSE of position over time

Note that the measurement noise does not change over time, the discount factor $\rho = 1 - \exp(-3)$ is near to 1 for both VB information filters. The sample time is 4 s, and simulation results were obtained by 100 Monte Carlo runs with 50 time steps for each run. Figures 4.1 and 4.2 show RMSE of estimated position and velocity over time. In the figures, RMSEs of SCIF, VB-ACIF, and VB-ASCIF are given. It can be seen that the RMSEs of two VB information filters are almost the same, and both of them are a little smaller than SCIF with true noise. The smaller RMSE indicates the better performance, so results mean that the performance of VB information filters similar and a little better than SCIF. The variance decreasing factor ρ may affect the performance of VB information filters, so the simulations with different ρ were processed. The parameter varies between [0.5, 1], and interval of increased step is 0.02. The corresponding average RMSEs of position and velocity are given in Fig. 4.3. In the figure, the bottom is average RMSE of velocity and the top is average RMSE of position, respectively. From the figure, we can see that the performance becomes better with the higher parameter ρ, although it is not exactly. It can be seen that the good performance can be achieved when the value of ρ is between [0.9, 1].

4.4.1.2 Sine Noise

Suppose that in the beginning, the measurement noise is known as $\sigma_p = 10$ m and $\sigma_\theta = 0.2°$. However, the measurement noise deviation becomes unknown for some reason and is like a sine curve after 50 s. The noise deviation curves are given in

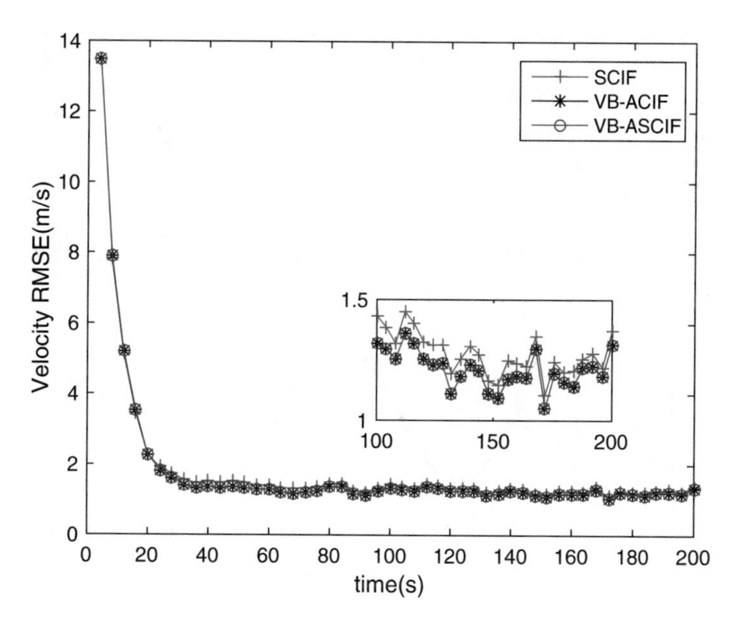

Fig. 4.2 RMSE of velocity over time

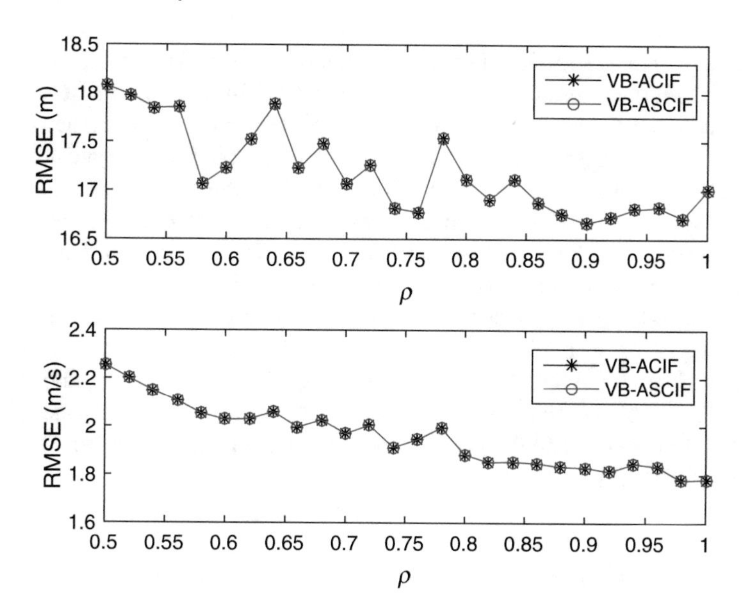

Fig. 4.3 Average RMSE changing with different ρ

Fig. 4.4, where the top is deviation of range and the bottom is deviation of angle. The measurement noise variance of conventional SCIF is $R_k = \text{diag}([(10\,\text{m})^2, (0.2°)^2])$, and the initial sufficient statistics for VB-ACIF and VB-ASCIF are the same as the

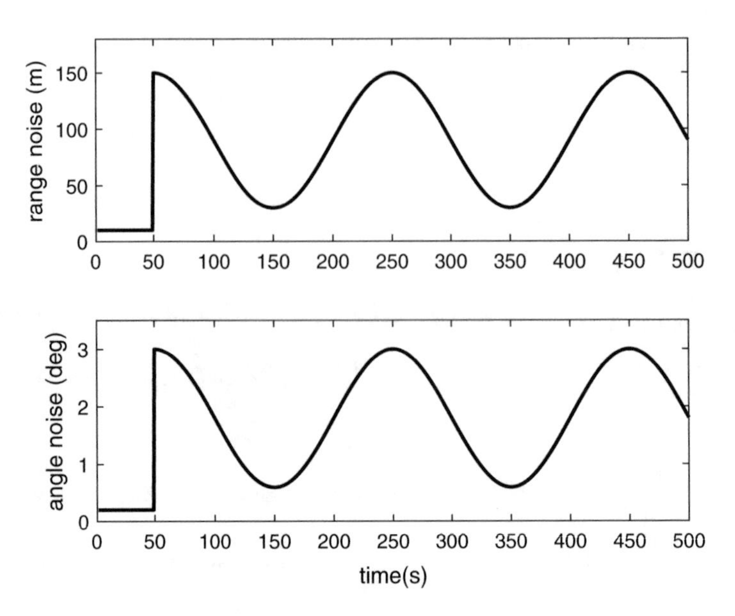

Fig. 4.4 The measurement noise deviation changing with time

first experiment. The discount factor is $\rho = 0.9$ for both VB information filters. And the SCIF with true measurement noise variance (SCIF-t) is used as a benchmark in order to compare the performance of proposed algorithms. Figures 4.5 and 4.6 show RMSE of estimated position and velocity over time. From Figs. 4.5 and 4.6, we can see that the performance of the two VB information filters is almost the same and is much better than the performance of conventional SCIF. Besides, the RMSE of two VB information filters follows the benchmark SCIF-t closely with a little delay and the peak errors of VB information filters are smaller than SCIF-t. The average RMSEs of position and velocity changing with variance decreasing factor ρ are given in Fig. 4.7 for both VB-ACIF and VB-ASCIF. The parameter settings are the same as before. In the figure, the bottom is average RMSE of velocity and the top is average RMSE of position, respectively. It can be seen that the performance changes with the parameter ρ, and the good performance can be achieved when the value is between [0.82, 0.9].

4.4.1.3 Rectangle Noise

Similar to the previous part, the measurement noise is known as $\sigma_p = 10$m and $\sigma_\theta = 0.2^\circ$ at the beginning. The measurement noise deviation becomes unknown with rectangle shape after 50 s and the curves are given in Fig. 4.8. In the figure, the top is deviation of range and the bottom is deviation of angle. The measurement noise variance of conventional SCIF is $R_k = \text{diag}([(10\,\text{m})^2, (0.2^\circ)^2])$. The initial parameters for VB-ACIF and VB-ASCIF are the same as before, and the discount

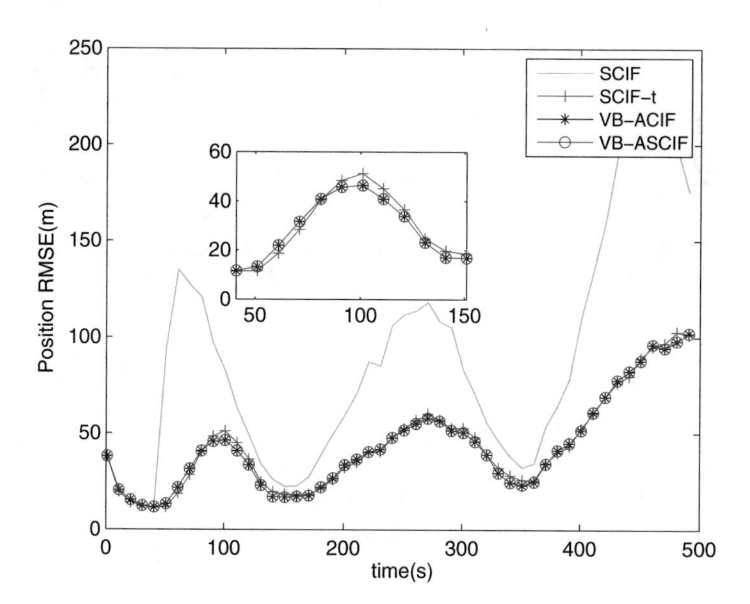

Fig. 4.5 RMSE of position over time

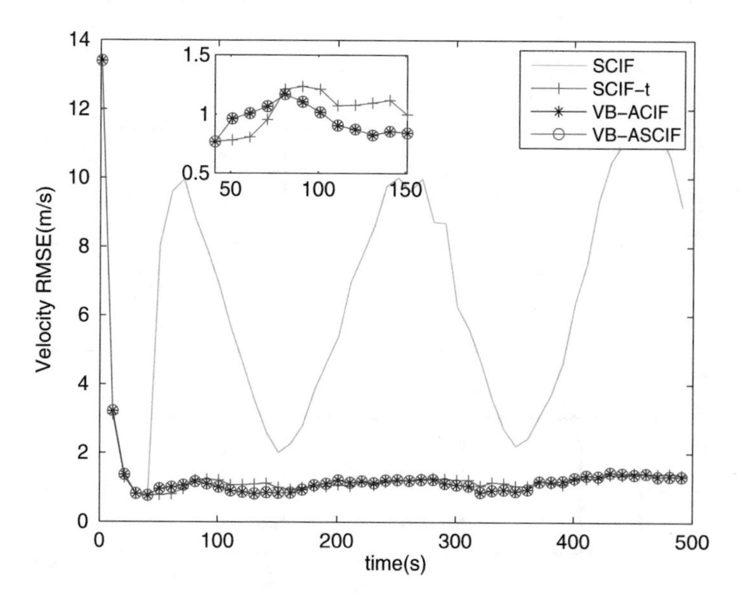

Fig. 4.6 RMSE of velocity over time

factor is $\rho = 0.82$ for both VB information filters. The SCIF with true measurement noise variance (SCIF-t) is used as a benchmark, too. Figures 4.9 and 4.10 show the RMSE of estimated position and velocity. The RMSE of position and velocity for conventional SCIF is much lager than the other filters when the noise ascends,

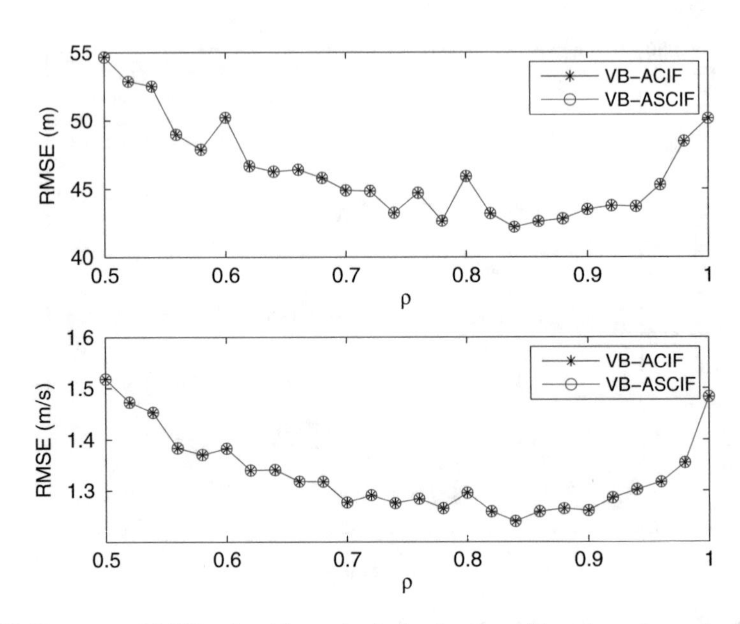

Fig. 4.7 The average RMSEs of position and velocity changing with variance decreasing factor

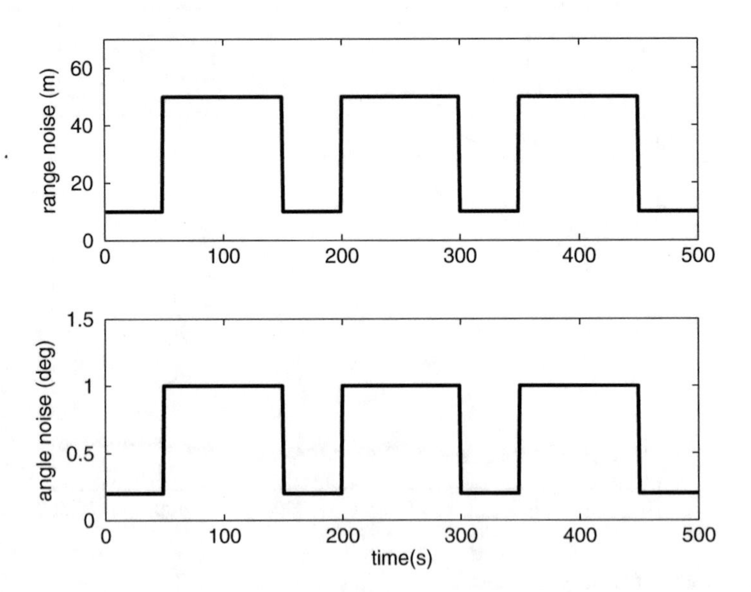

Fig. 4.8 The measurement noise deviation changing with time

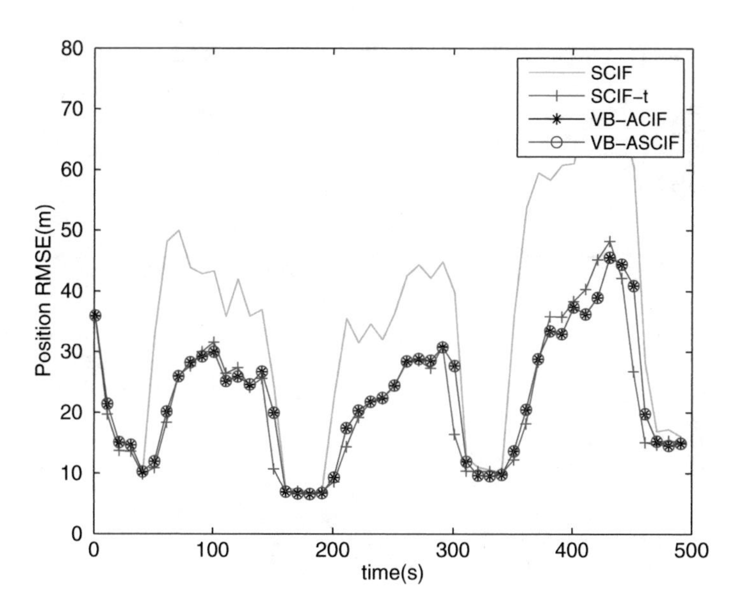

Fig. 4.9 RMSE of position over time

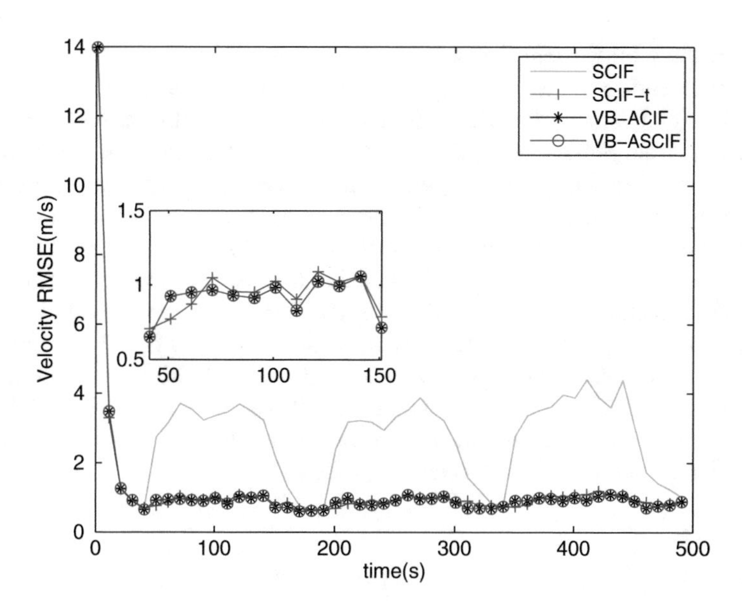

Fig. 4.10 RMSE of velocity over time

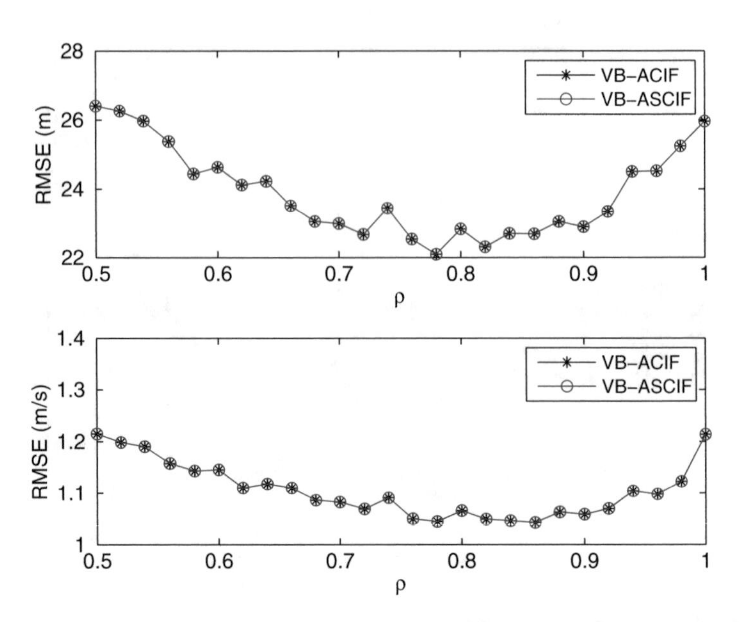

Fig. 4.11 The average RMSEs of position and velocity changing with variance decreasing factor

however, the RMSE become is smaller and is close to the other filters when the measurement noise decreases. This can be seen from Figs. 4.9 and 4.10. The RMSE of position and velocity for VB-ACIF and VB-ASCIF is almost the same, which can be seen from Figs. 4.9 and 4.10, too. As can be seen from Fig. 4.9, the position RMSE of VB information filters ascends and decreases a little slower compared to the benchmark SCIF-t when the true noise ascends and decreases. However, the velocity RMSE of VB information filters ascends slower and decreases quicker, which can be seen from Fig. 4.10. Meanwhile, the peak errors of VB information filters are smaller than SCIF-t. The average RMSEs of position and velocity changing with variance decreasing factor ρ are given in Fig. 4.7 for both VB-ACIF and VB-ASCIF, where the bottom is average RMSE of velocity and the top is average RMSE of position, respectively. It can be seen from Fig. 4.11 that the performance changes with the parameter ρ, and the good performance can be achieved when the value is between [0.74, 0.86].

4.4.2 *Multi-Sensor Fusion Tacking with Coordinated Turning Model*

A more challenging scenario is considered in this subsection. The target moves in a horizontal plane and undergoes a constant turn maneuver with unknown turn rate ω. The target dynamic can be modeled by a coordinated turning model with unknown turn rate

$$x_k = F_{ct} x_{k-1} + G_{ct} w_{k-1} \tag{4.67}$$

where

$$F_{ct} = \begin{bmatrix} 1 & \frac{\sin \omega T}{\omega} & 0 & \frac{\cos \omega T - 1}{\omega} & 0 \\ 0 & \cos \omega T & 0 & -\sin \omega T & 0 \\ 0 & \frac{1 - \cos \omega T}{\omega} & 1 & \frac{\sin \omega T}{\omega} & 0 \\ 0 & \sin \omega T & 0 & \cos \omega T & 0 \\ 0 & 0 & 0 & 0 & 1 \end{bmatrix}$$

$$G_{ct} = \begin{bmatrix} T^2/2 & T & 0 & 0 & 0 \\ 0 & 0 & T^2/2 & T & 0 \\ 0 & 0 & 0 & 0 & T \end{bmatrix}^T$$

where state is $x = [p_x, \dot{p}_x, p_y, \dot{p}_y, \omega]^T$, $w_k \sim N(0, Q_k)$, $Q_k = \text{diag}([w_x^2, w_y^2, w_\omega^2])$, $w_x^2 = w_y^2 = 0.1$, $w_\omega^2 = \exp(-5)$, sample time T = 1 s.

The target trajectory was generated by coordinated turning model (4.67) with the following true initial state

$$x_0 = [1000\,\text{m}, 200\,\text{m/s}, 1000\,\text{m}, 0\,\text{m/s}, 1.5°/\text{s}]^T$$

In the simulations, initial states for filters were chosen randomly from $N(x_0, P_0)$ in each turn, where

$$P_0 = \text{diag}([100^2\,\text{m}^2, 10^2\,\text{m}^2/\text{s}^2, 100^2\text{m}^2, 10^2\,\text{m}^2/\text{s}^2, (0.1°)^2])$$

There are three sensors which are randomly located in the area $[0, 7000]\,\text{m} \times [0, 7000]\,\text{m}$. Defining a measurement noise variance $R_0 = \text{diag}([(10\,\text{m})^2, (0.2°)^2])$, the true measurement noise variances can be modeled by a staircase function

$$R_k = \begin{cases} R_0 & if \quad t < t_m \\ \alpha R_0 & \text{else} \end{cases} \tag{4.68}$$

where the time parameters t_m for three sensors are 10 s, 15 s, and 20 s, respectively; the scale parameters α for three sensors are 10, 8, and 5, respectively.

The SCIF, VB-ACIF, and VB-ASCIF method for multi-sensor fusion were compared in the simulations, and benchmark is the SCIF for multi-sensor with true noise (SCIF-t). The estimation precision was assessed by RMSE of position, velocity, and turn rate, where RMSE in turn rate was defined similar to the previous two. In order to compare the numerical robustness of information filters, we introduced the divergence rate. The filter was declared to diverge when RMSE of position at a certain Monte Carlo run $\mathrm{RMSE}_p(n)$ was larger than 200 m. $\mathrm{RMSE}_p(n)$ is given by

$$\mathrm{RMSE}_p(n) = \sqrt{\frac{1}{K}\sum_{k=1}^{K}((p_{x,k}^n - \hat{p}_{x,k}^n)^2 + (p_{y,k}^n - \hat{p}_{y,k}^n)^2)}$$

Those diverged runs were removed from the final calculations of ARMSE and RMSE at time k. The measurement noise variance of conventional SCIF is R_0. The initial parameters for VB-ACIF are $\nu_0 = 1$ and $V_0 = 2R_0^{-1}$. The freedom of VB-ASCIF is the same as VB-ACIF, and $S_{V,0} = \mathrm{chol}(V_0)$. The number of fixed point iterations per time step was 3 and the discount factor is $\rho = 0.85$ for both VB information filters. Simulation results were obtained by 500 Monte Carlo runs with 50 time steps for each run, and the results were shown in Figs. 4.12, 4.13, 4.14, and Table 4.4.

Table 4.4 gives comparisons of the four filters in ARMSE and the number of divergences. It can be seen from Table 4.4 that the ARMSE of VB-ASCIF is the smallest and ARMSE of conventional SCIF is the largest. The ARMSE of VB-ACIF is smaller than SCIF-t, and this is because that the square root form has more numerical precision especially in the strong nonlinearity situation. The ARMSE of SCIF-t is larger than the VB information filters, this may be because that the fusion rule of VB information filters is more suitable for the changed measurement noise. The numbers of diverges for SCIF, SCIF-t, VB-ACIF, and VB-ASCIF are 80,

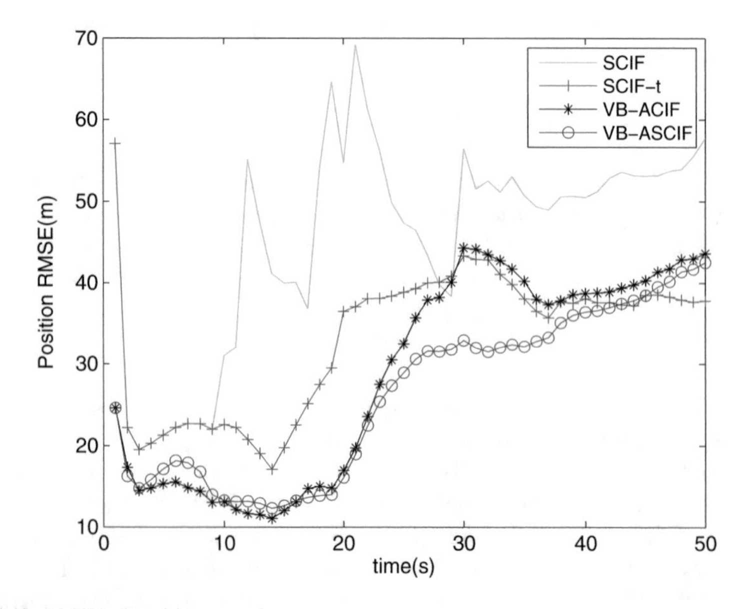

Fig. 4.12 RMSE of position over time

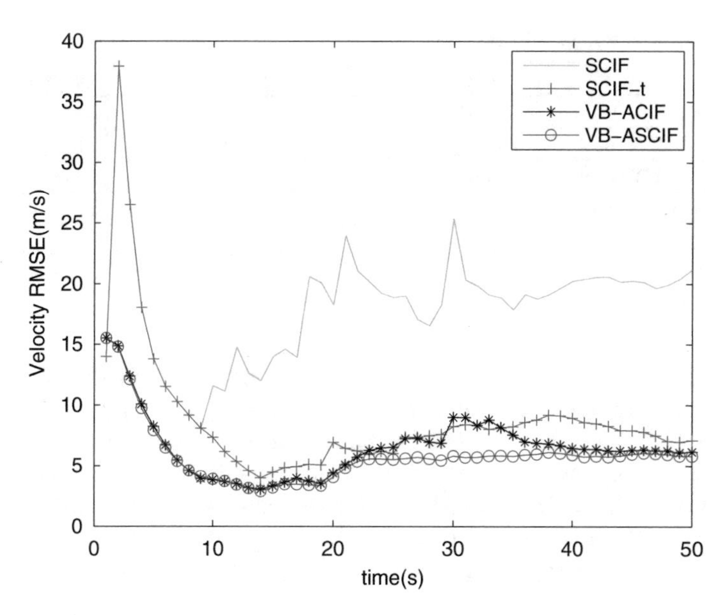

Fig. 4.13 RMSE of velocity over time

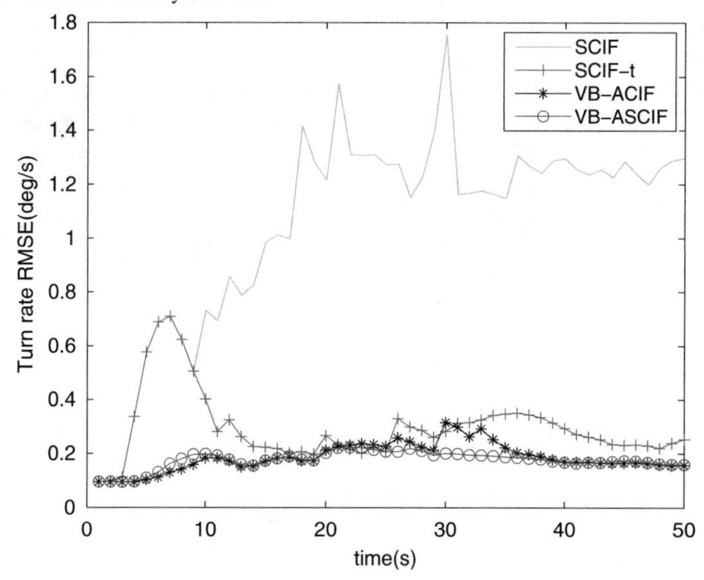

Fig. 4.14 RMSE of turn rate over time

25, 2, and 2. It means the robustness of VB information filters is much better than SCIF and SCIF-t, which benefits from the adaptive laws of VB. So the performance of VB-ASCIF is the best of the four methods both in estimated precision and robustness.

Table 4.4 Comparison of different methods

Method	SCIF	SCIF-t	VB-ACIF	VB-ASCIF
Position ARMSE (m)	45.65	33.10	28.66	26.23
Velocity ARMSE (m/s)	18.17	8.80	6.55	5.83
turn rate ARMSE (deg)	1.06	0.298	0.183	0.178
Number of divergences	80	25	2	2

4.5 Conclusion

In this chapter, we have introduced an adaptive cubature information filter based on cubature integration rule and variational Bayesian approximation for Gaussian nonlinear state space models. Unlike the traditional cubature information filter, the presented algorithms can recursively estimate both the information state and the measurement information matrix when the noise statistics is unknown. The consequent VB-ASCIF was derived for the improved numerical precision and stability. The proposed algorithms were tested on two types of dynamic models: one is the linear state model with nonlinear model and the other is the nonlinear state model with nonlinear measurement model. In the first scenario, the two proposed VB information filters performed much better than the traditional CIF for different unknown noise. Besides, the performance of the two VB information filters was almost the same and close to CIF-t. This is because that the VB information filters can estimate the unknown measurement information matrix. In the second scenario, three sensors were used to verify the performance of adaptive filters. In the simulations, the VB-SCIF has the best performance benefiting from propagation of the square root of both the state information matrix and the scale matrix. The two VB information filter performed better than the traditional CIF and CIF-t.

Appendix

The Gaussian and Wishart distributions have the form

$$
\begin{aligned}
N(x|\hat{x}, P) &= (2\pi)^{-d/2}|P|^{-1/2} \\
&\quad \times \exp\left(-\frac{1}{2}\mathrm{tr}((x-\hat{x})(x-\hat{x})^T P^{-1})\right)
\end{aligned} \tag{4.69}
$$

$$
\mathrm{W}(\Lambda|\nu, V) = C|\Lambda|^{(\nu-d-1)/2}\exp\left(-\frac{1}{2}\mathrm{tr}(V^{-1}\Lambda)\right) \tag{4.70}
$$

where $|\cdot|$ denotes the determinant, $\mathrm{tr}(\cdot)$ denotes the trace operator, and C is the normalization constant

$$C(\nu, V) = |\Lambda|^{-\nu/2} \times \left(2^{\nu d/2}\pi^{d(d-1)/4}\prod_{i=1}^{d}\Gamma\left(\frac{\nu+1-i}{2}\right)\right)^{-1} \tag{4.71}$$

where $\Gamma(\cdot)$ denotes the gamma function. It should noticed that the scale matrix $V > 0$ and $\nu > d - 1$ are well known for the normal Wishart distribution.

References

1. Agamennoni G, Nieto JI, Nebot EM (2012) Approximate inference in state-space models with heavy-tailed noise. IEEE Trans Signal Process 60(10):5024–5037
2. Arasaratnam I (2013) Sensor fusion with square-root cubature information filtering. Intell Control Autom 04(1):11–17
3. Arasaratnam I, Haykin S (2008) Square-root quadrature Kalman filtering. IEEE Trans Signal Process 56(6):2589–2593
4. Arasaratnam I, Haykin S (2009) Cubature Kalman filters. IEEE Trans Autom Control 54(6):1254–1269
5. Arasaratnam I, Haykin S, Elliott RJ (2007) Discrete-time nonlinear filtering algorithms using Gauss-Hermite quadrature. Proc IEEE 95(5):953–977
6. Beal MJ (2003) Variational algorithms for approximate Bayesian inference. University of London, London
7. Bishop CM (2006) Pattern recognition and machine learning (information science and statistics). Springer, New York
8. Chandra KPB, Gu DW, Postlethwaite I (2013) Square root cubature information filter. IEEE Sensors J 13(2):750–758
9. Christmas J, Everson R (2011) Robust autoregression: student-t innovations using variational Bayes. IEEE Trans Signal Process 59(1):48–57
10. Dong P, Jing Z, Leung H, Shen K (2017) Variational Bayesian adaptive cubature information filter based on Wishart distribution. IEEE Trans Autom Control 63(12):1–8
11. Gelman A, Carlin JB, Stern HS, Rubin DB (2014) Bayesian data analysis, vol 2. Taylor & Francis, Boca Raton
12. Huang Y, Zhang Y, Li N, Zhao L (2015). Improved square-root cubature information filter. Trans Inst Meas Control. https://doi.org/10.1177/0142331215608428
13. Ito K, Xiong K (2000) Gaussian filters for nonlinear filtering problems. IEEE Trans Autom Control 45(5):910–927
14. Julier SJ, Uhlmann JK (2004) Unscented filtering and nonlinear estimation. Proc IEEE 92(3):401–422
15. Kailath T, Sayed AH, Hassibi B (2000) Linear estimation, vol. 1. Prentice Hall, Upper Saddle River
16. Lee DJ (2008) Nonlinear estimation and multiple sensor fusion using unscented information filtering. IEEE Signal Process Lett 15:861–864
17. Liu G, Worgotter F, Markelic I (2011) Nonlinear estimation using central difference information filter. In: IEEE statistical signal processing workshop (SSP), pp 593–596
18. Liu G, Worgotter F, Markelic I (2012) Square-root sigma-point information filtering. IEEE Trans Autom Control 57(57):2945–2950
19. Mbalawata IS, Srkk S, Vihola M, Haario H (2015) Adaptive metropolis algorithm using variational Bayesian adaptive Kalman filter. Comput Stat Data Anal 83:101–115

20. Mutambara AGO (1998) Decentralized estimation and control for multisensor systems. CRC Press, Boca Raton
21. Nrgaard M, Poulsen NK, Ravn O (2000) New developments in state estimation for nonlinear systems. Automatica 36(11):1627–1638
22. Nummenmaa (2009) A Recursive noise adaptive Kalman filtering by variational Bayesian approximations. IEEE Trans Autom Control 54(3):596–600
23. Pakki K, Chandra B, Gu DW, Postlethwaite I (2011) Cubature information filter and its applications. In: Proceedings of the American control conference, pp 3609–3614
24. Safarinejadian B, Estahbanati ME (2016) A novel distributed variational approximation method for density estimation in sensor networks. Measurement 89:78–86
25. Sarkka S, Hartikainen J (2013) Non-linear noise adaptive Kalman filtering via variational Bayes. In: IEEE international workshop on machine learning for signal processing (MLSP), pp 1–6
26. Smidl V, Quinn, A (2006) The variational Bayes method in signal processing. Springer, Berlin
27. Vercauteren T, Wang X (2005) Decentralized sigma-point information filters for target tracking in collaborative sensor networks. IEEE Trans Signal Process 53(8):2997–3009
28. Wang S, Feng J, Chi KT (2014) A class of stable square-root nonlinear information filters. IEEE Trans Autom Control 59(7):1893–1898. https://doi.org/10.1109/tac.2013.2294619
29. Xu D, Shen C, Shen F (2014) A robust particle filtering algorithm with Non-Gaussian measurement noise using student-t distribution. IEEE Signal Process Lett 21(1):30–34
30. Zhu H, Leung H, He Z (2013) State estimation in unknown non-gaussian measurement noise using variational Bayesian technique. IEEE Trans Aerosp Electron Syst 49(49):2601–2614

Chapter 5
Multi-Target Tracking with Gaussian Mixture CPHD Filter and Gating Technique

5.1 Introduction

Multi-target tracking methods based on random finite set (RFS) [7, 12, 14, 15, 17, 19, 20, 23, 24] have drawn much attention in recent years. Multi-target state and measurement are modeled by random finite sets in the RFS formulation for multi-target tracking. Therefore, mathematical tools provided by finite set statistics (FISST)[7] can be used to extend the multi-target tracking problem of Bayesian inference. Compared with the traditional association-based multi-target tracking methods, the difficulties caused by data association can be avoided. However, the multi-target Bayesian recursive algorithm proposed by R. Mahler is hard to solve in practice [12]. The probability hypothesis density (PHD) is the first moment filter proposed in [12] to be associated with multi-target posteriori. PHD filter is an easier method to handle the optimal multi-target filtering method. Because the intensity function's domain is a single-target state space, its propagation is much computation less than multi-target posteriori. The PHD filter implementation includes sequential Monte Carlo [21] (SMC-PHD) and Gaussian mixture PHD [5, 20] (GM-PHD). In particular, the GM-PHD filter is a promising implementation which has an easy-to-realize peak extraction. The GM-PHD filter with nonlinear dynamical models was proposed in [6]. The so-called CPHD filter was presented in [13, 14, 16]. The CPHD filter propagates the entire probability distribution on target number while the PHD filter only propagates the first moment of target intensity. In addition, the CPHD filter can adapt more general false alarm models such as i.i.d. (independent, identically distribution) cluster processes [3, 24] rather than the Poisson process. It can provide more accurate estimates of target number and the states of the targets than the PHD filter. The cost of this additional capability is a higher amount of computation. References [22, 24] pointed out that the Gaussian mixture implementation of the CPHD (GM-CPHD) filter outperforms joint probabilistic data association (JPDA) filter in cluttered environment. It has successfully applied to track multiple ground moving targets using road-map information [19]. Besides,

Z. Jing et al., *Non-Cooperative Target Tracking, Fusion and Control*,
Information Fusion and Data Science, https://doi.org/10.1007/978-3-319-90716-1_5

for the single-target case the GM-CPHD filter is equivalent to the multihypothesis tracker with a sequential likelihood test for track extraction, which has been proven in [19].

In fact, computational cost of the CPHD filter can be reduced by reducing the number of measurements (cardinality) in measurement set. The gating techniques used in traditional tracking algorithms [1, 2, 4, 10, 18] can be utilized to reduce the cardinality of measurement set. In this chapter, the elliptical gating technique is incorporated in the GM-CPHD filter to reduce the computational cost. However, it is difficult to detect new born targets since elliptical gating may eliminate all measurements which are not associated with the detected targets. In order to solve this problem, the CPHD filter birth-target model accounting for the possibility of newly-appearing targets is used. Simulation results show that this method can greatly improve the computational efficiency without losing too much estimation accuracy.

This chapter is organized as follows. Section 5.2 gives an overview of the GM-CPHD recursion. The GM-CPHD filter with gating technique is proposed in Sect. 5.3. Demonstrations and numerical studies are given in Sect. 5.4. The presentation of this chapter is based on the work [25, 26].

5.2 Gaussian Mixture CPHD Filter

In the framework of the FISST, the multi-target state and multi-target measurement are modeled by random finite sets (RFS)

$$X_k = \left\{x_{k,1}, x_{k,2}, \ldots, x_{k,N_k}\right\} \subset F(\mathrm{X}), \tag{5.1}$$

$$Z_k = \left\{z_{k,1}, z_{k,2}, \ldots, z_{k,M_k}\right\} \subset F(\mathrm{Z}), \tag{5.2}$$

$$\mathrm{Z}_{1:k} = \bigcup_{i=1}^{k} Z_i, \tag{5.3}$$

where $F(\mathrm{X})$ is the collection of finite subset of X, $F(\mathrm{Z})$ is the collection of finite subset of Z, N_k is the number of targets at time k, and M_k is the number of measurement at time k.

In order to deal with limitations of the PHD filter, Mahler [13–15] proposed the CPHD filter. It propagates the entire probability distribution function on target number while the PHD filter only propagates the PHD. A closed-form solution to the CPHD recursion called the Gaussian mixture CPHD filter is proposed in [22, 24]. It requires additional linear Gaussian multi-target models besides the assumptions of the original CPHD filter. The class of linear Gaussian multi-target models consists of standard linear Gaussian assumptions of the transition and observation models of individual targets, as well as certain assumptions on the birth, death, and detection of targets.

Suppose that each target follows a linear Gaussian dynamical model, i.e.,

$$f_{k|k-1}(x|\zeta) = N\left(x; F_{k-1}\zeta, Q_{k-1}\right), \tag{5.4}$$

$$g_k(z|x) = N\left(z; H_k x, R_k\right), \tag{5.5}$$

where $N(\cdot; m, P)$ is a Gaussian density with mean m and covariance P, F_{k-1} denotes the state transition matrix, Q_{k-1} denotes the process noise covariance, H_k denotes the observation matrix, and R_k denotes the measurement noise covariance, k represents the time index.

The survival and detection probabilities are assumed to be independent of the state, i.e.,

$$p_{s,k}(x) = p_{s,k},$$

$$p_{D,k}(x) = p_{D,k}.$$

The intensity of the birth RFS at time k is a Gaussian mixture with the form

$$\gamma_k(x) = \sum_{i=1}^{J_{\gamma,k}} \omega_{\gamma,k}^{(i)} N\left(x; m_{\gamma,k}^{(i)}, P_{\gamma,k}^{(i)}\right), \tag{5.6}$$

where $\omega_{\gamma,k}^{(i)}$, $m_{\gamma,k}^{(i)}$, and $P_{\gamma,k}^{(i)}$ are the shape parameters of the Gaussian mixture intensity. Let $D_{k|k-1}$ and D_k be the intensities associated with the multi-target predicted density $p(X_k|Z_{1:k-1})$ and the posterior density $p(X_k|Z_{1:k})$ in the recursion Eqs. (5.1)–(5.2), respectively. Besides, we use $p_{k-1}(n)$ and $p_{k|k-1}(n)$ to denote the posterior cardinality distribution and predicted cardinality distribution, respectively. For the linear Gaussian multi-target model, the GM-CPHD filter [24] can be given by following two steps.

Prediction Given the posterior intensity $D_{k-1}(x)$ and posterior cardinality distribution $p_{k-1}(n)$ at time k, and $D_{k-1}(x)$ is a Gaussian mixture with the form

$$D_{k-1}(x) = \sum_{j=1}^{J_{k-1}} \omega_{\gamma,k-1}^{(i)} N\left(x; m_{\gamma,k-1}^{(i)}, P_{\gamma,k-1}^{(i)}\right). \tag{5.7}$$

Then, the predicted intensity $D_{k|k-1}(x)$ is also a Gaussian mixture, and the CPHD prediction is given by

$$p_{k|k-1}(n) = \sum_{i=0}^{n} p_{\Gamma,k}(n-i) \sum_{l=i}^{\infty} C_i^l p_{k-1}(l) p_{s,k}^i (1-p_{s,k})^{l-i}, \tag{5.8}$$

$$D_{k|k-1}(x) = D_{s,k|k-1}(x) + \gamma_k(x), \tag{5.9}$$

where $p_{\Gamma,k}(\cdot)$ is the cardinality distribution of births at time k, $\gamma_k(x)$ is the intensity of spontaneous births at time k,

$$D_{s,k|k-1}(x) = p_{s,k} \sum_{j=1}^{J_{k-1}} \omega_{k-1}^{(j)} N\left(x; m_{s,k|k-1}^{(j)}, P_{s,k|k-1}^{(j)}\right),$$

$$m_{s,k|k-1}^{(j)} = F_{k-1} m_{k-1}^{(j)},$$

$$P_{s,k|k-1}^{(j)} = Q_{k-1} + F_{k-1} P_{k-1}^{(j)} F_{k-1}^{\mathrm{T}}.$$

Update Given the predicted intensity $D_{k|k-1}(x)$ and predicted cardinality distribution $p_{k|k-1}(n)$ at time k, and $D_{k|k-1}(x)$ is a Gaussian mixture with the form

$$D_{k|k-1}(x) = \sum_{j=1}^{J_{k|k-1}} \omega_{k|k-1}^{(j)} N\left(x; m_{k|k-1}^{(j)}, P_{k|k-1}^{(j)}\right). \tag{5.10}$$

Then, the posterior intensity is also a Gaussian mixture, and the CPHD update is given by

$$p_k(n) = \frac{\Upsilon_k^0\left[\omega_{k|k-1}, Z_k\right](n) p_{k|k-1}(n)}{\left\langle \Upsilon_k^0\left[\omega_{k|k-1}, Z_k\right], p_{k|k-1}\right\rangle}, \tag{5.11}$$

$$D_k(x) = \frac{\left\langle \Upsilon_k^1\left[\omega_{k|k-1}, Z_k\right], p_{k|k-1}\right\rangle}{\left\langle \Upsilon_k^0\left[\omega_{k|k-1}, Z_k Z_k\right], p_{k|k-1}\right\rangle}(1 - p_{D,k}) D_{k|k-1}(x)$$

$$+ \sum_{z \in Z_k} \sum_{j=1}^{J_{k|k-1}} \omega_k^{(i)} N\left(x; m_k^{(i)}, P_k^{(i)}\right), \tag{5.12}$$

where

$$\Upsilon_k^u\left[\omega, Z\right](n) = \sum_{j=0}^{\min(|Z|,n)} (|Z| - j)! p_{K,k}\left(|Z| - j\right)$$

$$P_{j+u}^n \frac{(1 - p_{D,k})^{n-(j+u)}}{\langle 1, \omega\rangle^{j+u}} \sigma_j\left(\Lambda_k(\omega, Z)\right),$$

$$\Lambda_k(\omega, Z) = \left\{ \frac{\langle 1, \kappa_k\rangle}{\kappa_k(z)} p_{D,k} \omega^{\mathrm{T}} q_k(z) : z \in Z \right\}, \tag{5.13}$$

$$\omega_{k|k-1} = \left[\omega_{k|k-1}^{(1)}, \omega_{k|k-1}^{(2)}, \ldots, \omega_{k|k-1}^{(J_{k|k-1})}\right]^{\mathrm{T}},$$

$$q_k(z) = \left[q_k^{(1)}(z), q_k^{(2)}(z), \ldots, q_k^{(J_{k|k-1})}(z)\right],$$

$$q_k^{(j)}(z) \sim N(z; H_k m_{k|k-1}^{(j)}, S_k^{(j)}),$$

$$
\begin{aligned}
S_k^{(j)} &= H_k P_{k|k-1}^{(j)} H_k^{\mathrm{T}} + R_k, \\
\omega_k^{(j)}(z) &= p_{D,k}\omega_{k|k-1}^{(j)} q_k^{(j)}(z) \frac{\langle 1, \kappa_k\rangle}{\kappa_k(z)} \frac{\langle \Upsilon_k^1 \left[\omega_{k|k-1}, Z_k \setminus \{z\}\right], p_{k|k-1}\rangle}{\langle \Upsilon_k^0 \left[\omega_{k|k-1}, Z_k\right], p_{k|k-1}\rangle},
\end{aligned} \tag{5.14}
$$

$$
\begin{aligned}
m_k^{(j)}(z) &= m_{k|k-1}^{(j)} + K_k(z - H_k m_{k|k-1}^{(j)}), \\
P_k^{(j)} &= \left[I - K_k^{(j)} H_k\right] P_{k|k-1}^{(j)}, \\
K_k^{(j)} &= P_{k|k-1}^{(j)} H_k^{\mathrm{T}} \left[S_{k|k-1}^{(j)}\right]^{-1}.
\end{aligned}
$$

$\kappa_k(\cdot)$ is the intensity of clutter measurements at time k. $p_{K,k}(\cdot)$ denotes the cardinality distribution of clutter at time k. $P_j^n = n!/(n-j)!$ is the permutation coefficient. $\langle\alpha, \beta\rangle = \int \alpha(x)\beta(x)dx$ is the inner product defined between two real-valued functions α and β or $\langle\alpha, \beta\rangle = \sum_{l=0}^{\infty} \alpha(l)\beta(l)$ when α and β are real sequences. $\sigma_j(\cdot)$is the elementary symmetric function [14] of order j defined for a finite set Z of real number by $\sigma_j(Z) = \sum_{S\subseteq Z,|S|=j} \left(\Pi_{\zeta\in S}\zeta\right)$, with $\sigma_0(Z) = 1$. (e.g., if $Z = \{z_1, z_2, z_3\}$, then $\sigma_1(Z) = z_1 + z_2 + z_3$, $\sigma_2(Z) = z_1z_2 + z_1z_3 + z_2z_3$, and $\sigma_3(Z) = z_1z_2z_3$).

It should be noted that the GM-CPHD recursion can be extended to nonlinear models based on the extended Kalman filter or the unscented Kalman filter [22, 24].

5.3 GM-CPHD with Gating Technique

As addressed in the introduction, computational cost of the CPHD filter can be reduced by reducing the number of cardinality of the measurement set. The gating techniques in traditional tracking algorithms can be used to reduce the cardinality of measurement set. It can also be used for GM-PHD filter, which has been suggested in [24]. In this section, the elliptical gating technique is incorporated in the GM-CPHD filter to reduce the computational cost. Under Gaussian assumption, the validation region is an elliptical region [11] given by

$$
\Omega(k, \gamma) = \left\{z : \left[z - Hm_{k|k-1}\right]^{\mathrm{T}} S_k^{-1} \left[z - Hm_{k|k-1}\right] \le \gamma\right\}, \tag{5.15}
$$

where γ denotes a gate threshold. The volume of the validation region is $V_k = c_{n_z}\sqrt{|S_k|} \cdot \gamma^{n_z/2}$, where the coefficient c_{n_z} depends on the dimension of the measurement (it is the volume of the n_z-dimensional unit hypersphere: $c_1 = 2$, $c_2 = \pi$, $c_3 = 4\pi/3$, etc.). Suppose that p_G is the probability that a target-generated measurement falls in the validation region, then we have

$$
\gamma = -2\ln(1 - p_G), \quad \text{for } n_z = 2.
$$

Suppose the predicted intensity $D_{k|k-1}(x) = \sum_{j=1}^{J_{k|k-1}} \omega_{k|k-1}^{(j)} N\left(x; m_{k|k-1}^{(j)}, P_{k|k-1}^{(j)}\right)$ and the measurement set Z_k are known at time k. It can be seen from Eq. (5.9) that the intensity of RFS of the birth target is included in $D_{k|k-1}(x)$. Using this technique, the newborn target can be detected by the CPHD filter.

The GM-CPHD filter with the gating technique proposed here includes the following steps.

Step 1: Eliminating Unnecessary Gaussian Components
Given a weight threshold T_e $(0.1 > T_e \gg T_{tr})$, where T_{tr} is the truncation threshold for pruning procedure in the GM-CPHD filter [24]. After the pruning procedure, the remained Gaussian component set is given by

$$W_k = \left\{\omega_{k|k-1}^{(j)} N\left(x; m_{k|k-1}^{(j)}, P_{k|k-1}^{(j)}\right) | \omega_{k|k-1}^{(j)} > T_e, j = 1, \ldots, J_{k|k-1}\right\}. \tag{5.16}$$

If the weight of each Gaussian component of newborn targets is greater than a certain threshold (here it is 0.1), then the newborn targets can be concluded by W_k.

Step 2: Construction of New Measurement Set
Given a gate threshold $\gamma = -2\ln(1 - p_g)$, for $n_z = 2$. Denote

$$G_k(z) = \left\{d_k^{(1)}(z), \ldots, d_k^{(i)}(z), \ldots, d_k^{(\Gamma_{W_k})}(z) | z \in Z_k\right\}, \tag{5.17}$$

where Γ_{W_k} is the cardinality of set W_k at time k. And

$$\begin{aligned} d_k^{(i)}(z) &= \left[z - Hm_{k|k-1}^{(i)}\right]^{\mathrm{T}} \left(S_k^{(i)}\right)^{-1} \left[z - Hm_{k|k-1}^{(i)}\right], \\ S_k^{(i)} &= H_k P_{k|k-1}^{(i)} H_k^{\mathrm{T}} + R_k, \\ &\omega_{k|k-1}^{(i)} N\left(x; m_{k|k-1}^{(i)}, P_{k|k-1}^{(i)}\right) \in W_k. \end{aligned}$$

The combined validation region Ω_k corresponding to W_k can be denoted by

$$\Omega_k = \bigcup_{i=1}^{\Gamma_{W_k}} \Omega_k^i(\gamma), \tag{5.18}$$

where $\Omega_k^i(\gamma) = \left\{z : d_k^{(i)}(z) \le \gamma\right\}, i = 1, 2, \ldots, \Gamma_{W_k}$.

Measurements fallen in the combined validation region can be given by:

$$Z_k^{\text{new}} = \{z \in Z_k | \min(G_k(z)) \le \gamma\}. \tag{5.19}$$

Elliptical gating may eliminate all measurements which are not associated with the detected targets. Therefore it is difficult to detect newborn targets. In order to deal with this problem, the CPHD filter birth-target model accounting for the possibility of newborn targets is used here.

Step 3: Update $D_{k|k-1}(x)$ and $p_{k|k-1}(n)$
The predicted intensity $D_{k|k-1}(x)$ and the predicted cardinality density $p_{k|k-1}(n)$ can be updated using the new measurement set Z_k^{new}.

Since only measurements in the combined validation region Ω_k are remained, the clutter intensity must be modified to accommodate to the proposed method. The clutter in the combined validation area is also modeled as the Poisson RFS with intensity $K_k^{\text{new}}(z) = \lambda V_k^{\text{new}} u_k^{\text{new}}(z)$, where λ denotes the average clutter intensity, V_k^{new} represents the volume of the combined validation region, and $u_k^{\text{new}}(z)$ is the uniform probability density over the combined validation region. Therefore, Eqs. (5.13) and (5.14) of the update recursion of the GM-CPHD filter should be modified as follows:

$$\Lambda_k(w, Z) = \left\{ \frac{\langle 1, K_k^{\text{new}} \rangle}{K_k^{\text{new}}(z)} p_{D,k} w^{\mathrm{T}} q_k(z) : z \in Z \right\}, \tag{5.20}$$

$$\omega_k^{(j)}(z) = p_{D,k} \omega_{k|k-1}^{(i)} q_k^{(j)}(z) \frac{\langle 1, K_k^{\text{new}} \rangle}{K_k^{\text{new}}(z)} \frac{\Upsilon_k^1 \left[\omega_{k|k-1}, Z_k \setminus \{z\} \right], p_{k|k-1}}{\left\langle \Upsilon_k^0 \left[\omega_{k|k-1}, Z_k \right], p_{k|k-1} \right\rangle}, \tag{5.21}$$

where $\Lambda_k(w, Z)$ denotes a set for calculating the elementary symmetric function and $\omega_k^{(i)}$ is the weight of the Gaussian component in the updated step of the GM-CPHD filter.

$$\frac{\langle 1, K_k^{\text{new}} \rangle}{K_k^{\text{new}}(z)} = \frac{\lambda V_k^{\text{new}}}{\lambda V_k^{\text{new}} u_k^{\text{new}}(z)} = \frac{1}{u_k^{\text{new}}(z)} = V_k^{\text{new}}. \tag{5.22}$$

At each time step, the volume of the combined validation region can be given by

$$V_k^{\text{new}} \approx \begin{cases} V' = \sum_{i=1}^{\Gamma_{W_k}} V_k^{(i)}, & V' \leq V \\ V, & V' > V \end{cases}, \tag{5.23}$$

where V is the volume of the surveillance region, $V_k^{(i)} = c_{n_z} \sqrt{|\gamma S_k^{(i)}|}$ is the volume of validation region of index i, and Γ_{W_k} is the cardinality of set W_k. For the sake of simplification, the intersecting part of the validation region is not considered. Thus, the volume of the joint validation area is less than or equal to V.

The cardinality of the new measurement set is

$$m_{\text{new}} = n + \lambda V_k^{\text{new}}, \tag{5.24}$$

where n is the number of target.

Therefore, the computational complexity of the GM-CPHD filter using the proposed gating technology is $O(nm^3)$. And we can conclude that if the number of targets is small, the cost of the GM-CPHD filter can be significantly reduced by the proposed gating technology.

5.4 Simulations

5.4.1 Tracking for a Linear Gaussian Case

We use a two-dimensional simulation scenario to validate the proposed algorithm. The surveillance area is 2000 m × 2000 m wide, as shown in Fig. 5.1. The dynamical model of target motion is given by

$$x_k = F_k x_{k-1} + w_k, \tag{5.25}$$

where w_k is the Gaussian process noise with a zero mean and known covariance Q_k, and the target state x_k at time k consists of position and velocity

$$x = [x\ \dot{x}\ y\ \dot{y}]^{\mathrm{T}},$$

$$F_k = \begin{bmatrix} \tilde{F} & \\ & \tilde{F} \end{bmatrix}, \tilde{F} = \begin{bmatrix} 1 & T \\ 0 & 1 \end{bmatrix},$$

$$Q_k = \begin{bmatrix} \tilde{Q} & \\ & \tilde{Q} \end{bmatrix}, \tilde{Q} = \sigma_w^2 \begin{bmatrix} T^4/4 & T^3/2 \\ T^3/2 & T^2 \end{bmatrix},$$

where $\sigma_w = 5\,\mathrm{m/s^2}$, T is the sample time.

The measurement model is given by

$$z_k = H_k x_k + v_k, \tag{5.26}$$

where $H = \begin{bmatrix} 1 & 0 & 0 & 0 \\ 0 & 0 & 1 & 0 \end{bmatrix}$, the measurement noise v_k is the Gaussian noise with zero mean and the known covariance $R = \sigma_v^2 \begin{bmatrix} 1 & 0 \\ 0 & 1 \end{bmatrix}$, where $\sigma_v = 5\,\mathrm{m}$.

Target births appear from four different locations according to a Poisson RFS with intensity

$$\gamma_k(x_k) = \sum_{i=1}^{4} w_i N\left(x_k; \bar{m}_i, \bar{P}_{\gamma,i}\right), \tag{5.27}$$

where $\bar{m}_1 = [-980, 0, 10, 0]^{\mathrm{T}}$, $\bar{m}_2 = [980, 0, 480, 0]^{\mathrm{T}}$,

$\bar{m}_3 = [-480, 0, -980, 0]^{\mathrm{T}}$, $\bar{m}_4 = [480, 0, -480, 0]^{\mathrm{T}}$,
$P_{\gamma,i} = \mathrm{diag}([100, 50, 100, 50]), i = 1, 2, 3,$
$P_{\gamma,4} = \mathrm{diag}([100, 400, 100, 400]),$
$w_i = 0.1, i = 1, 2, 3, 4.$

Gaussian components of new birth target will not be eliminated by T_e, since $w_i > T_e$.

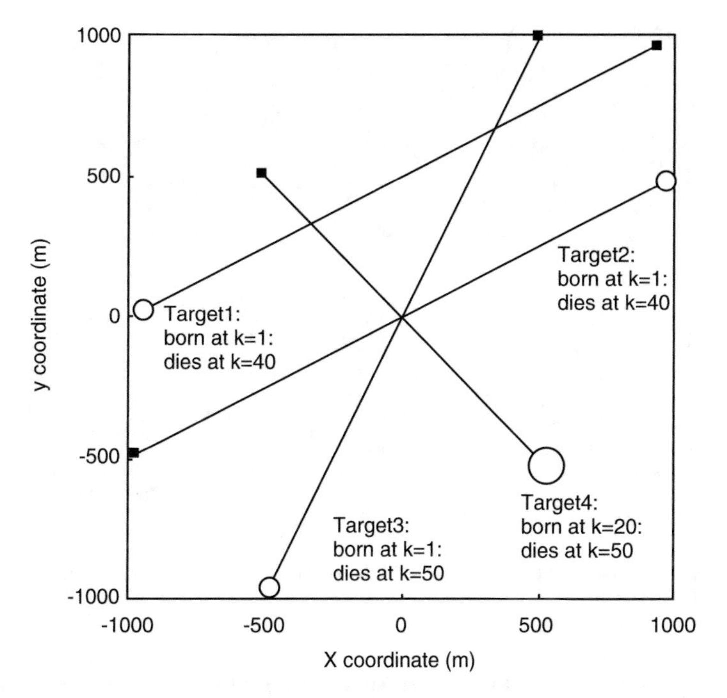

Fig. 5.1 The start/end points for each track are denoted by hollow circles and solid squares, respectively

The target birth regions of a target means the target will appear in such regions with probability P_a. Suppose $P_a = 0.9$, then the target birth regions of those four targets can be pictured in Fig. 5.1 by four circles with radius r_i centered at points $o_i = H\bar{m}_i$ $(i = 1, 2, 3, 4)$, respectively. r_i can be calculated by the elliptical gating technique, where $r_i \approx 28.9,\ i = 1, 2, 3,\ r_4 \approx 49.5$.

The probability of the detection is set to $p_{D,k} = 0.98$, and the probability of survival is set to $p_{s,k} = 0.99$. The sampling time is set to 1 s. The clutter can be modeled by a Poisson RFS with intensity

$$K_k(z) = \lambda V u(z), \tag{5.28}$$

where $\lambda = 1.25 \times 10^{-5}\ \mathrm{m}^{-2}$ is the average number of clutter returns per unit volume, V is the volume of the surveillance region, and $u(z)$ is the uniform density over the surveillance region. Figure 5.1 gives trajectories of true targets, while Fig. 5.2 shows the measurements including clutter over time.

For all simulations, the gating parameters are set to $T_e = 100 \cdot T_{tr}$ and $p_G = 1 - 10^{-4}$ to ensure that measurements generated by target can fall in the combined validation region. The pruning truncation threshold is set to $T_{tr} = 10^{-5}$, merging threshold is set to $U = 10\,\mathrm{m}$, and maximum of Gaussian components is set to $J_{\max} = 200$. The state estimates are extracted from the GM-CPHD filter by taking the Gaussian components of which the weights are more than 0.5.

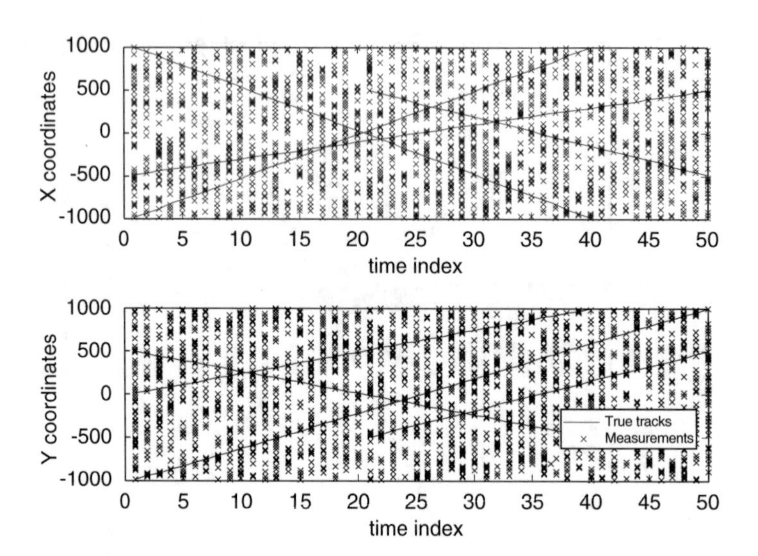

Fig. 5.2 Measurements and true target tracks in x- and y-coordinates versus time ($\lambda = 1.5 \times 10^{-5}\,\mathrm{m}^{-2}$)

If the gating technique is used at every time step, simulation results show that the GM-CPHD filter may achieve poor performance. It is because the gating technique may impair tracks initiation function of the GM-CPHD filter in the initialization of the GM-CPHD recursion. Therefore, we suggest that the gating technique should be used after the initialization of the recursion of the GM-CPHD filter. The GM-CPHD filter with gating technique using this method can perform successfully, which can be shown in Fig. 5.3. It can be seen from Fig. 5.4 that the cardinality of measurement set can be reduced significantly. Therefore the gating method can reduce the computational cost of the GM-CPHD filter significantly.

Since the elliptical gating may eliminate all measurements unrelated to the detected targets, newborn targets may not be detected in simulations. In order to deal with this problem, the birth-target model accounting for the possibility of newly-appearing targets is used in the CPHD filter.

The Wasserstein Distance (WD) was proposed as a useful metric for performance evaluation of multi-target tracking algorithms in [8, 9]. The WD between true multi-target states X $(|X| = m)$ and estimated multi-target states $\hat{X}$ $\left(|\hat{X}| = n\right)$ is defined by

$$d_p^W(X, \hat{X}) = \inf_C \left[\sum_{i=1}^{m} \sum_{j=1}^{n} C_{ij} \| X_i - \hat{x}_j \|^p \right]^{1/p}, \quad p < \infty, \tag{5.29}$$

where the infimum is taken over all $m \times n$ "transportation matrices" $C = \{C_{ij}\}$, $C_{ij} \geq 0$, and for all $i = 1, 2, \ldots, m$ and $j = 1, 2, \ldots, n$,

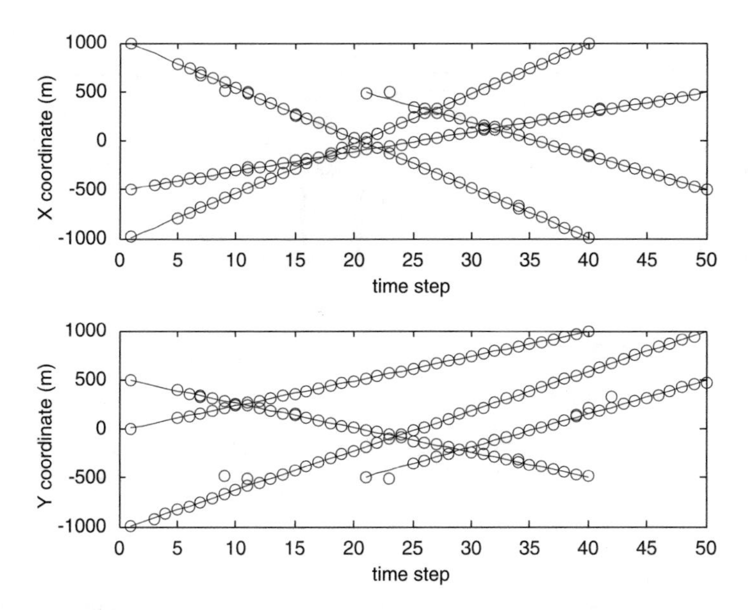

Fig. 5.3 GM-CPHD filter with the proposed gating technique estimates and true target tracks in x and y coordinates versus time. The true tracks and estimates for each tacks are denoted by solid line/open circle, respectively

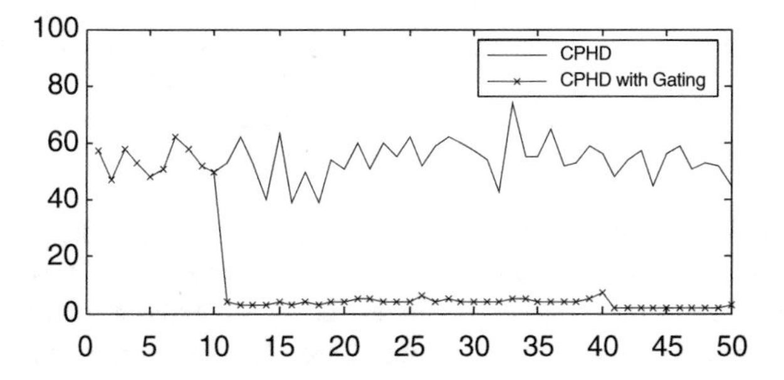

Fig. 5.4 Cardinality comparison to original measurement set and new measurement set versus time ($\lambda = 1.25 \times 10^{-5}\,\mathrm{m}^{-2}$)

$$\sum_{i=1}^{m} C_{ij} = \frac{1}{n}, \quad \sum_{j=1}^{n} C_{ij} = \frac{1}{m}.$$

It should be noted that both the errors of the estimated state and cardinality are considered in WD. As pointed out by Hoffman and Mahler [9], "the WD provides useful and intuitively sensible measures of the over-all ability of a multi-target tracker to correctly estimate the numbers and states of the targets in a multi-target scenario at any given time." In this part WD d_2^W is utilized to assess the performance of the GM-CPHD filter and the proposed method.

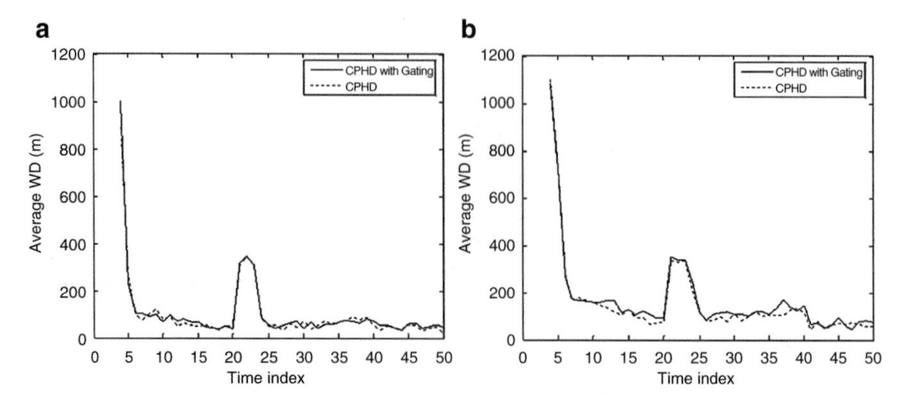

Fig. 5.5 Performance measures (MC average WD) comparison between the GM-CPHD with and without our proposed gating technique versus time for two different clutter intensities (**a**): ($\lambda = 1.0 \times 10^{-5}\,\mathrm{m}^{-2}$), (**b**): ($\lambda = 5.0 \times 10^{-5}\,\mathrm{m}^{-2}$)

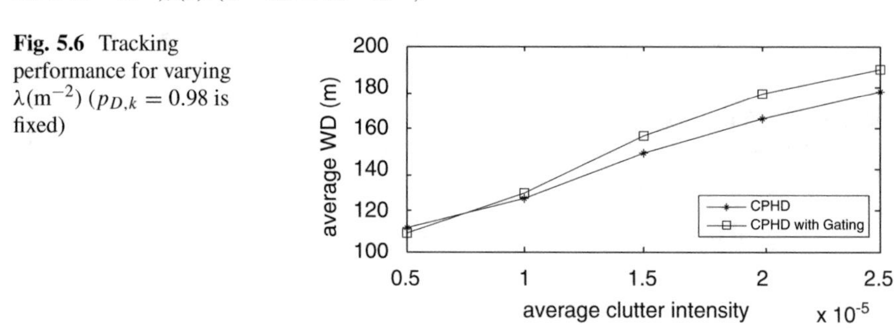

Fig. 5.6 Tracking performance for varying $\lambda(\mathrm{m}^{-2})$ ($p_{D,k} = 0.98$ is fixed)

200 Monte Carlo (MC) runs are performed in order to verify the performance of the GM-CPHD filter and the proposed method. The average WD for the standard GM-CPHD filter and the proposed method over 200 MC runs are shown in Fig. 5.5. 200 MC runs are also performed for both filters with the varying clutter rates. The results are shown in Fig. 5.6. It can be seen from Fig. 5.7 (implemented in MATLAB on a standard notebook computer-IBM X-60) that the computational cost of the our proposed gating technique is much lesser than that of the GM-CPHD filter.

In another example, the performance of the fixed number GM-CPHD filter (i.e., a special case of the GM-CPHD recursion for tracking a fixed number of targets [24, section V]) incorporating gating technique is compared with the JPDA filter. The results are shown in Fig. 5.8. Simulations over 200 Monte Carlo runs are performed. Simulation results are shown in Figs. 5.9 and 5.10. It can be seen from these figures that the fixed number GM-CPHD filter incorporating our proposed gating technique can achieve similar performance to the fixed number GM-CPHD filter, and outperforms the JPDA filter. Therefore, we can conclude that the GM-CPHD filter with our proposed gating technique can achieve similar performance of tracking with a lesser computational cost compared with the standard GM-CPHD filter.

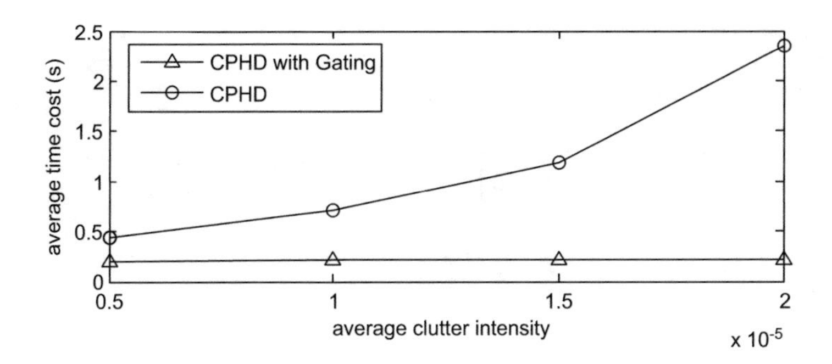

Fig. 5.7 Average time cost of one time step

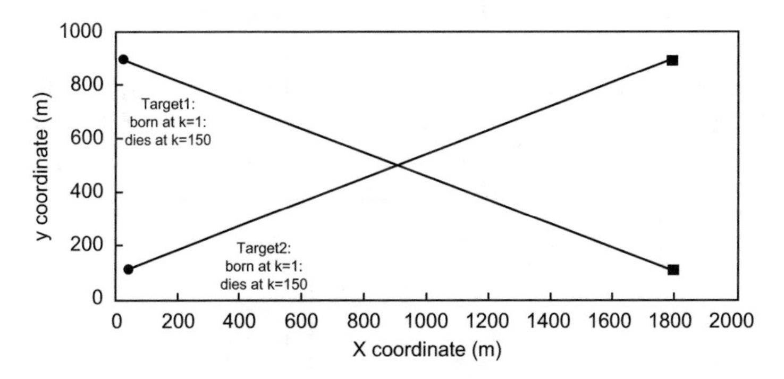

Fig. 5.8 True target tracks in the x y-plane. The start/end points for each track are denoted by filled circle/filled square respectively

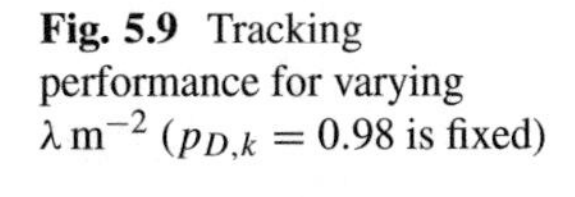

Fig. 5.9 Tracking performance for varying $\lambda\,\mathrm{m}^{-2}$ ($p_{D,k} = 0.98$ is fixed)

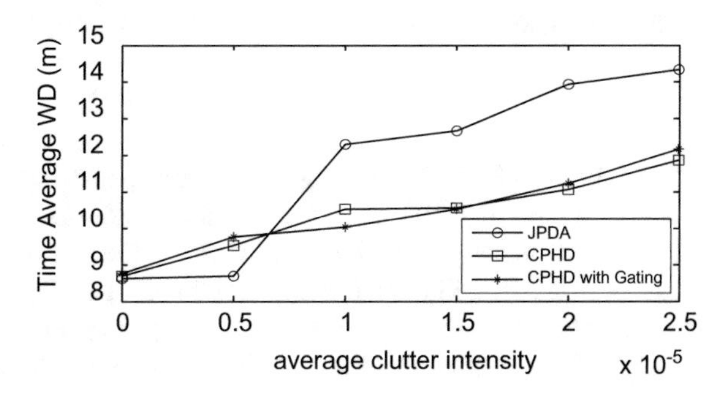

Remark 5.1 It is important that each true measurement must be used to update the 246 GM-CPHD filter with gating technique. Therefore, the parameter T_e used in Eq. (5.16) is the key coefficient for the proposed method. If it is too large, tracking performance may become bad since the real measurement of a target at time $k + 1$ may be blocked by the combined validation region. Thus, we suggest that T_e should not be greater than 0.001. In addition, the Gaussian component of birth target with sufficiently large variance should be used at every time step in order to make sure every measurement of birth target not be blocked.

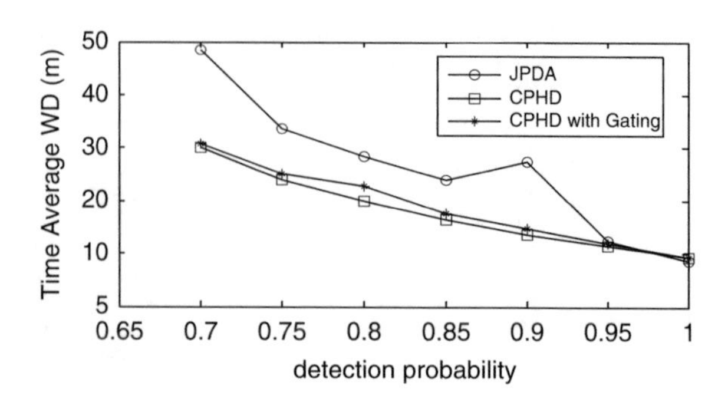

Fig. 5.10 Tracking performance for varying $p_{D,k}$($\lambda = 1.25 \times 10^{-5}\,\mathrm{m}^{-2}$is fixed)

5.4.2 *Tracking for a Nonlinear Gaussian Case*

In this example, there are ten targets with five maneuvering targets in the surveillance region. The true trajectories of the targets are shown in Fig. 5.11. Every target has a survival probability $p_{s,k} = 0.99$ and the detected probability $p_{D,k} = 0.98$. The state dynamics is linear with $\sigma_w = 10\,\mathrm{m/s}^2$ while the measurement consists of range and bearing

$$z_k = \begin{bmatrix} \sqrt{x_k^2 + y_k^2} \\ a\sin\frac{y_k^2}{\sqrt{x_k^2+y_k^2}} \end{bmatrix} + \varepsilon_k, \tag{5.30}$$

where $\varepsilon_k \sim N(\cdot, 0, R_k)$ with $R_k = \mathrm{diag}([\sigma_R^2, \sigma_\alpha^2])$, $\sigma_R = 10\,\mathrm{m}$, $\sigma_\alpha = \pi/180\,\mathrm{rad}$. The clutter can be modeled as a Poisson RFS with intensity

$$K_k(z) = \lambda V u(z), \tag{5.31}$$

where $\lambda = 3.5 \times 10^{-7}\,\mathrm{m}^{-2}$ is the average number of clutter returns per unit volume, $V = 6 \times 10^6\,\mathrm{m}^2$ is the volume of the surveillance region, and $u(z)$ is the uniform density over the surveillance region. The pruning parameters of the Extended Kalman Gaussian Mixture CPHD (EK-GM-CPHD) filter are same as aforementioned in Sect. 5.4.1. It can be seen from Fig. 5.12 that all the targets can be tracked well enough by the CPHD filter with our proposed gating technique.

5.5 Conclusion

In this chapter, a method for reducing the computational cost of GM-CPHD filter using elliptical gating techniques is proposed. However, elliptical gating may eliminate all measurements that are unrelated to the detected target, making it difficult to detect newborn targets. To solve this problem, we use the CPHD birth

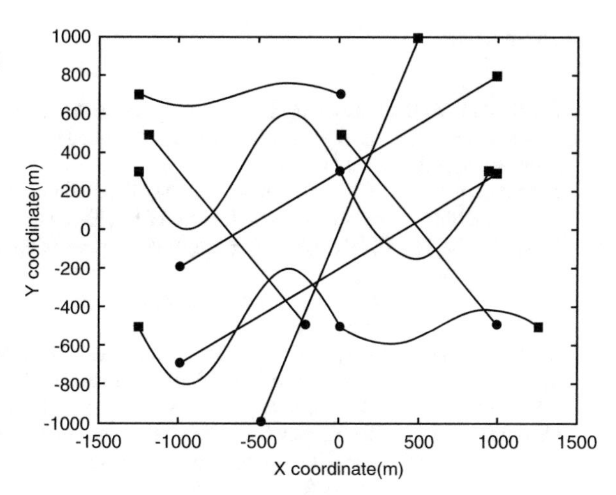

Fig. 5.11 True target tracks in the x y-plane. The start/end points for each track are denoted by filled circle/filled square, respectively

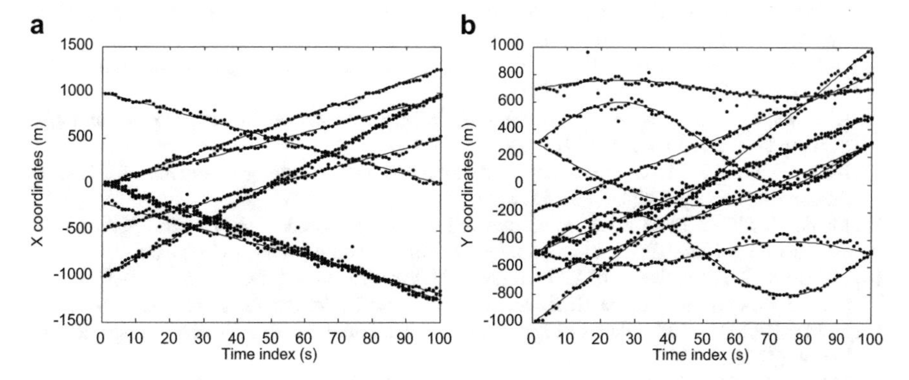

Fig. 5.12 GM-CPHD filter with our proposed gating technique estimates and true target tracks in x and y coordinates versus time. The true tracks and estimates for each tacks are denoted by solid line/filled circle, respectively

model for newborn targets. Simulation results show that the computational cost is reduced and the loss of tracking performance is not significant. Further simulation results show that a fixed number of the GM-CPHD filter with gating also performed better than the JPDA filter. In addition, the gating technique should be used in the GM-CPHD recursion to avoid initialization of the tracking initiation function that impairs the GM-CPHD filter, and the weight threshold T_e should not be greater than 0.001, otherwise it will cause poor tracking performance.

References

1. Bar-Shalom Y (1987) Tracking and data association. Academic, New York
2. Bar-Shalom Y, Tse E (1975) Tracking in a cluttered environment with probabilistic data association. Automatica 11(5):451–460
3. Bar-Shalom Y, Li XR, Kirubarajan T (2004) Estimation with applications to tracking and navigation: theory algorithms and software. Wiley, New York
4. Blackman SS (1986) Multiple-target tracking with radar applications. Artech House, Inc. Dedham, MA
5. Clark D, Vo BN (2007) Convergence analysis of the gaussian mixture PHD filter. IEEE Trans Signal Process 55(4):1204–1212
6. Clark D, Vo BT, Vo BN, Godsill S (2008) Gaussian mixture implementations of probability hypothesis density filters for non-linear dynamical models. In: IET seminar on target tracking and data fusion: algorithms and applications, pp 21–28. IET, Stevenage
7. Goodman I, Mahler RP, Nguyen HT (1997) Mathematics of data fusion. Springer, Netherlands
8. Hoffman JR, Mahler RP (2002) Multitarget miss distance and its applications. In: Proceedings of the fifth international conference on information fusion, vol 1, pp 149–155. IEEE, New York
9. Hoffman JR, Mahler RP (2004) Multitarget miss distance via optimal assignment. IEEE Trans Syst Man Cybern Part A Syst Humans 34(3):327–336
10. Houles A, Bar-Shalom Y (1989) Multisensor tracking of a maneuvering target in clutter. IEEE Trans Aerosp Electron Syst 25(2):176–189
11. Kirubarajan T, Bar-Shalom Y (2004) Probabilistic data association techniques for target tracking in clutter. Proc IEEE 92(3):536–557
12. Mahler RP (2003) Multitarget bayes filtering via first-order multitarget moments. IEEE Trans Aerosp Electron Syst 39(4):1152–1178
13. Mahler R (2006) PHD filters of second order in target number. In Defense and security symposium, p 62360. International Society for Optics and Photonics, Bellingham
14. Mahler R (2006) A theory of PHD filters of higher order in target number. In: Defense and security symposium, p 62350. International Society for Optics and Photonics, Bellingham
15. Mahler R (2007) PHD filters of higher order in target number. IEEE Trans Aerosp Electron Syst 43(4):1523–1543
16. Mahler R (2007) Unified sensor management using CPHD filters. In: 10th international conference on information fusion, pp 1–7. IEEE, New York
17. Mahler RP (2007) Statistical multisource-multitarget information fusion. Artech House Inc., Norwood
18. Musicki D, Evans R, Stankovic S (1994) Integrated probabilistic data association. IEEE Trans Autom Control 39(6):1237–1241
19. Ulmke M, Erdinc O, Willett P (2007) Gaussian mixture cardinalized PHD filter for ground moving target tracking. In: 10th international conference on information fusion, pp 1–8. IEEE, New York
20. Vo BN, Ma WK (2006) The gaussian mixture probability hypothesis density filter. IEEE Trans Signal Process 54(11):4091–4104
21. Vo BN, Singh S, Doucet (2005) A Sequential monte carlo methods for multitarget filtering with random finite sets. IEEE Trans Aerosp Electron Syst 41(4):1224–1245
22. Vo BT, Vo BN, Cantoni A (2006) The cardinalized probability hypothesis density filter for linear gaussian multi-target models. In: 40th annual conference on information sciences and systems, pp 681–686. IEEE, Princeton
23. Vo BT, Vo BN, Cantoni A (2006) Performance of PHD based multi-target filters. In: 9th international conference on information fusion, pp 1–8. IEEE, Florence
24. Vo BT, Vo BN, Cantoni A (2007) Analytic implementations of the cardinalized probability hypothesis density filter. IEEE Trans Signal Process 55(7):3553–3567
25. Zhang H (2009) Finite-set statistics based multiple target tracking. Ph.D. thesis, Shanghai Jiao Tong University
26. Zhang H, Jing Z, Hu S (2009) Gaussian mixture CPHD filter with gating technique. Signal Process 89(8):1521–1530

Chapter 6
Bearing-Only Multiple Target Tracking with the Sequential PHD Filter for Multi-Sensor Fusion

6.1 Introduction

It is well known that a single target can be located in space R^2 by the intersection of multiple line-of-sight angles [4, 6, 14]. For just two sensors, measurements of multiple targets may create a number of false triangulations, which is called *ghosts* (see Fig. 6.1). Therefore, additional sensors should be employed to resolve the ambiguous target association of the two sensors. The minimum number of spatially distinct sensors to uniquely localize N emitters is $N + 1$, which has been proven in [2].

It is a widely studied problem to locate multiple emitters from passive angle measurements. In [15], the method of using frequency and bearing estimation and the corresponding frequency range relation to locate the transmitter is considered. However, this method will fail when the frequencies of emitters are the same. And the method proposed in [2] will fail when measurements are contaminated by clutter [1]. Data association is the central problem for traditional multiple targets tracking. Pattipati et al. [1, 5, 17, 19] treated data association as an S-D assignment problem. For assignment algorithm, data association is formulated as constrained optimization problem, and the maximum cost function is usually used to estimate the global likelihood ratio from the result of state estimator. However, even under the assumption of zero false alarm and unity detection probabilities, the matching problem for $S > 3$ is NP hard [5, 17]. The computational complexity analysis of the SD assignment programming is given in Appendix. It can be seen that NP hard problem cannot be solved by an algorithm with complexity of polynomial time [1] for all possibilities. Multistage Lagrangian relaxation (with computational complexity $O(kM^3)$, where k is the number of relaxation iterations and M is the number of reports from each sensor) is a practical solution to solve

Z. Jing et al., *Non-Cooperative Target Tracking, Fusion and Control*,
Information Fusion and Data Science, https://doi.org/10.1007/978-3-319-90716-1_6

Fig. 6.1 Ghost problems

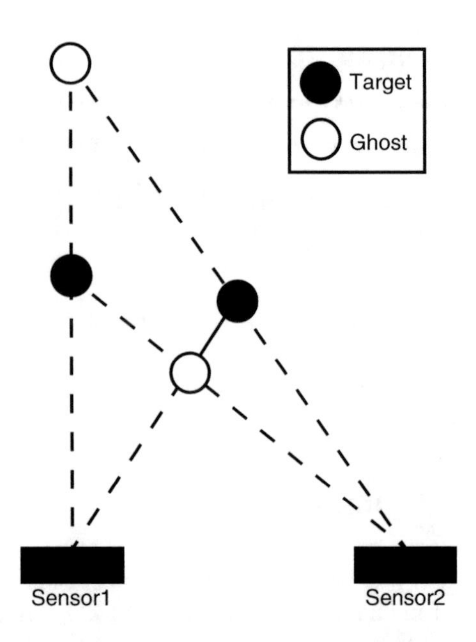

the multidimensional assignment problem. As the number of sensors increases the position estimation error decreases. However, its computational complexity rapidly increases with the sensor's growth. In addition, satisfactory results cannot be achieved in dense clutter.

The Random Finite Set (RFS) approach [7, 10, 13, 16, 25] to multi-target tracking is different from traditional multi-target tracking methods, because multi-target states and multi-target measurements are naturally modeled by random finite sets in the RFS formulation. The mathematical tools provided by Finite Set Statistics (FISST) can apply to extend the Bayesian inference to multi-target tracking problems. Multi-target tracking based on FISST is a technique for the existence of association uncertainty. Therefore, it is possible to avoid multi-target tracking difficulties caused by data association. Rao-Blackwellised particle filters have been introduced in RFS-based multi-target tracking and the bearings-only tracking problem has been successfully solved [20]. One of the RFS-based methods, called the sequential Probability Hypothesis Density (PHD) filter, was used to track multiple speakers in an acoustic environment [11, 18, 23]. The Gaussian Mixture PHD (GM-PHD) filter has been employed to solve the passive sensor correlation problem for three sensors case in [28]. The measurements of each sensor are used in a circular manner in this case. In this way, the location of these ghosts can be as chaotic as real clutter. The GM-PHD filter can be used to remove ghosts like removing clutter. However, only one sensor pair is used to report targets in each time step. Therefore, the computational complexity of this method is $O(M^2N)$, where N is the number of emitters and M is the number of measurements from each sensor.

In this chapter, the sequential PHD filter, which uses two different ways to locate multiple emitters for passive sensors, is introduced. One uses a different combination of sensors in a circular manner, and the other uses the measuring group of each sensor, respectively. Simulation results show that in the presence of clutter, the proposed sequential PHD filter can achieve better performance with smaller computational complexity than the method based on S-D assignment programming. In addition, since the report of each sensor is used at any time step, it can get better performance than the method proposed in [28]. The presentation of this chapter is based on the work in [27, 29].

This chapter is organized as follows. Section 6.2 provides an overview of multi-target states estimation based on FISST and the sequential PHD Filter is also introduced. Deghosting methods based on the sequential PHD filter are introduced in Sect. 6.3. Simulations are given in Sects. 6.4, and 6.5 shows the conclusion.

6.2 Sequential PHD Filter for Multi-Sensor Fusion

The multi-target state and multi-target observation based on the FISST are RFS given by

$$\begin{aligned} X_k &= \left\{x_{k,1}, x_{k,2}, \ldots, x_{k,M_k}\right\} \subset F(X), \\ Z_k &= \left\{z_{k,1}, z_{k,2}, \ldots, z_{k,N_k}\right\} \subset F(Z) \\ Z_{1:k} &= \bigcup_{i=1}^{k} Z_i, \end{aligned}$$

where $F(X)$ and $F(Z)$ are the respective collections of all finite subset of X and Z. M_k and N_k are respectively the number of targets and measurement at time k.

Multiple set integrals involved in the multi-target Bayes recursion are computationally intractable. The PHD filter [24] is a more tractable method. It recursively propagates the first-order moment of the multiple targets state. The domain of the intensity function is on the single-target state space ($\mathbb{R}^2$ or $\mathbb{R}^3$). Therefore, much less computational cost is required for it than the multi-target posterior. The recursions of PHD filter are given by

$$D_{k|k-1}(x) = \int \phi_{k|k-1}(x, \zeta) D_{k-1}(\zeta) d\zeta + \gamma_k(x), \tag{6.1}$$

$$D_k(x) = [1 - p_D(x)]\, D_{k|k-1}(x) + \sum_{z \in Z_k} \frac{\Psi_{k,z}(x) D_{k|k-1}(x)}{K_k(z) + \int \Psi_{k,z}(\xi) D_{k|k-1}(\xi) d\xi}, \tag{6.2}$$

where $\phi(x, \zeta) = p_{S,k|k-1}(\zeta) f_{k|k-1}(x|\zeta) + \beta_{k|k-1}(x|\zeta)$; $\Psi_{k,z}(x) = p_D(x) g_k(z|x)$; $g_k(z|x)$ is the measurement likelihood; $\gamma_k(x)$ is the intensity of the birth RFS; $\beta_{k|k-1}(x|\zeta)$ is the intensity of the RFS spawn by a target previous state ζ;

$p_{S,k|k-1}(\zeta)$ is the probability that a target still exists given that its previous state is ζ; $p_D(x)$ is the probability of detection given a state x; $K_k(z)$ is the intensity of the clutter RFS.

The SMC-PHD filter [22, 24] and the GM-PHD filter [3, 21, 26] are two typical implementations of PHD filter. The GM-PHD filter is a more promising implementation because it has an easy state extraction procedure.

If each target follows linear Gaussian dynamics

$$f_{k|k-1}(x|\zeta) = N\left(x; F_{k-1}\zeta, Q_{k-1}\right), \tag{6.3}$$

$$h_k(z|x) = N\left(z; H_k x, R_k\right), \tag{6.4}$$

where $N(\cdot; m, P)$ are the Gaussian density with mean m and covariance P, F_{k-1} denotes the state matrix, Q_{k-1} is the covariance of the process noise, H_k is the measurement matrix, and R_k is the covariance of the measurement noise. The GM-PHD filter can be represented by the following two propositions [21]

Proposition 6.1 *Given the posterior intensity in a Gaussian mixture form*

$$D_{k-1}(x) = \sum_{i=1}^{J_{k-1}} w_{k-1}^{(i)} N\left(x; \hat{x}_{k-1}^{(i)}, P_{k-1}^{(i)}\right). \tag{6.5}$$

Then, the predicted intensity at time k is also a Gaussian mixture. It is given by

$$D_{k|k-1}(x) = D_{k|k-1}^{(s)}(x) + D_{k|k-1}^{(\beta)}(x) + \gamma_k(x), \tag{6.6}$$

where $D_{k|k-1}^{(s)}(x) = e_k \sum_{i=1}^{J_{k-1}} w_{k-1}^{(i)} N(x; F_{k-1}x_{k-1}^{(i)}, Q_{k-1} + F_{k-1}P_{k-1}^{(i)}F_{k-1}^T)$ *is the PHD of survival targets;* $D_{k|k-1}^{(\beta)}(x) = \sum_{i=1}^{J_{k-1}} \sum_{j=1}^{J_k^{(\beta)}} w_{k-1}^{(i)} w_k^{(\beta,j)} N(x; \hat{x}_{k|k-1}^{(i,j)}, P_{k|k-1}^{(i,j)})$ *is the PHD of spawn targets;*

$$\hat{x}_{k|k-1}^{(i,j)} = \hat{x}_{k-1}^{(i)} + \hat{x}_k^{(\beta,j)};$$

$$P_{k|k-1}^{(i,j)} = P_{k-1}^{(i)} + P_k^{(\beta,j)};$$

$$\gamma_k(x) = \sum_{i=1}^{J_k^{\gamma}} w_k^{\gamma,i} N\left(x; m_k^{\gamma,i}, P_k^{(\gamma,i)}\right).$$

Proposition 6.2 *Given the predicted intensity* $D_{k|k-1}(x)$ *in a Gaussian mixture form*

$$D_{k|k-1}(x) = \sum_{i=1}^{J_k} w_{k|k-1}^{(i)} N\left(x; \hat{x}_{k|k-1}^{(i)}, P_{k|k-1}^{(i)}\right). \tag{6.7}$$

Then, the posterior intensity at time k is also a Gaussian mixture. It is given by

$$D_k(x) = (1 - p_D) D_{k|k-1}(x) + \sum_{z \in Z_k} D_k^{(v)}(x; z), \tag{6.8}$$

$$D_k^{(v)}(x; z) = \sum_{i=1}^{J_{k|k-1}} \frac{p_D w_{k|k-1}^{(i)} q_k^{(i)}(z) N\left(x; \hat{x}_k^{(i)}(z), P_k^{(i)}\right)}{K_k(z) + p_D \sum_{j=1}^{J_{k|k-1}} w_{k|k-1}^{(j)} q_k^{(j)}(z)},$$

where

$$\begin{aligned}
q_k^{(i)}(z) &= N\left(z; H_k x_{k|k-1}^{(j)}, R_k + H_k P_{k|k-1}^{(j)} H_k^T\right); \\
\hat{x}_k^{(i)}(z) &= \hat{x}_{k|k-1}^{(i)} + K_k^{(i)}\left(z - H_k \hat{x}_{k|k-1}^{(i)}\right); \\
P_k^{(i)} &= \left(I - K_k^{(i)} H_k\right) P_{k|k-1}^{(i)}; \\
K_k^{(i)} &= P_{k|k-1}^{(i)} H_k^T \left(H_k P_{k|k-1}^{(i)} H_k^T + R_k\right)^{-1}.
\end{aligned}$$

It should be noted that the two propositions can also be extended to nonlinear situation [21].

As R. Mahler pointed out in [12, p. 1169], there are some difficulties to generalize single-sensor PHD filter to multisensor case. The rigorous formula of PHD corrector for the multisensor is too complicated for practical use. Therefore, some heuristic approach [11, 23] was proposed to deal with multiple speakers tracking problem. The PHD corrector step of the multisensor is given by

$$D_k(x) = \Psi_k^{(l)}\left(Z_k^{(l)}|x\right) \cdots \Psi_k^{(1)}\left(Z_k^{(1)}|x\right) D_{k|k-1}(x), \tag{6.9}$$

where

$$\begin{aligned}
\Psi_k^{(j)}\left(Z_k^{(j)}|x\right) &= \sum_{z^{(j)} \in Z_k^{(j)}} \frac{p_{D,j}(x) h_{k,j}(z^{(j)}|x)}{K_k(z^{(j)}) + D_{k|k-1}^{(j-1)}\left[p_{D,j} h_{k,j}(z^{(j)}|x)\right]} \\
&\quad + 1 - p_{D,j(x)}; \\
D_{k|k-1}^{(j-1)}\left[p_{D,j} h_{k,j}(z^{(j)}|x)\right] &= \int p_{D,j} h_{k,j}(z^{(j)}|x) D_{k|k-1}^{(j-1)}(x) \mu(dx); \\
D_{k|k-1}^{(0)}(x) &= D_{k|k-1}(x); \\
Z_k &= Z_k^{(1)} \cup Z_k^{(2)} \cup \cdots \cup Z_k^{(l)}.
\end{aligned}$$

Since $\Psi_k^{(j)}(Z_k^{(j)}|x)$, $j = 1, 2, \ldots, l$ cannot be commuted, the filter output depends on the order of the sensors in the sequential PHD filter. The Gaussian mixture form of Eq. (6.9) is proposed in [18]. The computational complexity of the sequential PHD filter is $O(lMN)$, where l is the number of sensors, M is the cardinality of measurement set of each sensor, N is the number of detected targets.

6.3 Multiple Target Tracking Based on Sequential PHD Filter

This section describes a sequential PHD filter that uses a passive sensor to locate multiple emitters in two different ways.

6.3.1 Multiple Target Tracking with Sensors in Pairs

Suppose the sensors used in the sequential PHD filter are in pairs (a sensor pair includes two passive sensors with different locations). There are $\frac{l(l-1)}{2}$ sensor pairs for l sensors. At least

$$\lceil 0.5l \rceil = \begin{cases} 0.5l & l \text{ is even} \\ 0.5(l+1) & l \text{ is odd} \end{cases}$$

sensor pairs should be used by the sequential PHD filter if all the sensors are used in this way. The minimum sensor pairs which include all those l sensors are called sensor combination. It can be seen that there are total $\frac{l(l-1)}{2\lceil 0.5l \rceil}$ sensor combinations if the number of passive sensors is l. If those passive sensors are utilized in pairs, at least three sensor combinations should be used in the sequential PHD filter.

Besides, the conception of *sensor combination* is also introduced by an example. It can be seen from Fig. 6.2 that T is a target in the surveillance region. There are four passive sensors that are located along with the x axis. α, β, γ, and θ are the measurements for sensors O_1, O_2, O_3, and O_4, respectively. The measurement model of each sensor is nonlinear

$$\alpha = \arctan \frac{x}{y}, \ \beta = \arctan \frac{x-L}{y},$$

$$\gamma = \arctan \frac{x-2L}{y}, \ \theta = \arctan \frac{x-3L}{y},$$

where $L_1 = L_2 = L_3 = L$.

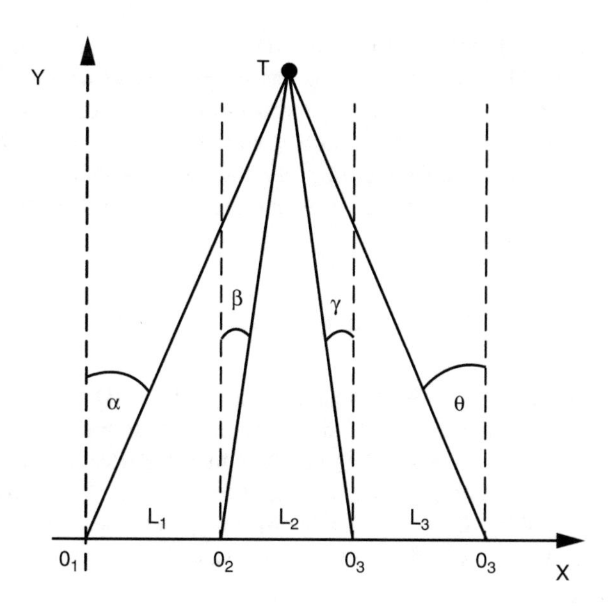

Fig. 6.2 A target in the surveillance region (four passive sensors)

The likelihood functions of two sensor pairs are given by

$$p\left(Z_k^1|X_k\right) = N\left(Z_k^1;\, \psi_1(X_k),\, R_k\right), \tag{6.10}$$

$$p\left(Z_k^2|X_k\right) = N\left(Z_k^2;\, \psi_2(X_k),\, R_k\right), \tag{6.11}$$

where $R_k = \sigma_\omega^2 \begin{bmatrix} 1 & 0 \\ 0 & 1 \end{bmatrix}$, σ_ω^2 denotes the measurement noise variance, $\psi_1(X_k) = \begin{bmatrix} \alpha \\ \gamma \end{bmatrix}$ denotes the measurement equation for sensor pair o_1o_3, and $\psi_2(X_k) = \begin{bmatrix} \beta \\ \theta \end{bmatrix}$ for sensor pair o_2o_4. The cardinality of the measurement set of a sensor pair is $n_in_j(i \neq j)$, where n_i is the cardinality of the measurement set for sensor i. In the sequential PHD filter, those two sensor pairs should be used sequentially. Thus, a *sensor combination* $\{o_1o_3, o_2o_4\}$ is established in case of four sensors when these sensors are used in pairs. Other two *sensor combinations* are $\{o_1o_2, o_3o_4\}$ and $\{o_1o_4, o_2o_3\}$. There are three sensor combinations ($\{o_1o_2, o_2o_3\}$, $\{o_1o_3, o_1o_2\}$, and $\{o_2o_3, o_1o_3\}$) in case of three passive sensors.

The sequential PHD filter [11, 18, 23] can be applied here, if the sensor combination has been constructed. However, when locating two crossing targets with three sensors (see Fig. 6.2), the performance is not as ideal as expected if only the same one sensor combination is used at each time step (see Fig. 6.2). If those sensor combinations are used in a cyclic fashion (i.e., sensor combination $\{o_1o_2, o_2o_3\}$ is used at time $k-1$, sensor combination $\{o_1o_3, o_1o_2\}$ at time k, and sensor combination $\{o_2o_3, o_1o_3\}$ at time $k+1$), the ghosts can be disordered more like real clutters. Therefore, these ghosts can be removed by the sequential PHD filter like removing clutter. The EKF-based sequential GM-PHD filter is employed for nonlinear models. The computational complexity of this filter is $O(\lceil 0.5l \rceil M^2 N)$. This is because the cardinality of the measurement set for each sensor pair is M^2, where l is the number of passive sensors, M is the cardinality of measurement set for each sensor, and N is the number of detected targets.

6.3.2 *Multiple Target Tracking with Sensors Separately*

Since only one sensor can be used in the PHD filter at each time step, sensors are used in pair in [28]. However, in the sequential PHD filter there is no need to use the passive sensors in pairs. The measurement set of each passive sensor can be used separately in the sequential PHD filter, and the computational complexity is $O(lMN)$. It can be seen that the computational complexity will be greatly reduced compared with the method where sensors are used in pairs.

6.4 Simulations

The target dynamic is given by

$$x_k = F_k x_{k-1} + v_k, \tag{6.12}$$

where v_k denotes the Gaussian process noise with zero-mean and known covariance Q_k, x_k denotes the state of target at time k, and

$$\begin{aligned} x_k &= [x\ \dot{x}\ y\ \dot{y}]^{\mathrm{T}}, \\ F_k &= \begin{bmatrix} \tilde{F} & \\ & \tilde{F} \end{bmatrix}, \tilde{F} = \begin{bmatrix} 1 & T \\ 0 & 1 \end{bmatrix}, \\ Q_k &= \begin{bmatrix} \tilde{Q} & \\ & \tilde{Q} \end{bmatrix}, \tilde{Q} = \sigma_v^2 \begin{bmatrix} T^4/4 & T^3/2 \\ T^3/2 & T^2 \end{bmatrix} \end{aligned}$$

where $\sigma_v = 100\,\mathrm{m/s^2}$ is the standard deviation of the process noise, $T = 1\,\mathrm{s}$ is the sample time.

Suppose that the spontaneous birth RFS is a Poisson RFS with intensity

$$\gamma_k(x_k) = 0.2N\left(x_k; \bar{m}_1, \bar{P}\right) + 0.2N\left(x_k; \bar{m}_2, \bar{P}\right), \tag{6.13}$$

Fig. 6.3 Estimation tracks in the x-y plane, if only one sensor combination is used by the sequential PHD filter at each time index, the true tracks and estimates for each track are denoted by dots and circles, respectively ($L = 1000\,\text{m},\ \lambda_k = 0\,\text{rad}^{-1}$)

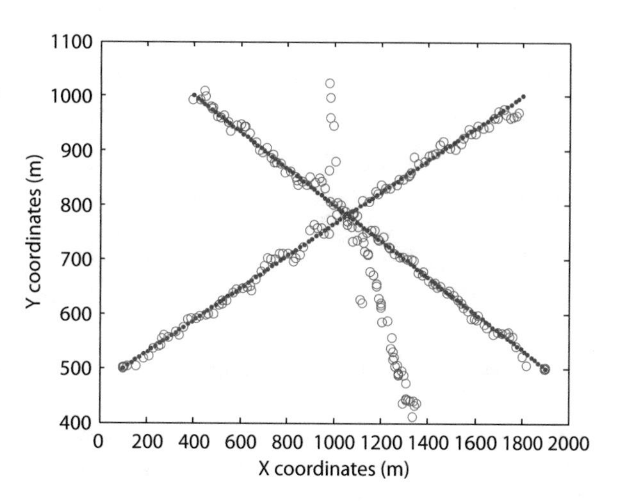

where

$$\bar{m}_1 = [317, 0, 1500, 0]^T,$$

$$\bar{m}_2 = [317, 0, 1072, 0]^T,$$

$$\bar{P} = \text{diag}\left([100, 50, 100, 50]^T\right).$$

The clutter RFS is modeled by a Poisson RFS with intensity

$$K_k(z_k) = \lambda_k V u(z_k), \tag{6.14}$$

where $\lambda_k(\text{rad}^{-1})$ denotes the clutter rate, $V = \pi$ denotes the volume of the surveillance region, and $u(z_k) = \pi^{-1}$ denotes the uniform density over the surveillance region. Assume $\sigma_w = \frac{\pi}{180}$ rad, $p_{S,k|k-1} = 0.99$, $p_{D,i} = 0.98$, $i = 1, 2, 3, 4$, truncation threshold $Tr = 10^{-5}$, and merging threshold $U = 5$. The state estimates can be obtained by taking the means of the Gaussian components of which the weights greater than 0.5.

For all simulations, we assume that sensors are located on the x axis (Fig. 6.2). Suppose $L = 1000\,\text{m}$ and $\lambda_k = 0\,\text{rad}^{-1}$, simulations demonstrated when locating two crossing targets with three passive sensors, the performance is not as ideal as expected, as shown in Fig. 6.3. The circles rising vertically in the figure represent ghosts that cannot be eliminated, if only same one sensor combination is utilized in the sequential PHD filter at each time step. Since the ghosts are similar to targets, they cannot be eliminated in the sequential PHD filter using the passive sensors in pairs. On the other hand, it can be seen from Fig. 6.4 that the sequential PHD filter can achieve better performance when different sensor combinations are used in a cyclic fashion or the sensors are used separately. When the sensor combinations are used in a cyclic way, the ghosts can be disordered more like clutter. Thus, the sensors combination should be used in a cyclic way if the passive sensors must be used in pairs.

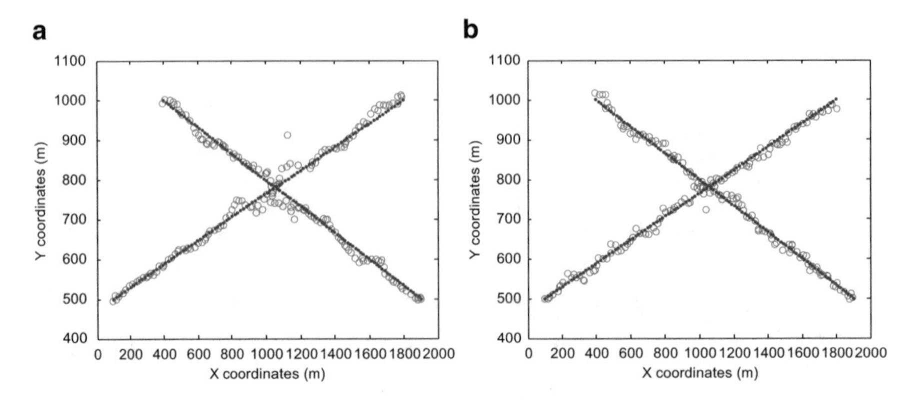

Fig. 6.4 Simulation result of the sequential PHD filter, (**a**) estimation tracks in the x-y plane, if several sensor combinations are used in a cyclic fashion, (**b**) estimation tracks in the x-y plane, if those sensors are used separately. The true tracks and estimates for each track are denoted by dots and circles, respectively ($L = 1000\,\text{m},\ \lambda_k = 0\,\text{rad}^{-1}$)

The Wasserstein distance (WD) [8, 9] is used for performance evaluation of multi-target tracking algorithms here. In the simulations, WD(d_2^W) is employed to evaluate the performance of the proposed method.

The sensor number of the method proposed in [17] is three. As pointed out by Bishop and Pathirana [2], at most two targets can be located when there are three passive sensors. The huge number of candidate associations makes the association process more difficult for more sensors and more targets case in dense clutter environment [5]. Since the association graph is strongly connected, it cannot be decomposed into subproblems. It can be seen from appendix that the S-D assignment problem is extremely difficult due to the huge number of candidate associations. Its computational complexity rapidly increases with the sensor's growth. For purposes of simplification, only two crossing emitters and three passive sensors are used here.

For comparison, the sequential PHD filter, the method based on S-D assignment programming, and the method proposed in [28] are used here. Simulations are obtained by 100 Monte Carlo (MC) runs. The average WDs of these methods are plotted in Fig. 6.5 for each time step. Figure 6.5 shows that the sequential PHD filter using the sensors pairwise does not work well while the sequential PHD filter using those sensors single-wise and the S-D assignment approach have similar performance when those emitters are closely spaced. The WDs over 100 MC runs for these methods with different clutter intensity are shown in Fig. 6.6. It can be seen from these figures that sequential PHD filter outperforms the S-D assignment method proposed in [17] and the method proposed in [28] in dense clutter environment.

The traditional way to track multiple targets is to track each target separately and use a separate filter. It requires the correct association of each target with its measurement. To a certain extent, the performance of the traditional tracking

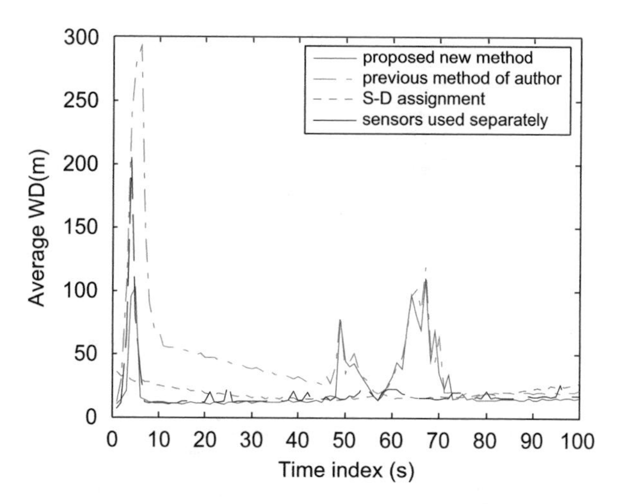

Fig. 6.5 Performance measures (MC average WD) comparison of four methods versus time ($\lambda_k = 0.15\,\text{rad}^{-1}$ is fixed)

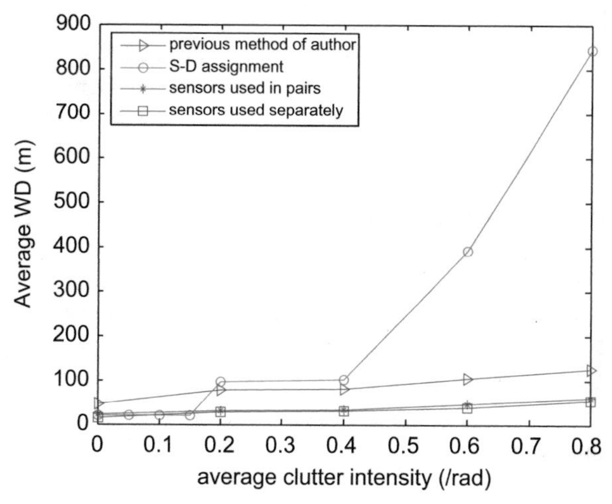

Fig. 6.6 Tracking performance measure of four methods versus clutter intensity ($p_{D,k} = 0.98$ is fixed)

algorithms depends on the correctness of the data association. However, it is not easy to obtain the correct correlation results in a dense clutter environment. Multi-target tracking based on FISST is a method for the existence of association uncertainty. Therefore, it is possible to avoid the difficulties caused by data association. The simulation in many literature shows that the PHD filter is insensitive to the clutter density. Therefore, in a dense and chaotic environment, the sequential PHD filter is superior to the S-D allocation programming for multiple-emitter locations.

The sequential PHD filter is used for each time step compared with our previous method in [28]. However, our previous method uses only one sensor pair for each time step. Therefore, it is obvious that the deghosting method based on the sequential PHD filter outperforms our previous method.

Figure 6.7 plots the tracking performance of the three methods for different distances between sensors. It can be seen from Fig. 6.7 that performance of the three

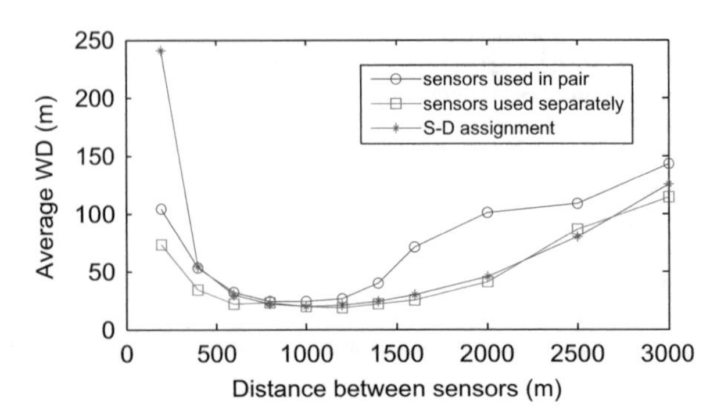

Fig. 6.7 Tracking performance (MC average WD) of three methods for different distance between sensors ($p_{D,k} = 0.98$ and $\lambda_k = 0\,\text{rad}^{-1}$, 100 runs)

approaches varies with the distances between sensors. Single-wise-sensor sequential PHD filter and S-D assignment approach perform well and have similar performance while the pairwise-sensor sequential PHD filter does not perform well. Thus, we conclude that the two methods that sensors are used by the sequential PHD filter in two different manners are not mathematically equivalent. When those passive sensors are used in pairs, the sequential PHD filter does not perform well when distances between sensors are too big. This is because the distance between these ghosts is very small. For example, when L is equal to 3000 m, the ghosts are dense in a region of the surveillance area. Thus the intensity of the clutter will increase in this area, because these ghosts are considered to be clutter. Therefore, it is difficult to eliminate these ghosts by using the sequential PHD filters with sensors in pairs.

In addition, the sequential PHD filter is used to locate five emitters with six sensors. Figure 6.8 plots the true tracks of the three targets. The six sensors are positioned along with the x axis as shown in Fig. 6.2. Suppose $L = 800$ m, simulation results show that the sequential PHD filter can achieve satisfactory performance, when emitters are subject to birth, death, spawning, and merging situation(see Fig. 6.9).

6.5 Conclusion

In this chapter, the sequential PHD filter, which uses two ways to locate multiple emitters with the passive sensors, is introduced. Compared with the method based on S-D assignment planning in dense clutter environment, the sequential PHD filter can achieve better performance and smaller computational complexity. The simulation results show that the sequential PHD filter can obtain better performance and the computational complexity can be greatly reduced when passive sensors are used separately. In addition, the sequential PHD filter can obtain satisfactory performance when multiple targets are subject to birth, death, spawning, and merging situation.

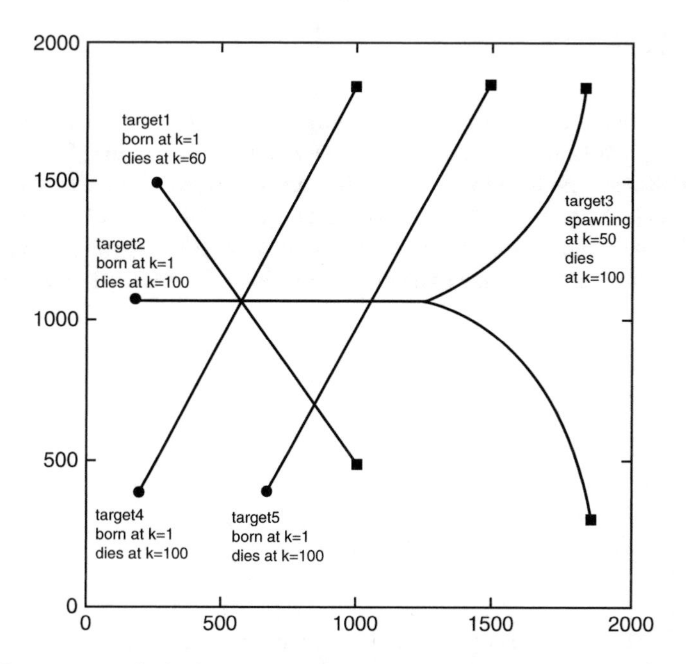

Fig. 6.8 True target tracks in the x-y plane for another example. The start/end points for each track are denoted by test respectively

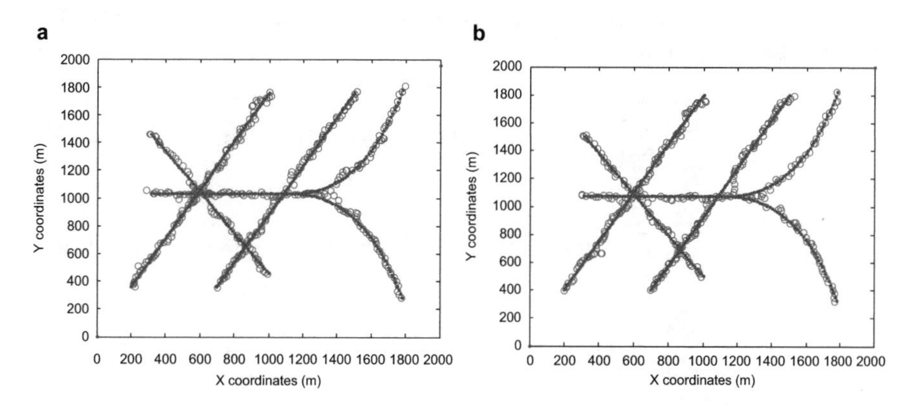

Fig. 6.9 Simulation result of the sequential PHD filter for two different clutter intensities, estimation tracks in the x-y plane, those sensors are used separately, the true tracks and estimates for each track are denoted by dots and circles, respectively $L = 800\,\text{m}$, (**a**) $\lambda_k = 0\,\text{rad}^{-1}$, (**b**) $\lambda_k = 0.4\,\text{rad}^{-1}$

Appendix

The computational complexity of the bearings-only multiple-emitter location in R^2 using S-D assignment programming is discussed in this appendix.

For simplicity, the notations used in this appendix are the same as the notations used in [17]. As pointed by Pattipati et al. [17], the formulation of the data association problem of the multiple-emitter location using three passive sensors, mathematically, leads to a generalization of the three-dimensional (3D) assignment problem.

$$J^* = \min_{\rho \in P} J(\rho), \tag{6.15}$$

where

$$J(\rho) = \sum_{i_1=0}^{n_1} \sum_{i_2=0}^{n_2} \sum_{i_3=0}^{n_3} c_{i_1 i_2 i_3} \cdot \rho_{i_1 i_2 i_3}, \tag{6.16}$$

$$c_{i_1 i_2 i_3} = \sum_{s=1}^{3} \left[(1 - \delta_{0 i_s}) \left(\ln \left(\frac{\sqrt{2\pi} \sigma_s}{P_{D_s} \Phi_s} \right) + \frac{1}{2} \left(\frac{z_{s i_s} - \hat{\theta}_{sj}}{\sigma_s} \right)^2 - \delta_{0 i_s} \in \left(1 - P_{D_s}\right) \right) \right]. \tag{6.17}$$

The set P of constraints, denoting the set of all feasible partitions, is formulated as the set of linear equalities

$$\sum_{i_1=0}^{n_1} \sum_{i_2=0}^{n_2} \rho_{i_1 i_2 i_3} = 1, \quad \text{for all } i_3 = 1, 2, \ldots, n_3, \tag{6.18}$$

$$\sum_{i_1=0}^{n_1} \sum_{i_3=0}^{n_3} \rho_{i_1 i_2 i_3} = 1, \quad \text{for all } i_2 = 1, 2, \ldots, n_2, \tag{6.19}$$

$$\sum_{i_2=0}^{n_2} \sum_{i_3=0}^{n_3} \rho_{i_1 i_2 i_3} = 1, \quad \text{for all } i_1 = 1, 2, \ldots, n_1. \tag{6.20}$$

Suppose there are no missed detection and false measurements, and $n_1 = n_2 = n_3 = 2$, i.e., the number of targets is two. Denote

$$\rho = [\rho_{111} \ \rho_{112} \ \rho_{121} \ \rho_{122} \ \rho_{211} \ \rho_{212} \ \rho_{221} \ \rho_{222}]^{\mathrm{T}}$$
$$c = [c_{111} \ c_{112} \ c_{121} \ c_{122} \ c_{211} \ c_{212} \ c_{221} \ c_{222}]$$

then $J(\rho) = c \cdot \rho$.

Therefore, the 3D assignment problem can be solved by binary integer programming as follows.

$$J^* = \min J(\rho),$$

where constraint Eqs. (6.18), (6.19), and (6.20) can be reformulated as follows:

$$\begin{bmatrix} 1\,0\,1\,0\,1\,0\,1\,0 \\ 0\,1\,0\,1\,0\,1\,0\,1 \\ 1\,1\,0\,0\,1\,1\,0\,0 \\ 0\,0\,1\,1\,0\,0\,1\,1 \\ 1\,1\,1\,1\,0\,0\,0\,0 \\ 0\,0\,0\,0\,1\,1\,1\,1 \end{bmatrix} \begin{pmatrix} \rho_{111} \\ \rho_{112} \\ \rho_{121} \\ \rho_{122} \\ \rho_{211} \\ \rho_{212} \\ \rho_{221} \\ \rho_{222} \end{pmatrix} = \begin{pmatrix} 1 \\ 1 \\ 1 \\ 1 \\ 1 \\ 1 \\ 1 \\ 1 \end{pmatrix} \tag{6.21}$$

Then function 'bintprog' of 'Matlab' can be employed to solve the 3-D assignment problem.

However, this problem may become infeasible in clutter environment, for example.

$$Z_1 = \{0.1906\,\text{rad}\ 1.3389\,\text{rad}\ -1.1345\,\text{rad}\}$$

$$Z_2 = \{-1.0653\,\text{rad}\ 1.0650\,\text{rad}\}$$

$$Z_3 = \{-1.3011\,\text{rad}\ -0.1967\,\text{rad}\}$$

The reason is that since constraint Eqs. (6.18), (6.19), and (6.20) are used, false measurements must be included in the optimal solution partition.

Therefore, the number of targets must be known before the S-D assignment programming is used to solve the bearings-only location problem in R^2. In order to find the optimal partition ρ, the computational complexity will be greatly increased.

Suppose the detection probability is one, the number of targets is two and there are three passive sensors. Then the number of the feasible partitions is

$$Num_{\text{partition}} = \prod_{i=1}^{3} \frac{n_i(n_i - 1)}{2}. \tag{6.22}$$

Then the computation complexity of this problem is $O(n^6)$, if $n_i = n,\ i = 1, 2, 3$.

Suppose detection probability is also unit one, the number of targets becomes three and there are four passive sensors. Then the number of the feasible partitions is

$$Num_{\text{partition}} = \prod_{i=1}^{4} C_{n_i}^3, \tag{6.23}$$

where $C_{n_i}^3$ is the $C_i^l = \frac{l!}{i!(l-i)!}$ is the binomial coefficient.

Then the computation complexity is $O(n^12)$, if $n_i = n,\ i = 1, 2, 3, 4$.

Then the number of the feasible partitions for m targets and l sensor($m < l$) is

$$Num_{\text{partition}} = \prod_{i=1}^{l} C_{n_i}^{m}. \tag{6.24}$$

Then the computation complexity of this problem is $O(n^{lm})$, if $n_i = n, i = 1, 2, \ldots, l$.

References

1. Bar-Shalom Y (1990) Multitarget-multisensor tracking: advanced applications. Artech House, Norwood
2. Bishop AN, Pathirana PN (2007) Localization of emitters via the intersection of bearing lines: a ghost elimination approach. IEEE Trans Veh Technol 56(5):3106–3110
3. Clark D, Vo BN (2007) Convergence analysis of the gaussian mixture phd filter. IEEE Trans Signal Process 55(4):1204–1212
4. Daniels H, Wishart J (1951) The theory of position finding. J R Stat Soc Ser B Methodol 13:186–207
5. Deb S, Yeddanapudi M, Pattipati K, Bar-Shalom Y (1997) A generalized sd assignment algorithm for multisensor-multitarget state estimation. IEEE Trans Aerosp Electron Syst 33(2): 523–538
6. Gavish M, Weiss AJ (1992) Performance analysis of bearing-only target location algorithms. IEEE Trans Aerosp Electron Syst 28(3):817–828
7. Goodman IR, Mahler RP, Nguyen HT (2013) Mathematics of data fusion, vol 37. Springer Science & Business Media, New York
8. Hoffman JR, Mahler RP (2002) Multitarget miss distance and its applications. In: Proceedings of the fifth international conference on information fusion, vol 1. IEEE, New York, pp 149–155
9. Hoffman JR, Mahler RP (2004) Multitarget miss distance via optimal assignment. IEEE Trans Syst Man Cybern Syst Hum 34(3):327–336
10. Lin L, Bar-Shalom Y, Kirubarajan T (2006) Track labeling and phd filter for multitarget tracking. IEEE Trans Aerosp Electron Syst 42(3):778–795
11. Ma WK, Vo BN, Singh SS, Baddeley A (2006) Tracking an unknown time-varying number of speakers using tdoa measurements: a random finite set approach. IEEE Trans Signal Process 54(9):3291–3304
12. Mahler RP (2003) Multitarget bayes filtering via first-order multitarget moments. IEEE Trans Aerosp Electron Syst 39(4):1152–1178
13. Mahler RP (2007) Statistical multisource-multitarget information fusion. Artech House, Norwood
14. Nardone S, Lindgren A, Gong K (1984) Fundamental properties and performance of conventional bearings-only target motion analysis. IEEE Trans Autom Control 29(9):775–787
15. Naus H, Van Wijk C (2004) Simultaneous localisation of multiple emitters. IEE Proc Radar Sonar Navig 151(2):65–70
16. Panta K, Vo BN, Singh S (2007) Novel data association schemes for the probability hypothesis density filter. IEEE Trans Aerosp Electron Syst 43(2):556–570
17. Pattipati KR, Deb S, Bar-Shalom Y, Washburn RB (1992) A new relaxation algorithm and passive sensor data association. IEEE Trans Autom Control 37(2):198–213

18. Pham NT, Huang W, Ong SH (2007) Multiple sensor multiple object tracking with gmphd filter. In: Proceedings of the 10th international conference on information fusion. IEEE, New York, pp 1–7
19. Popp RL, Pattipati KR, Bar-Shalom Y (2001) m-best sd assignment algorithm with application to multitarget tracking. IEEE Trans Aerosp Electron Syst 37(1):22–39
20. Vihola M (2007) Rao-blackwellised particle filtering in random set multitarget tracking. IEEE Trans Aerosp Electron Syst 43(2):689–705
21. Vo BN, Ma WK (2006) The gaussian mixture probability hypothesis density filter. IEEE Trans Signal Process 54(11):4091–4104
22. Vo BN, Singh S, Doucet A (2003) Sequential monte carlo implementation of the phd filter for multi-target tracking. In: Proceedings of the sixth international conference on information fusion, pp 792–799
23. Vo BN, Singh S, Ma WK (2004) Tracking multiple speakers using random sets. In: IEEE international conference on acoustics, speech, and signal processing, vol 2. IEEE, New York, pp 357–360
24. Vo BN, Singh S, Doucet A (2005) Sequential monte carlo methods for multitarget filtering with random finite sets. IEEE Trans Aerosp Electron Syst 41(4):1224–1245
25. Vo BT, Vo BN, Cantoni A (2007) Analytic implementations of the cardinalized probability hypothesis density filter. IEEE Trans Signal Process 55(7):3553–3567
26. Vo BN, Vo BT, Mahler RP (2010) A closed form solution to the probability hypothesis density smoother. In: 13th conference on information fusion (FUSION). IEEE, New York, pp 1–8
27. Zhang H (2009) Finite-set statistics based multiple target tracking. Ph.D. thesis, Shanghai Jiao Tong University
28. Zhang H, Jing Z, Hu S (2007) Bearing-only multi-target location based on gaussian mixture phd filter. In: 10th international conference on information fusion. IEEE, New York, pp 1–5
29. Zhang H, Jing Z, Hu S (2010) Localization of multiple emitters based on the sequential phd filter. Signal Process 90(1):34–43

Chapter 7
Joint Detection, Tracking, and Classification Using Finite Set Statistics and Generalized Bayesian Risk

7.1 Introduction

Multi-target joint detection, tracking, and classification (JDTC) is a critical problem in airborne surveillance systems, which consists of two estimation and one decision subproblems: (1) estimate the number of the targets; (2) estimate their kinematic states; (3) determine their classes. These three problems are usually interdependent, for example, tracking may provide flight envelop and dynamic feature from observation sequence to distinguish the target type; according to the class, appropriate dynamic models are chosen for accurate tracking; besides, the change of target number implies a modification of tracking and classification procedures [21]. Actually, multi-target JDTC is a joint decision and estimation (JDE) problem.

Most traditional JDTC algorithms can be classified into three categories. (1) Estimation-then-decision (ETD) strategy: In this category, the target states are estimated first, then the classification is derived based on the kinematic estimates [13–16]. The drawback of this strategy is that, because the classification is significantly dependent on the estimates, the classifying is affected by the estimation error obviously. As shown in [6], the classification performance is deteriorated due to the inaccurate state estimates calculated with the error data association. (2) Decision-then-estimation (DTE): In this category, the decision is made first, and the estimates are then calculated based on the decisions made before. The disadvantage of this strategy is that the error of the decision is not considered. In [1], the state estimates are calculated with classification-aided data association, however, the classification is done without regarding the quality of the estimation it would lead to. (3) Based on the joint density-probability: the target state and class are inferred by the joint density-probability. Recently, the probability hypothesis density (PHD) filter and multi-Bernoulli filter [11, 12] were proposed for multi-target tracking based on the random finite set (RFS) theory. In [9, 21], the class dependent multi-target density is calculated using the particle implementation of PHD/MeMBer filter [14] with the corresponding motion model set. The target density is represented by a clustering of

Z. Jing et al., *Non-Cooperative Target Tracking, Fusion and Control*,
Information Fusion and Data Science, https://doi.org/10.1007/978-3-319-90716-1_7

the particles, and the class is then inferred by the maximum percentage of particles in each cluster. However, the state and class of each target are not explicitly derived in this category. Furthermore, the overall performance may not be necessarily good, because they do not reach the final goal directly and the approximation is needed to derive the final result [5].

Recently, Li et al. proposed a new approach for the problems involving interdependent decision and estimation based on a new generalized Bayesian risk [7]. The decision and estimation costs are converted to a unified measure by additional weight coefficients, and the optimal JDE solution is derived to minimize the risk. This method is inherently superior to the conventional strategies, especially when decision and estimation problems are highly correlated. In [10], the recursive JDE (RJDE) algorithm was developed to fit the dynamic JDE problems and was used to solve single-target JTC problem. Moreover, a joint performance measure (JPM) was proposed for evaluating the overall performance. In [2], the RJDE algorithm was extended to multi-sensor scenario. However, these algorithms are computationally intensive because the JDE solution is derived over the whole data space. In [5], the conditional JDE (CJDE) algorithm was proposed, new Bayes risk is defined conditioned on data. Because the optimal estimates and the costs are computed directly by the measurements once the decision is made, the computation is simplified greatly. In [4], a multiple-model recursive CJDE (RCJDE) method was proposed to solve target JTC problem. In [3], multi-target joint detection and tracking JDT problem were handled by using CJDE with the risk inspired by optimal subpattern assignment (OSPA) [17, 18].

In this chapter, a recursive solution is proposed to solve multi-target joint detection, tracking, and classification (JDTC) problem based on the finite set statistics and generalized Bayesian risk. A new Bayesian risk is defined involving the costs of multi-target cardinality estimation (detection), state estimation (tracking), and classification for the labeled random finite set variables. In addition, the estimates and costs are calculated using the labeled multi-Bernoulli (LMB) filter [15–20] conditioned on different hypotheses and decisions of target classes. For the explicit expression of the posterior density involving the measurement-target-associations (MTA), the costs are exactly computed by summing up all the costs for each track. In addition, the Gaussian mixture implementation of the proposed algorithm is also provided. Because the optimal solution is then derived to minimize the new Bayesian risk, the interdependence of detection, tracking, and classification is considered. Therefore, the performance of the proposed algorithm is better than traditional methods. Simulations demonstrate the effectiveness and superiority of the proposed multi-target JDTC algorithm. The presentation of this chapter is based on the work in [8].

7.2 Conditional Joint Decision and Estimation

In the joint decision and estimation (JDE) framework, a new risk is defined to represent the interdependence of decision and estimation.

$$\bar{R} \triangleq \sum_i \sum_j (\alpha_{ij} c_{ij} + \beta_{ij} E[C(x, \hat{x})|D_i, H_j]) P(D_i, H_j) \tag{7.1}$$

where c_{ij} is the cost of decision on D_i while the true hypothesis is H_j, and conditional expected estimation cost $E[C(x, \hat{x})|D_i, H_j] = mse(\hat{x}|D_i, H_j)$ is the mean square error given the corresponding decision and hypothesis, and $P\{D_i, H_j\}$ is the joint probability of decision and hypothesis. These two costs are converted to a unified measurement by introducing additional weight coefficient $\{\alpha_{ij}, \beta_{ij}\}$.

The optimal solution is derived to minimize this new intermediate risk. For any given $E[C(x, \hat{x})|D_i, H_j]$, the optimal decision is

$$D = D_i \qquad if \qquad C^i(z) \leq C^l(z) \quad \forall l \tag{7.2}$$

where the posterior cost is given by

$$C^i = \sum_j (\alpha_{ij} c_{ij} + \beta_{ij} E[C(x, \hat{x})|D_i, H_j]) P(H_j) \tag{7.3}$$

Actually, an optimal Bayes estimator is a function of observations z that minimizes the Bayes risk $E[C(x, \hat{x})|D_i, H_j]$. Therefore, the decision D_i is equivalent to the event $\{z \in \mathscr{D}_i\}$, where $\mathscr{D}_i$ is the decision region for D_i in the data space. Assume that $\{\mathscr{D}^1, \mathscr{D}^2, \ldots, \mathscr{D}^M\}$ is the decision partition of the measurement data space, the optimal estimator is only defined if $z \in \mathscr{D}_i$. The optimal state estimates is:

$$\check{x}_{ij} = \sum_{i,j} \hat{x}_{ij} P^i\{D_i, H_j|z\} \tag{7.4}$$

where

$$\hat{x}_{ij} = E[\hat{x}_j|z, H_j] \tag{7.5}$$

$$\bar{P}(H_j|Z) = \frac{\beta_{ij} P(H_j|Z)}{\sum_h \beta_{ih} P(H_h|Z)} \tag{7.6}$$

Therefore, in the JDE method, the estimation cost ϵ_{ij} is computed by the currently available partition of Z_k. Repeat the E-step and D-step until the algorithm converges, the optimal partition is then derived, and the JDE solution is obtained.

The flexible and powerful framework of JDE is represented as:

1. The decision candidates and hypothesis are not necessary one-to-one;

2. In the cost $C(x, \hat{x})$; $\hat{x}$ and x can be of different dimensions;
3. The coefficient $\{\alpha_{ij}, \beta_{ij}\}$ can be chosen and the two strategy E-then-D, D-then-E can be represented by this framework.

The JDE method is computationally intensive since the solution is derived over the whole data space. To overcome this defect, the CJDE algorithm [7] is proposed. The solution is derived based on a risk that depends on the particular received measurement z other than over all the realizations of the data, that is

$$\bar{R}(z) \triangleq \sum_i \sum_j (\alpha_{ij} c_{ij} + \beta_{ij} E[C(x, \hat{x})|D^i, H^j, z]) P\{D^i, H^j|z\} \tag{7.7}$$

where $P\{D^i, H^j|z\}$ is the joint probability of decision and hypothesis, c_{ij} is the cost of decision D^i when the true hypothesis is H^j, and the conditional expected estimation cost $E[C(x, \hat{x})|D^i, H^j] = mse(\hat{x}|D^i, H^j)$ is the mean square error. The optimal solution is derived to minimize this new Bayes risk and the optimal decision D is:

$$D = D^i \quad \text{if} \quad C_C^i(z) \leq C_C^n(z) \quad \forall n \tag{7.8}$$

where the posterior cost is

$$C_C^i(z) = \sum_j (\alpha_{ij} c_{ij} + \beta_{ij} E[C(x, \hat{x})|D^i, H^j, z]) P\{H^j|z\} \tag{7.9}$$

To calculate $C_C^i(z)$ with $C(x, \hat{x}) = \tilde{x}'\tilde{x}$, the key is to obtain the estimation cost ϵ_{ij}. Assume that the optimal target estimate is:

$$\hat{x}_n = \sum_j E(\hat{x}^{(j)}|H^j, z) P\{H^j|z\} \tag{7.10}$$

then the estimation cost is:

$$\begin{aligned} \epsilon^{ij}(z) &\triangleq E[\tilde{x}'\tilde{x}|D^i, H^j, z] \\ &= mse(\hat{x}^{(ij)}|D^i, H^j, z) + E[(\hat{x}^{(ij)} - \hat{x})'(\cdot)|D^i, H^j, z] \\ &= mse(\hat{x}^{(j)}|H^j, z) + (\hat{x}^{(j)} - \check{x}^{(i)})(\hat{x}^{(j)} - \check{x}^{(i)}), \forall z \in D^i \end{aligned} \tag{7.11}$$

where $(\cdot)$ denotes the same term right before it.

The recursive CJDE algorithm is shown as follows:

1. Initialize the parameters: $\hat{x}_{k-1}^{(j)}$, $P\{H^j|Z^{k-1}\}$ and so on.
2. Predict the state based on dynamics of x_k; update $\hat{x}_k^{(j)}$ and $P\{H^j|Z^k\}$ by z_k; then compute $\check{x}_k^{(i)}$ for decision i.

3. Compute $\epsilon^{ij}(Z^k)$ and get cost $C_C^i(Z^k)$; then $D_k^i : C_C^i(Z^k) \leq C_C^n(Z^k), \forall n$.
4. Output the CJDE solution for time k; let $D_k = D_k^i$ and $\hat{x}_k = \check{x}_k^{(i)}$.

CJDE inherits the virtue of JDE, the coupling between decision and estimation is considered. In addition, the optimal estimate and corresponding cost are directly calculated using the measurement located inside the region of the decision D. Therefore, the CJDE is more applicable for the computational complexity is largely reduced.

7.3 Recursive JDTC with FISST and Generalized Bayesian Risk

7.3.1 Recursive JDTC Algorithm

Suppose that the class of a target is a time-invariant attribute, which can be distinguished according to the dynamic behavior. The class-dependent target kinematic state and measurement at time k can be modeled as:

$$x_k = f_{k|k-1,c}(x_{k-1}, w_{k,c}) \tag{7.12}$$

$$z_k = h_k(x_k, v_k) \tag{7.13}$$

where $f_{k|k-1,c}$ is the class-dependent state transition function, h_k is the measurement function, $w_{k,c}$ and v_k are the uncorrelated Gaussian process and measurement noise, respectively. Assume that, at time k, $X_k = \{x_{k,1}, \ldots, x_{k,n}\}$ is the set of multi-target states, and $Z_k = \{z_{k,1}, \ldots, z_{k,m}, c_1, \ldots, c_i\}$ is the set of noisy and cluttered measurements, where $\{z_{k,1}, \ldots, z_{k,m}\}$ are the measurements of the targets and $\{c_1, \ldots, c_i\}$ is the set of clutter. The multi-target JDTC algorithm aims to estimate the target number, their kinematic states, and determine their classes from the sequence of noisy and cluttered measurement sets.

Because multi-target JDTC is a dynamic problem and the measurements are usually obtained sequentially, a new recursive Bayesian risk is firstly defined for the labeled RFS variables. Suppose that $\mathscr{C} = \{\mathrm{c}_j\}_{j=1}^J$ is the class set which contains J possible target classes, $\mathbf{X}_k$ is the multi-target state RFS at time k, H_ℓ^j and $D_{k,\ell}^i$ are the class hypothesis and decision of track ℓ, $H^m = \{H_\ell^j\}_{\ell \in \mathscr{L}(\mathbf{X})}$ and $D_k^n = \{D_{k,\ell}^i\}_{\ell \in \mathscr{L}(\mathbf{X})}$ are then the class hypothesis and decision sets of all the targets, respectively. The new Bayesian risk is then given by

$$\begin{aligned}\bar{R}_C(Z_k) = \sum_{m,n} (\alpha_{mn} c_{mn} + \beta_{mn} E[C(\mathbf{X}, \hat{\mathbf{X}}) | D_k^n, H^m, Z_k] \\ + \gamma_{mn} E[(|I_{mn} - \hat{I}|) | D_k^n, H^m, Z_k]) P\{D_k^n, H^m | Z_k\}\end{aligned} \tag{7.14}$$

where c_{mn} is the cost of deciding on D^n when the hypothesis H^m is true, $C[(\mathbf{X}, \hat{\mathbf{X}}) | D^n, H^m, Z_k]$ is the conditional expected estimation cost of multi-target states, and $E[(|I_{mn} - \hat{I}|)|D^n, H^m, Z_k]$ is the conditional expected multi-target cardinality estimation error; $P\{D_k^n, H^m|Z_k\}$ is the posterior probability of decision and hypothesis set. α_{mn}, β_{mn}, and γ_{mn} are the nonnegative weights used to unify the costs.

To minimize $\bar{R}_C(Z_k)$, the optimal decision D_k is:

$$D_k = D_k^n \qquad if \qquad C_n(Z_k) \leq C_i(Z_k), \forall i \tag{7.15}$$

where the cost $C_n(Z_k)$ for the decision n is given by:

$$\begin{aligned} C_n(Z_k) = \sum_m \Big(& \alpha_{mn} c_{mn} + \beta_{mn} E[C(\mathbf{X}, \hat{\mathbf{X}})|D_k^n, H^m, Z_k] \\ & + \gamma_{mn} E[(|I_{mn} - \hat{I}|)|D_k^n, H^m, Z_k] \Big) P_n(H^m|Z_k) \end{aligned} \tag{7.16}$$

To calculate the costs in $C_n(Z_k)$, the multi-target states and cardinality estimates given the class decision set are proposed as follows:

1. Prediction: Suppose that the density of target ℓ can be given by

$$p_{k-1}(x, \ell) = \sum_j f_{k-1}\left(x, \ell|H_\ell^j\right) P\left(H_\ell^j\right) \tag{7.17}$$

where $P(H_\ell^j)$ is the probability of the class hypothesis, and $f_{k-1}(x, \ell|H_\ell^j)$ is the class dependent target density. Suppose that the multi-target prior is the LMB RFS, the class-dependent multi-target predicted density given hypothesis set $H^m = \{H_\ell^j\}_{\ell \in \mathcal{L}(\mathbf{X})}$ can then be represented as

$$\begin{aligned} \pi_{k|k-1}(X_+) = & \Delta(X_+) \sum_{(I_+, \xi) \in \mathcal{F}(\mathbb{L}_+) \times \Xi} \omega_+^{(I_+, \xi)} \delta_{I_+}(\mathcal{L}(X_+)) \\ & \left[\sum_j p_{k|k-1}^{(\xi)}\left(x, \ell|H_\ell^j\right) P^{(\xi)}\left(H_\ell^j\right) \right]^{X_+} \end{aligned} \tag{7.18}$$

where

$$\omega_+^{(I_+, \xi)} = \omega_B(I_+ \cap \mathbb{B}) \omega_S^{(\xi)}(I_+ \cap \mathbb{L}) \tag{7.19}$$

$$p_{k|k-1}^{(\xi)}\left(x, \ell|H_\ell^j\right) = 1_{\mathbb{L}}(\ell) p_S^{(\xi)}\left(x, \ell|H_\ell^j\right) + (1 - 1_{\mathbb{L}}(\ell)) p_B\left(\cdot|H_\ell^j\right) \tag{7.20}$$

$$p_S^{(\xi)}\left(x, \ell|H_\ell^j\right) = \frac{\langle p_S(\cdot, \ell) f_c(x|\cdot, \ell), p_{k-1}^{(\xi)}(\cdot, \ell)\rangle}{\eta_S^{(\xi)}(\ell)} \tag{7.21}$$

$$\eta_S^{(\xi)}(\ell) = \int \langle p_S(\cdot, \ell) f_c(x|\cdot, \ell), p_{k-1}^{(\xi)}(\cdot, \ell)\rangle dx \tag{7.22}$$

Here, $f_c(x|\cdot, H_\ell^j)$ is the class hypothesis dependent state transition function, $p_s(\cdot, \ell)$ is the target survival probability. For the new birth targets, $p_B(\cdot|H_\ell^j)$ is the prior existence probability and density of the new birth target. $\omega_+^{(I_+,\xi)}$ is the product of the weight $\omega_S^{(\xi)}(I_+ \cap \mathbb{L})$ of survival targets and the weight $\omega_B(I_+ \cap \mathbb{B})$ of new birth targets.

2. Update: Conditioned on the class decision set $D_k^n = \{D_{k,\ell}^i\}_{\ell \in \mathscr{L}(\mathbf{X})}$, whether the measurement used to update the multi-target states lies inside the region of the decision needs to be determined. Suppose that $\mathscr{D}_\ell^i$ is the decision region for $D_{k,\ell}^i$ in the measurement data space, $\{\mathscr{D}_\ell^i\}_{i=1}^n$ is then the set of regions which forms a partition of the measurement data space [7]. The inclusion function $1_{\mathscr{D}_\ell^i}(z)$ is defined here to indicate whether the measurement z lies inside the region $\mathscr{D}_\ell^i$. If $z \in \mathscr{D}_\ell^i$, $1_{\mathscr{D}_\ell^i}(z) = 1$; if $z \notin \mathscr{D}_\ell^i$, $1_{\mathscr{D}_\ell^i}(z) = 0$. When the measurement set Z_k is collected at time k, the posterior multi-target density conditioned on the class decision set $D_k^n = \{D_{k,\ell}^i\}_{\ell \in \mathscr{L}(\mathbf{X})}$ is:

$$\pi_k^n\left(\mathbf{X}|D_k^n, Z_k\right) = \Delta(\mathbf{X}) \sum_{(I_+,\theta) \in \mathscr{F}(\mathbb{L}_+) \times \Theta} \omega_n^{(I_+,\theta)}(Z_k)\delta_{I_+}(\mathscr{L}(\mathbf{X})) \times \left[p_n^{(\theta)}\left(\cdot, \ell|H_\ell^j, D_{k,\ell}^i, Z_k\right) P_n^{(\theta)}\left(H_\ell^j|D_{k,\ell}^i, Z_k\right)\right]^{\mathbf{X}} \tag{7.23}$$

where Θ is the space of mappings $\theta : \mathbb{L} \to \{0, 1, \ldots, |Z|\}$.

$$\omega_n^{(I_+,\theta)}(Z_k) \propto \omega_+^{(I_+)}\left[\eta_Z^{(\theta)}\left(\cdot|D_{k,\ell}^i, H_\ell^j\right)\right]^{I_+} \tag{7.24}$$

$$p_n^{(\theta)}\left(x, \ell|H_\ell^j, D_{k,\ell}^i, Z_k\right) = \frac{1_{\mathscr{D}_\ell^i}(z_{\theta(\ell)})\psi_Z(x, \ell; \theta)p_{k|k-1}^{(\ell)}\left(x|H_\ell^j\right)}{\eta_Z^{(\theta)}\left(\ell|D_{k,\ell}^i, H_\ell^j\right)} \tag{7.25}$$

$$P_n^{(\theta)}\left(H_\ell^j|D_{k,\ell}^i, Z_k\right) = \frac{\eta_Z^{(\theta)}\left(\ell|D_{k,\ell}^i, H_\ell^j\right) P_{k|k-1}\left(H_\ell^j\right)}{\sum_j \eta_Z^{(\theta)}\left(\ell|D_{k,\ell}^i, H_\ell^j\right) P_{k|k-1}\left(H_\ell^j\right)} \tag{7.26}$$

$$\eta_Z^{(\theta)}\left(\ell|D_{k,\ell}^i, H_\ell^j\right) = \left\langle\psi_Z(x, \ell; \theta), p_{k|k-1}^{(\ell)}\left(x|H_\ell^j\right)\right\rangle \tag{7.27}$$

$$\psi_Z(x, \ell; \theta) = \begin{cases} q_d(x, \ell), & z_{\theta(\ell)} = \varnothing \\ \dfrac{p_d(x, \ell)g(z_{\theta(\ell)}|x, \ell)}{\kappa(z_{\theta(\ell)})}, & \text{other} \end{cases} \tag{7.28}$$

Because the components propagated by the GLMB filter grow exponentially, the computational complexity of the method in [20] is intensive. To reduce the computational load, the GLMB density can be approximated as [15] with exactly preserving the first moment of the multi-target posterior density. The approximated density can be represented using the parameter set

$$\pi_k^n(x|Z_k) = \left\{ \left(r_n^{(\ell)}, p_n^{(\ell)}\left(x|H_\ell^j\right) P_n\left(H_\ell^j\right) \right) \right\}_{\ell \in \mathbb{L}_+} \tag{7.29}$$

where

$$r_n^{(\ell)} = \sum_{(I_+,\theta)\in\mathscr{F}(\mathbb{L}_+)\times\Theta} \omega_n^{(I_+,\theta)}(Z_k) 1_{I_+}(\ell) \tag{7.30}$$

$$p_n^{(\ell)}\left(x|H_\ell^j\right) = \frac{1}{r^{(\ell)}} \sum_{(I_+,\theta)\in\mathscr{F}(\mathbb{L}_+)\times\Theta} \omega_n^{(I_+,\theta)}(Z_k) 1_{I_+}(\ell) p_n^{(\theta)}\left(x, \ell|H_\ell^j\right) \tag{7.31}$$

$$P_n\left(H_\ell^j\right) = \frac{1}{r^{(\ell)}} \sum_{(I_+,\theta)\in\mathscr{F}(\mathbb{L}_+)\times\Theta} \omega_n^{(I_+,\theta)}(Z_k) 1_{I_+}(\ell) P_n^{(\theta)}\left(H_\ell^j\right) \tag{7.32}$$

In this approximation, the density of each track is represented using the sum of class dependent components, and only one component represents the density of each target will be propagated, and the computational load will be largely reduced. As given in [15], the computational complexity of the approximated GLMB filter is at worst cubic in the number of measurements, i.e., $O(m^3)$. By comparison, the proposed multi-target JDTC algorithm involves predicting the target state according to the class hypothesis, and updating the state estimates given the class decision. Therefore, the increase in computational load is linear in the number J of the possible target classes.

For the exact calculation of the multi-target posterior density involving MTA, the CJDE cost can be explicitly computed as

$$C_n(Z_k) = \sum_m \left(\sum_c \omega_n^c(\alpha_{mn} c_{mn} + \beta_{mn}\varepsilon_X) + \gamma_{mn}\varepsilon_I \right) P_n(H^m|Z_k) \tag{7.33}$$

where c represents $(I, \theta) \in \mathscr{F}(\mathbb{L}) \times \Theta$, the term ε_X denotes the estimation cost of multi-target state, which can be calculated by summing up the state estimation costs of each target

$$\begin{aligned} \varepsilon_X &= \sum_{\ell\in\mathscr{L}(\mathbf{X})} E\left[C(x_\ell, \hat{x}_\ell)|D_{k,\ell}^i, H_\ell^j, Z_k\right] \\ &= \sum_{\ell\in\mathscr{L}(\mathbf{X})} mse\left(\hat{x}_{k,\ell}^{ij}\right) + \left(\hat{x}_{k,\ell}^{ij} - \check{x}_{k,\ell}^{i}\right)' \left(\hat{x}_{k,\ell}^{ij} - \check{x}_{k,\ell}^{i}\right) \end{aligned} \tag{7.34}$$

where $\hat{x}^{ij}_{k,\ell}$ is the class dependent state estimate, and $\check{x}^{i}_{k,\ell}$ is the optimal estimate for the decision $D^{i}_{k,\ell}$, which can be calculated as:

$$\check{x}^{i}_{k,\ell} = \sum_{j} \hat{x}^{ij}_{k,\ell} P_n\left(H^{j}_{\ell}\right), \quad z_{\theta(\ell)} \in \mathscr{D}^{i}_{k,\ell} \tag{7.35}$$

If no measurements lie inside the region of the decision $D^{i}_{k,\ell}$, the estimation cost can be computed by replacing the estimate $\hat{x}^{ij}_{k,\ell}$ with the prediction. Similarly, the decision cost of track ℓ can be calculated as:

$$c_{mn} = \sum_{\ell \in \mathscr{L}(\mathbf{X})} c^{ij}_{k,\ell} \tag{7.36}$$

where $c^{ij}_{k,\ell}$ is the cost of deciding on $D^{i}_{k,\ell}$ when hypothesis H^{j}_{ℓ} is true for track ℓ.

Because the posterior cardinality estimate $\hat{I}_k$ of the original labeled multi-Bernoulli filter is unbiased [15], it is used as the optimal cardinality estimate here. The estimation cost of multi-target cardinality can be calculated as:

$$\begin{aligned} \epsilon^{ij}_{k,I} &= E\left[\left(|I^{ij}_{k} - \hat{I}_k|\right) | D^{i}_{k}, H^{j}, Z^{k}\right] = \sum_{n} n(\rho(n) - \rho_{ij}(n)) \\ &= \sum_{n} n \left(\sum_{c} \omega^{(I,\xi,\theta)} - \sum_{c} \omega^{(I,\xi,\theta)}_{ij}\right) \end{aligned} \tag{7.37}$$

where c denotes $(I, \xi) \in \mathscr{F}_n(\mathbb{L}) \times \Xi$ and $\theta \in \Theta$. Because $\sum_j P^{(\xi,\theta)}_{i}(H^{j}) = 1$. As given in (7.24)–(7.26), the posterior multi-target cardinality estimates is mainly dependent on the decision. Therefore, the coefficients γ_{ij} are reasonable set to be equal for all class hypotheses, i.e., $\gamma_{mn} = \gamma_m$ for all H^m when γ_{ij} is equal for all H^j given D^i:

$$\begin{aligned} \epsilon^{i}_{k,I} &= \sum_{j} \gamma_{ij} E\left[\left(|I^{ij}_{k} - \hat{I}_k|\right) | D^{i}_{k}, H^{j}, Z^{k}\right] P_i(H^{j}) \\ &= \gamma_i \sum_{n} n \left(\sum_{c} \omega^{(I,\xi,\theta)} - \sum_{c} \omega^{(I,\xi,\theta)}_{i}\right) \end{aligned} \tag{7.38}$$

Moreover, because the classes of different targets are independent, the hypothesis probability

$$P_n(H^{m}|Z_k) = \prod_{\ell} P_n\left(H^{j}_{\ell}\right) \tag{7.39}$$

Calculating the cost using (7.34)–(7.39), then the optimal decision is $D_k^n : C_n(Z_k) \leq C_i(Z_k), \forall i$, and the corresponding target state estimates are derived using the conditional LMB filter. The proposed recursive multi-target JDTC algorithm is summarized as follows:

Algorithm 1 The recursive CJDE-LMB algorithm

1: Predict prior multi-target density using the class-dependent dynamic model according to the hypothesis.
2: Update $\check{x}_{k,\ell}^i$, $P_n(H_\ell^j)$ and $\omega_n^{(I,\theta)}$ for decision D_k^n using the conditional LMB filter.
3: Calculate the joint detection, tracking and classification cost $C_n(Z^k)$ using (7.34)–(7.39), and the optimal decision is then $D_k^n : C_n(Z_k) \leq C_i(Z_k), \forall i$.
4: Output the CJDE solution for time k: the optimal decision $D_k = D_k^n$, the target existence probability $r_n^{(\ell)}$ and the state estimate $\check{x}_{k,\ell}^i$.

7.3.2 *Gaussian Mixture Implementation*

In this subsection, the Gaussian mixture implementation of the proposed recursive JDTC algorithm is developed.

1. Prediction: Suppose that at time $k-1$, $p_{k-1}^{(\ell)}(x|H_\ell^j)$ is the density of track ℓ that can be typically modeled by a Gaussian mixture (GM):

$$p_{k-1}^{(\ell)}\left(x|H_\ell^j\right) = \sum_{n=1}^{N_{k-1}^j} \omega_{k-1,j}^{(n)} \mathcal{N}\left(x, m_{k-1,j}^{(n)}, P_{k-1,j}^{(n)}\right) \tag{7.40}$$

where $m_{k-1,j}^{(n)}$ and $P_{k-1,j}^{(n)}$ are the mean value and covariance of the state vector, the predicted multi-target density can then be represented as (7.18), and the predicted density of multi-target is

$$\pi_+(X_+) = \Delta(X_+) \sum_{(I_+,\xi)\in\mathcal{F}(\mathbb{L}_+)\times\Xi} \omega_+^{(I_+,\xi)} \delta_{I_+}(\mathcal{L}(X_+)) \left[\sum_j p_+^{(\xi)}\left(x,\ell|H_\ell^j\right) p^{(\xi)}\left(H_\ell^j\right)\right]^{X_+} \tag{7.41}$$

the density of track ℓ is

$$p_{k|k-1,s}^{(\ell)}\left(x|H_\ell^j\right) = \sum_{n=1}^{N_{k-1,\ell}^j} \omega_{k-1,j}^{(n)} \mathcal{N}\left(x; m_{k|k-1,j}^{(n)}, P_{k|k-1,j}^{(n)}\right)$$
$$p_b^{(\ell)}\left(x|H_\ell^j\right) = \sum_{n=1}^{N_{b,\ell}^j} \omega_{b,k,j}^{(n)} \mathcal{N}\left(x; m_{b,k,j}^{(n)}, P_{b,k,j}^{(n)}\right) \tag{7.42}$$

where

$$m_{k|k-1,j}^{(n)} = F_{k|k-1}^j m_{k-1,j}^{(n)} \tag{7.43}$$

$$P_{k|k-1,j}^{(n)} = F_{k|k-1}^j P_{k-1,j}^{(n)} \left(F_{k|k-1}^j\right)^T + Q_{k|k-1}^j \tag{7.44}$$

Here, $F_{k|k-1}^j$ and $Q_{k|k-1}^j$ are the class dependent state transition matrix and process noise covariance matrix, and $m_{b,k,j}^{(n)}$ and $P_{b,k,j}^{(n)}$ are the mean and covariance of the state vector of new birth target.

2. Update: When the measurement set Z_k is collected at time k, the posterior multi-target density conditioned on the decision $\{D_k^n\}$ is:

$$\pi_k^n(\mathbf{X}|Z_k) = \Delta(\mathbf{X}) \sum_{(I_+,\theta)\in\mathcal{F}(\mathbb{L}_+)\times\Theta} \omega_n^{(I,\theta)}(Z_k)\delta_{I_+}(\mathcal{L}(\mathbf{X})) \tag{7.45}$$
$$\times \left[p_n^{(\theta)}\left(\cdot, \ell|H_\ell^j, D_{k,\ell}^i, Z_k\right) P_n^{(\theta)}\left(H_\ell^j|D_{k,\ell}^i, Z_k\right)\right]^{\mathbf{X}}$$

where the weight is:

$$\omega_n^{(I,\theta)}(Z_k) \propto \omega_+^{(I_+)} \left[\eta_Z^{(\theta)}\left(\ell|D_{k,\ell}^i, H_\ell^j\right)\right]^{I_+} \tag{7.46}$$

$$\eta_Z^{(\theta)}\left(\ell|D_{k,\ell}^i, H_\ell^j\right) = 1_{\mathcal{D}_\ell^i}(z_{\theta(\ell)})\Bigg((1-p_d) + p_d\frac{1}{\lambda c(k)} \tag{7.47}$$
$$\times \sum_{n=1}^{N_{k|k-1}^j} \omega_{k|k-1,j}^{(n)} \mathcal{N}\left(z; H_k m_{k|k-1,j}^{(n)}, H_k P_{k|k-1,j}^{(n)} H_k^T + R_k\right)\Bigg)$$

and the posterior density of each target is:

$$p_n^{(\theta)}(x, \ell|Z_k) = \sum_{n=1}^{N_{k|k-1}^j} \omega_{k|k-1}^{(n)}\Bigg((1-p_d)\mathcal{N}\left(x; m_{k|k-1,j}^{(n)}, P_{k|k-1,j}^{(n)}\right)$$
$$+p_d q_{k,j}^{(n)}(z_{\theta(\ell)})\mathcal{N}\left(x; m_{k,j}^{(n)}, P_{k,j}^{(n)}\right)\Bigg) \tag{7.48}$$

where

$$q_{k,j}^{(n)}(z_{\theta(\ell)}) = \mathcal{N}\left(z_{\theta(\ell)}; H_k m_{k|k-1,j}^{(n)}, P_{k|k-1,j}^{(n)}\right) \tag{7.49}$$

$$m_{k,j}^{(n)} = m_{k|k-1,j}^{(n)} + K_{k,j}^{(n)}\left(z_{\theta(\ell)} - H_k m_{k|k-1,j}^{(n)}\right) \tag{7.50}$$

$$P_{k,j}^{(n)} = \left(I - K_{k,j}^{(n)} H_k\right) P_{k|k-1,j}^{(n)} \tag{7.51}$$

$$K_{k,j}^{(n)} = P_{k|k-1,j}^{(n)} H_k^T \left(H_k P_{k|k-1,j}^{(n)} H_k^T + R_k\right)^{-1} \tag{7.52}$$

Then the approximated target density can be derived using (7.30)–(7.32).

3. Calculate the risk: As given in (7.35), the class dependent posterior estimate and associated covariance are:

$$\begin{aligned} \hat{x}_{k,\ell}^{ij} &= \sum_{n=1}^{N_k^j} \omega_{k,ij}^{(n)} m_{k,ij}^{(n)} \\ P_{k,\ell}^{ij} &= \sum_{n=1}^{N_k^j} \omega_{k,ij}^{(n)} \left(P_{k,ij}^{(n)} + \left(m_{k,ij}^{(n)} - \hat{x}_k^{ij}\right)\left(m_{k,ij}^{(n)} - \hat{x}_k^{ij}\right)^T\right) \end{aligned} \tag{7.53}$$

and the optimal estimates of track ℓ is:

$$\check{x}_{k,\ell}^{i} = \sum_{j=1}^{J} \hat{x}_{k,\ell}^{ij} P_k^i(H_j) \tag{7.54}$$

For the explicit Gaussian mixture implementation of the conditioned LMB filter, the estimation cost in the risk can be given by

$$\varepsilon_X = \sum_c \sum_\ell \omega_n \left(tr\left(P_{k,\ell}^{ij}\right) + \left(\hat{x}_{k,\ell}^{ij} - \check{x}_{k,\ell}^{i}\right)'\left(\hat{x}_{k,\ell}^{ij} - \check{x}_{k,\ell}^{i}\right)\right) \tag{7.55}$$

Finally, compute the CJDE cost for decision D_k^n, then the optimal solution can be derived.

7.3.3 Analysis of Risk Parameters

As defined in (7.7), the coefficients $\alpha_{ij} \geq 0$, $\beta_{ij} \geq 0$, and $\gamma_{ij} \geq 0$ in the new risk $\bar{R}_C(z)$ are used as relative weights to combine the decision and estimation costs. These coefficients play important roles on the risk because their different choices may result in different $\bar{R}_C$, thereby leading to different JDTC results. In this section, we analyze their effects.

The values of α_{ij} and β_{ij} modify the classification and tracking costs, and the effects of these two parameters are discussed in [5] for the target JTC problem. Because the existence of the target needs to be inferred in the JDTC problem, herein we discuss the effect of $\gamma_i = \sum_j \gamma_{ij}$. For simplicity, we confine $\alpha_{ij} = 1$, $c_{ii} = 0$, $c_{ij} = 1$, $\beta_{ii} = \beta_{ij}$, and $\gamma_{ii} = \gamma_{ij}$.

If γ_i is large, the estimation cost of target number will be predominant. As the data regions for all decisions form a partition of the measurement space, different decisions may lead to different target detection results. Therefore, when γ_i is large, this JDE problem is solved with a certain target number derived using an optimal Bayesian estimator. However, if γ_i is relatively too large, the CJDE cost $C_i(Z) \approx \gamma_i \epsilon_p^i$. In this case, the results of tracking and classification do not affect the detection, and the interaction between the decision and estimation is weakened.

If γ_i is small, the CJDE cost $C_{ij}(Z) \approx \alpha_{ij} c_{ij} + \beta_{ij} \epsilon_x^{ij}$. In this case, the JDTC result may not be correct, particularly when the dynamic behavior of the target changes. For example, consider two air target classes: commercial and military aircraft. Suppose that the target class is distinguished by the kinematic data, the classification results tend to favor class 1 if no maneuvers are performed. When a maneuver is performed suddenly at time k, the true measurement may not belong to the data region for original decision D_{k-1}^i, and the target tends to be judged as missed for less tracking and classification costs. As a result, both the state estimation and classification results are likely to be incorrect because the target is undetected. Actually, all costs need to contribute to risk $\bar{R}_C$ significantly. There is a relative balance between the target detection, tracking, and classification. Because of the uncertainties of the target existence, if γ_i is small, the cost of the target detection can be ignored, and the solution with less state estimation and classification costs will be chosen, but the tracking and classification results are meaningless when the target is undetected. On contrary, when γ_i is too large, the risk $\bar{R}_C$ is dominated by the detection cost, and the interaction of the detection, tracking, and classification is weakened.

Appropriate parameters should be chosen to balance the target detection, tracking, and classification. As given in [5], the parameters α_{ij} and β_{ij} can be chosen to make $\alpha_{ij} \cdot 1 \approx \beta_{ij} \cdot \max(\varepsilon_{ij})$, where $\max(\varepsilon_{ij})$ denotes the maximum expected estimation cost, and $\varepsilon_{ij} = mse(\hat{x}^{ij}|Z, D^i, H^j) + E[(\hat{x}^{ij} - \hat{x})'(\hat{x}^{ij} - \hat{x})|Z, D^i, H^j]$ can be calculated using the simulated measurements that lie inside the decision region. As analyzed before, the tracking and classification results are meaningless when the target is undetected. Therefore, γ_i can be chosen to make the maximum cost of the target detection approximate equal to the sum of the maximum costs of estimation and classification. When no measurements lie inside the decision region $\mathscr{D}^i$, the maximum detection cost for decision D^i can be computed as $\gamma_i \cdot (1 - \bar{p})$, where $\bar{p}$ is the existence probability of one target calculated with an empty set of measurements. Therefore, to balance these costs, the parameters can be chosen to make $\gamma_i \cdot (1 - \bar{p}) \approx \sum_j (\alpha_{ij} \cdot 1 + \beta_{ij} \cdot \max(\varepsilon_{ij})) P(H^j)$.

7.4 Simulations

In this section, numerical examples are presented to illustrate the effectiveness and superiority of the proposed CJDE-LMB algorithm. In addition, the results derived with different parameters are also compared.

7.4.1 Example 1: Comparison with Different Methods

Suppose that there are several targets with two possible classes move in a two-dimensional scenario. The classes differ from each other in terms of the dynamic behaviors, each class has a corresponding set of possible motion models. The ith model for class j is:

$$x_k = F_{k,i} x_{k-1} + w_{k,i} \tag{7.56}$$

where $F_{k,i}$ is the model-dependent state transition matrix, and $w_{k,i}$ is the Gaussian noise with covariance $Q_{k,i}$. The target of class 1 only has the constant velocity (CV) model with the following parameters:

$$\begin{aligned} F_{k,1} &= \text{diag}\left(\begin{bmatrix} 1 & T \\ 0 & 1 \end{bmatrix}, \begin{bmatrix} 1 & T \\ 0 & 1 \end{bmatrix}\right) \\ Q_{k,1} &= \text{diag}\left(\begin{bmatrix} T^2 & T \\ T & 1 \end{bmatrix}, \begin{bmatrix} T^2 & T \\ T & 1 \end{bmatrix}\right)\sigma_v^2 \end{aligned} \tag{7.57}$$

where σ_v is the process noise with the covariance $\sigma_v^2 = 1\,\text{m}^2/\text{s}^2$.

The target of class 2 has two possible dynamic models, the CV model as before, and the constant accelerate (CA) model with parameters

$$\begin{aligned} F_{k,2} &= \text{diag}\left(\begin{bmatrix} 1 & T & \frac{1}{2}T^2 \\ 0 & 1 & T \\ 0 & 0 & 1 \end{bmatrix}, \begin{bmatrix} 1 & T & \frac{1}{2}T^2 \\ 0 & 1 & T \\ 0 & 0 & 1 \end{bmatrix}\right) \\ Q_{k,2} &= \text{diag}\left(\begin{bmatrix} \frac{1}{4}T^4 & \frac{1}{2}T^3 & \frac{1}{2}T^2 \\ \frac{1}{2}T^3 & T^2 & T \\ \frac{1}{2}T^2 & T & 1 \end{bmatrix}, \begin{bmatrix} \frac{1}{4}T^4 & \frac{1}{2}T^3 & \frac{1}{2}T^2 \\ \frac{1}{2}T^3 & T^2 & T \\ \frac{1}{2}T^2 & T & 1 \end{bmatrix}\right)\sigma_a^2 \end{aligned} \tag{7.58}$$

where σ_a is the process noise with the covariance $\sigma_a^2 = 10\,\text{m}^2/\text{s}^4$. The model transition probability matrix is set as

$$\pi = \begin{bmatrix} 0.7 & 0.3 \\ 0.3 & 0.7 \end{bmatrix} \tag{7.59}$$

The kinematic measurement is $z_k = [x_k, y_k]' + w_k$, where $[x_k, y_k]$ is the position of the target, and w_k is the Gaussian measurement noise with the covariance $R_k = \text{diag}[\sigma_x^2, \sigma_y^2]$, $\sigma_x = \sigma_y = 2\,\text{m}$. The target detection probability $p_d = 0.98$, and the intensity of the Poisson distributed clutter is 6×10^{-5}.

In the scenario, there are two non-maneuvering targets and one maneuvering target move within the two-dimensional scenario. Target 1 moves straight from the beginning to the end, with the initial location $[-200, 700]\,\text{m}$ and velocity $[50, 0]\,\text{m/s}$. Target 2 appears at $k = 5$ with the initial location $[-200, 1000]\,\text{m}$, and moves straight with constant velocity $[40, 30]\,\text{m/s}$ until it disappears at $k = 25$. The maneuvering target 3 appears at $k = 3$ and disappears at $k = 27$. It moves straight from location $[0, 1900]\,\text{m}$ with a constant acceleration of $[4, -3]\,\text{m/s}^2$.

The multi-target detection, tracking, and classification performance of the CJDE-LMB algorithm is compared with the traditional methods in terms of the target cardinality estimates, optimal subpattern assignment (OSPA) distance, and the probability of correct classification, respectively. Moreover, the overall performance is evaluated by the joint performance metric (JPM), which is calculated with the costs of target detection, tracking, and classification.

The compared methods are as follows:

1. Estimation-then-Decision: The target state is first estimated using the GNN approach, and the decision is then made based on the ratio of current measurement likelihoods of the predicted states conditioned on different hypotheses.
2. Decision-then-Estimation: The target class is first determined, which minimizes the Bayes decision risk, and the target state is then estimated given the decided class.
3. Estimate the joint density-probability of target states and class: As proposed in [21], the class-dependent posterior density is firstly calculated using the particle implementation of the PHD filter with the corresponding dynamic models. Then, the target state and class probabilities are obtained by clustering the particles. This method is referred to as YW-JDTC here.

In the simulation, the target survival probability is $p_s = 0.98$, and the target birth probability is $p_b = 0.02$. The density of the new birth target is $b_k = \mathcal{N}(x; m_b, Q_b)$, where the parameters $m_{\gamma,k}^1 = [-200, 50, 0, 700, 0, 0]^T$, $m_{\gamma,k}^2 = [-200, 40, 0, 1000, 30, 0]^T$, and $m_{\gamma,k}^3 = [0, 20, 4, 1900, -15, -3]^T$, while the state covariances are $P_{\gamma,k}^1 = P_{\gamma,k}^2 = P_{\gamma,k}^3 = \text{diag}([100, 10, 1, 100, 10, 1])$. All the classes have an equal initial probability, and the initial probabilities of the two models for the maneuvering hypothesis are equal to 0.5.

According to the guidance of parameter choice provided before, the parameters in the new CJDE risk are set to be $\alpha_{mn}^1 = 20$, $\beta_{mn}^1 = 1$, $\gamma_{mn}^1 = 100$. The simulation results are obtained over 1000 Monte Carlo trials.

Figure 7.1 illustrates the estimate of the multi-target cardinality. The targets are correctly detected by the proposed CJDE-LMB filter. The reason is the coefficient γ in the new CJDE risk is relatively large, the cost of target detection plays a more important role. The tracking performance is shown in Fig. 7.2. As illustrated, the

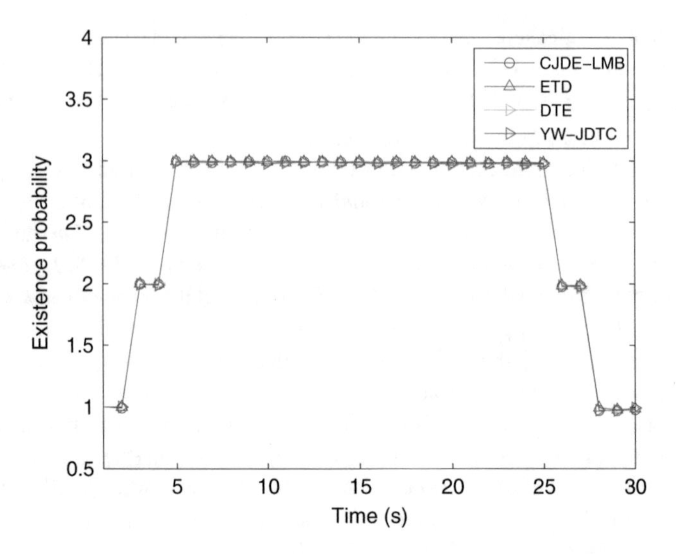

Fig. 7.1 Real target number and cardinality estimates

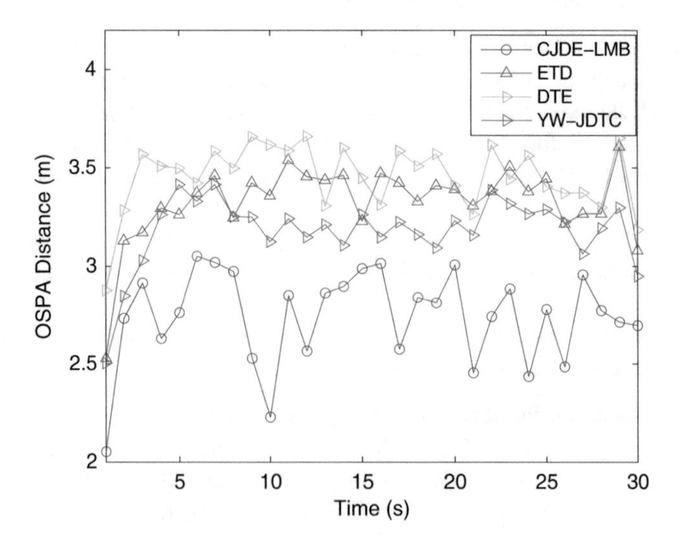

Fig. 7.2 OSPA distance

CJDE-LMB is the best in terms of the OSPA distance. The explanation of this result is that the interdependence between the decision and the estimation is considered, and the multi-target states are updated with reasonable MTA. On contrary, the decision of the target class is not regarded in the tracking when using ETD and YW-JDTC methods, while the error of the decision is not considered in the DTE method. Figure 7.3 shows the classification results. The CJDE-LMB algorithm also performs best while the ETD method is the worst. The reason for this phenomenon is that the

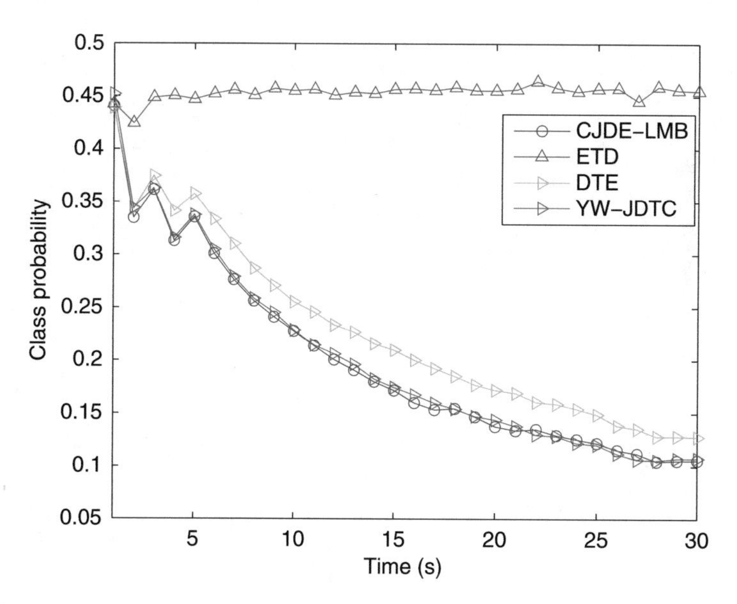

Fig. 7.3 Average incorrect decision rate

decision is only dependent on the current state estimation in the ETD method. In addition, although the superiority of the proposed CJDE-LMB algorithm over the YW-JDTC method is not very obvious, CJDE-LMB provides explicit decisions of the target classes, whereas YW-JDTC only computes the class probabilities.

Summing up all the costs and the overall performance is evaluated in terms of the JPM. As depicted in Fig. 7.4, the performance of the CJDE-LMB algorithm is better than that of the other methods.

This example shows that the performance of estimation and decision are improved because the interdependence between them are considered. Moreover, the proposed algorithm achieves the final goal directly and the explicit estimation and classification result are derived.

7.4.2 Example 2: Analysis of Different Risk Parameters

Consider the scenario similar to Example 1, three targets move in the two-dimensional scenario. The motion of target 1 and target 2 are the same as Example 1. For the maneuvering target, it moves straight with the constant velocity $[-20, 15]$ m/s for 7 steps. Then, it executes constant acceleration movement during $k = 8 - 27$ with $[-8, 6]\,\mathrm{m/s}^2$.

In order to illustrate the importance of the coefficients in the new Bayesian risk, the JDTC results are derived with different parameters in this example. Suppose that the coefficients are set to be $\alpha_{ij}^1 = 20, \beta_{ij}^1 = 1, \gamma_{ij}^1 = 100$, and $\alpha_{ij}^2 = 20, \beta_{ij}^2 =$

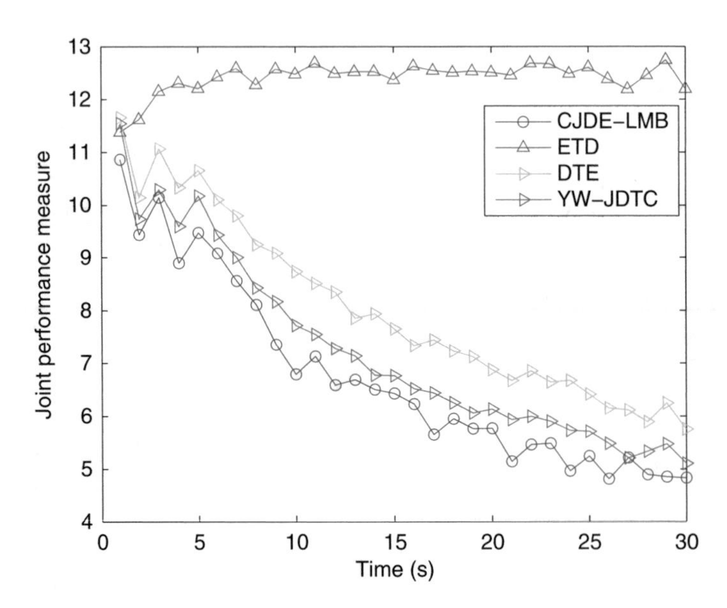

Fig. 7.4 Joint performance measure

$1, \gamma_{ij}^2 = 10$, respectively. The values of α and β make the costs of state estimation and classification balance. When $\gamma = 100$, the target detection plays a dual role as before, on contrary, when $\gamma = 10$, the cost of target miss detection contributes to $\bar{R}_C$ less significantly.

The performance of target detection, tracking, and classification under different parameters is illustrated in Figs. 7.5, 7.6, and 7.7, respectively. As shown in Fig. 7.5, when all the targets keep their motion modes, the tracks are detected correctly. After the target 3 executes constant acceleration, the tracks of all targets are maintained for $\gamma = 100$, whereas there exists target miss detection on some trials when $\gamma = 10$. The reason for this phenomenon is that after the target 3 performs maneuver, the optimal Bayesian decision converts to maneuvering, both the costs of estimation and decision increase due to the transition of dynamic model and the change of optimal Bayesian decision, respectively. In this case, the targets can be correctly detected when $\gamma = 100$ because the penalization is heavier on target miss detection. On contrary, the decision with less state estimation and classification costs is chosen when $\gamma = 10$.

Due to the incorrect target detection results, the average tracking and classification performance given $\gamma = 10$ is worse than $\gamma = 100$ as illustrated in Figs. 7.6 and 7.7. Moreover, the overall performance given $\gamma = 100$ is also better as shown in Fig. 7.8.

This example shows that because target detection is the prerequisite for accurate tracking and correct classification in the multi-target JDTC problem, the penalization on target miss detection need to be heavier.

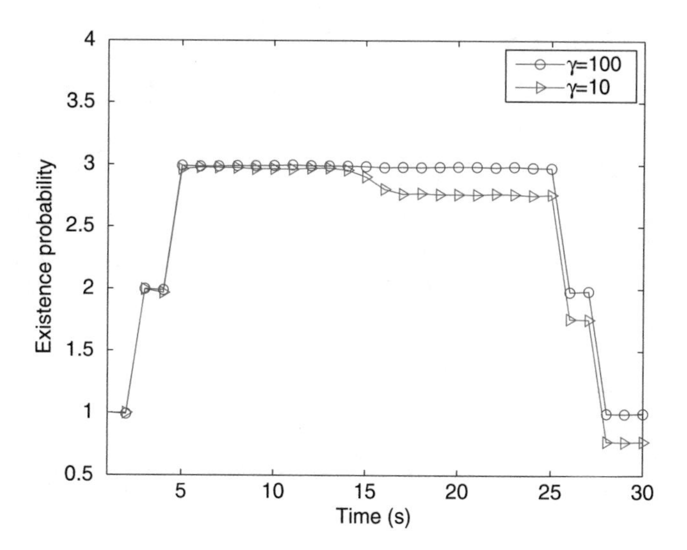

Fig. 7.5 Target existence probability estimates

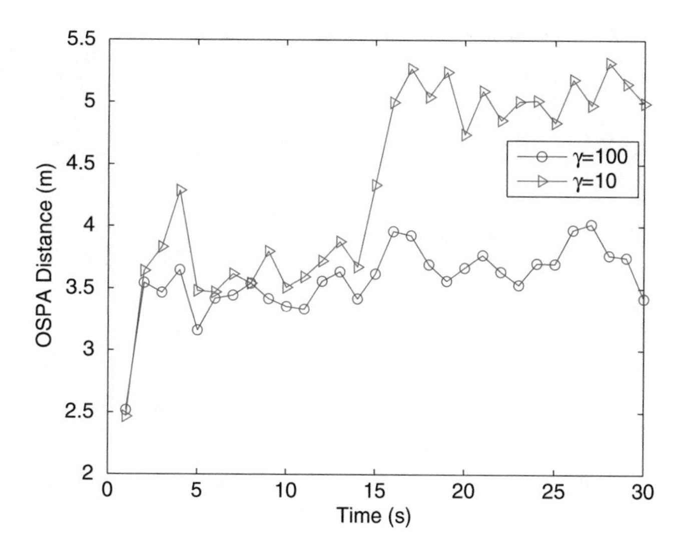

Fig. 7.6 OSPA distance

7.5 Conclusion

In this chapter, a recursive CJDE-LMB algorithm was proposed to solve the target joint detection, tracking, and classification problem. The optimal solution is derived based on a new generalized Bayesian risk defined for the label RFS variables involving the costs of target existence probability estimation (detection), state estimation (tracking), and classification. Furthermore, the density of the target

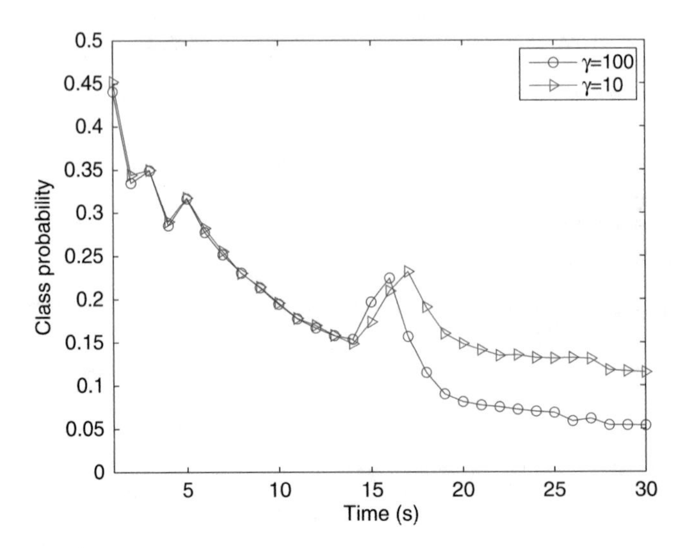

Fig. 7.7 Classification results

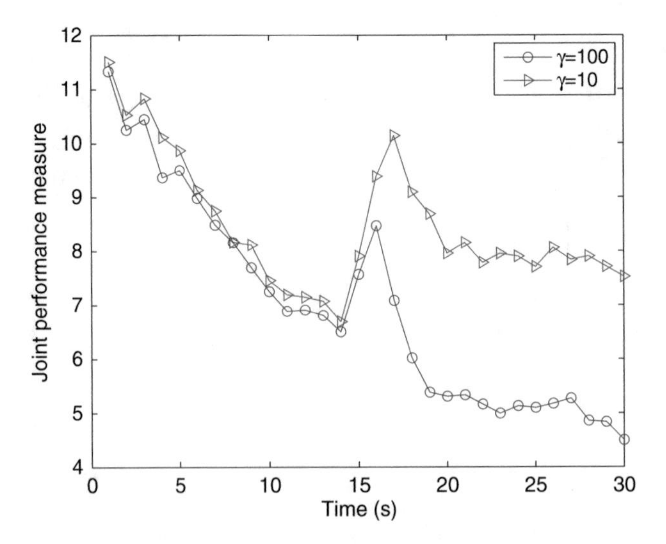

Fig. 7.8 Joint performance metric

is approximated using a sum of components and the computational complexity is largely reduced. Moreover, the method of the parameter selection for the new Bayesian risk is provided in order to derive reasonable solutions. In the simulations, the proposed algorithm was applied to the JDTC problem of a target with two possible classes, and the performance was compared with traditional methods. The results show that our method has the best performance, and the guidance of the parameter choice was also proven to be correct.

References

1. Bar-Shalom Y, Kirubarajan T, Gokberk C (2005) Tracking with classification-aided multiframe data association. IEEE Trans Aerosp Electron Syst 41(3):868–878
2. Cao W, Lan J, Li XR (2014) Joint tracking and classification based on recursive joint decision and estimation using multi-sensor data. In: Proceedings of the 17th international conference on information fusion, pp 1–8
3. Cao W, Lan J, Li XR (2015) Joint multi-target detection and tracking using conditional joint decision and estimation with ospa-like cost. In: Proceedings of the 18th international conference on information fusion, pp 1740–1747
4. Cao W, Lan J, Li XR (2015) Joint tracking and classification based on conditional joint decision and estimation. In: Proceedings of the 18th international conference on information fusion, pp 1764–1771
5. Cao W, Lan J, Li XR (2016) Conditional joint decision and estimation with application to joint tracking and classification. IEEE Trans Syst Man Cybern Syst 46(4):459–471
6. Hachour S, Delmotte F, Mercier D, Lefèvre E (2014) Object tracking and credal classification with kinematic data in a multi-target context. Inf Fusion 20:174–188
7. Li XR (2007) Optimal bayes joint decision and estimation. In: Proceedings of the 10th international conference on information fusion, pp 1–8
8. Li M, Jing Z, Dong P, Pan H (2017) Joint dtc based on fisst and generalised bayesian risk. IET Signal Proc 11(7):796–804
9. Lin G, Sun W, Wei P (2016) Extensions of the cbmember filter for joint detection, tracking, and classification of multiple maneuvering targets. Digital Signal Process 1(56):35–42
10. Liu Y, Li XR (2011) Recursive joint decision and estimation based on generalized bayes risk. In: Proceedings of the 14th international conference on information fusion, pp 1–8
11. Mahler RPS (2003) Multitarget bayes filtering via first-order multitarget moments. IEEE Trans Aerosp Electron Syst 39(4):1152–1178
12. Mahler R (2007) Statistical multisource-multitarget information fusion. Artech House, Norwood
13. Powell G, Marshall D, Smets P, Ristic B, Maskell S (2006) Joint tracking and classification of airbourne objects using particle filters and the continuous transferable belief model. In: Proceedings of the 9th international conference on information fusion, pp 1–8
14. Punithakumar K, Kirubarajan T, Sinha A (2008) Multiple-model probability hypothesis density filter for tracking maneuvering targets. IEEE Trans Aerosp Electron Syst 44(1):87–98
15. Reuter S, Vo BT, Vo BN, Dietmayer K (2014) The labeled multi-bernoulli filter. IEEE Trans Signal Process 62(12):3246–3260
16. Ristic B, Gordon N, Bessell A (2004) On target classification using kinematic data. Inf Fusion 5:15–21
17. Ristic B, Vo BN, Clark D, Vo BT (2011) A metric for performance evaluation of multi-target tracking algorithms. IEEE Trans Signal Process 59(7):3452–3457
18. Svensson D, Wintenby J, Svensson L (2009) Performance evaluation of mht and gm-cphd in a ground target tracking scenario. In: Proceedings of the 12th international conference on information fusion, pp 300–307
19. Vo BT, Vo BN (2013) Labeled random finite sets and multi-object conjugate priors. IEEE Trans Signal Process 61(13):3460–3475
20. Vo BN, Vo BT, Phung D (2014) Labeled random finite sets and the bayes multi-target tracking filter. IEEE Trans Signal Process 62(24):6554–6567
21. Wei Y, Yaowen F, Jianqian L, Xiang L (2012) Joint detection, tracking, and classification of multiple targets in clutter using the phd filter. IEEE Trans Aerosp Electron Syst 48(4):3594–3609

Chapter 8
Redundant Adaptive Robust Tracking for Active Satellite

8.1 Introduction

In this chapter, we present a switching-based H_∞ robust tracking method with application to tracking an active target satellite. The method is built on differential-orbital-element-based kinematic models, describing the orbital tracking motion, and a filtering algorithm named the redundant adaptive robust extended Kalman filter (RAREKF).

The RAREKF is developed because of the conservativeness of the traditional H_∞ robust filter and the existence of the model error. On the one hand, conservative design of the H_∞ robust filter results in the excessive loss of filtering optimality and decreases the filtering precision. On the other hand, for describing orbital relative motion, nonlinear modeling error is inevitable and leads to failure of the switching scheme of a previous switching-structured filter so that the filtering conservativeness cannot be improved. Hence, we introduced an error-redundant switching scheme in an H_∞ robust EKF, generating the RAREKF. Stability analysis and error evaluation are both given to discuss the presented filter.

We use a numerical example to show the advantage of the RAREKF to other typical filters. By using the designed evaluation method, the RAREKF has the lowest error level and the best performance among all the compared filters.

This chapter is organized as follows. We first give the satellite tracking models to describe relative motion between two Keplerian orbits. Then, we present the detailed filtering algorithm and discussions of the RAREKF and the RAREKF-based satellite tracking algorithm. Finally, some numerical simulations are demonstrated. The presentation of this chapter is primarily based on the works [15, 17].

Z. Jing et al., *Non-Cooperative Target Tracking, Fusion and Control*,
Information Fusion and Data Science, https://doi.org/10.1007/978-3-319-90716-1_8

8.2 Active Satellite Tracking

Inter-satellite tracking is a type of space operation, aiming to provide real-time motion information of the target satellite for a chaser satellite. It is one of the most significant techniques for many space missions, including spacecraft rendezvous, debris capture, space robot on-orbit service, etc.

8.2.1 *Traditional Problem*

Previously, the research of inter-satellite tracking is primarily focused on GPS tracking methods [5, 6]. Due to the innovative information provided by GPS system, satellite orbit can be identified with high precision [14, 28]. After that, Carrier-Phase Differential GPS (CDGPS) appeared and used in the navigation of formation flying [2, 26], achieving cm-level position precision for LEO satellites [20]. Currently, the application range has been extended to GEO and HEO that are higher than the GPS constellation [22].

Practically, for GPS-based tracking, we have to face potential problems like communication link jam, signal loss, system fault and so on, and it is questionable to use onboard GPS receiver only. So, we need to build the capability of autonomous inter-satellite tracking, realizing target motion determination by using direct relative measurements. The inter-satellite tracking scheme was first studied by Markley [19], Chow and Culp [3]. Psiaki [21] researched the effects from orbital parameters and J2 perturbation to observability. Kawase [12] adopted a relative motion model for the first time for relative orbit determination. With the improvement of relative orbital motion theory and detective techniques, inter-satellite relative tracking shows great importance in many areas such as rendezvous [9, 10], formation [8], and, in particular, the space operations to noncooperative space objects [25, 27].

8.2.2 *Noncooperative Maneuvering Satellite Tracking*

In past works, tracking targets are usually assumed passive, but it is not always true if the target is noncooperative with potential maneuver, or if the chaser needs formation reconfiguration. Here we consider a general case of relative tracking. The chaser satellite gets relative range and azimuth from onboard detective devices and the target satellite is active, having potential unknown maneuvers during tracking.

Compared with traditional inter-satellite tracking, the problem of the relative tracking stated above has two aspects affected greatly.

First is the orbital motion model. To describe the orbital motion, it requires a dynamical model of relative motion of using an unknown term for target maneuver, but such a model makes it quite hard to get a closed-form solution for target tracking and to estimate the exact target maneuver because of the noncooperative properties. The inherent nonlinear error is evitable so that the tracking errors of relative position and velocity will be accumulated. Second is the target tracking algorithm. The target maneuver is usually regarded as external disturbance, under which the traditional Kalman filters with optimal filtering gains may diverge due to poor robustness.

That necessitates the development of the adaptive and robust filtering methods such as those based on online adjusting noise covariance [30] or filter gains [24]. Xiong provided an adaptive robust extended Kalman filter (AREKF) [29] by using a self-switching attenuation factor to acquire adaptive filtering gains that changed alternatively between optimal and H_∞ robust filtering modes.

However, due to the existence of unavoidable modeling errors, the switching scheme of the AREKF fails to work properly, remaining at the robust mode with excessive loss of optimality, and leading to increase of the filtering precision. Therefore, an error redundant method is required to reactivate the switching scheme of the AREKF.

8.2.3 Active Satellite Tracking Method

The method provided in this chapter has the following considerations:

First, introducing the concept of orbit osculation. Actually, active satellite runs on an osculating Keplerian orbit at each time point. So, the dynamic relative motion is a combination of a cluster of kinematic motions and we may use kinematic model to approximate the real dynamic motion. Due to the lower propagation to modeling error [11], the kinematic model based on differential orbital elements (DOE) [7] is adopted to describe the relative motion to an active target satellite.

Second, considering active satellite maneuver as an external disturbance. Together with the internal modeling error, both cause decrease of tracking precision. Therefore, we designed a new version of AREKF with error-redundancy, that is, the redundant AREKF (RAREKF). By introducing a redundancy factor, the RAREKF not only has adaptability to the target maneuver but also loosens the switching condition, yielding a redundancy to the modeling error.

Third, using the unbiased converted measurement (UCM) technique given in [18] to form the measurement equation [4] so that the residue of the measurement noises can be minimized and the unbiased state estimates can be obtained.

To judge the effectiveness of the tracking methods based on different models and filtering algorithms, we need tracking error evaluation. Clearly, tracking error is resulted by both model and filter, but in previous studies, only stochastic noises [11] or initial states [1] are considered in evaluating orbital models. The evaluation index designed in [11] neglects the effect from modeling error, whereas the index in [1] is not a well representation of the error vector with lack of physical meaning. Thus, we here give an index considering both relative model and filtering algorithm, making it more suitable for tracking error evaluation.

8.3 Relative Motion Models

Consider two satellites called Chaser and Target. The Chaser is supposed flying freely and the Target is active with potential maneuver. This section presents the kinematic models of relative motion, including state equation and measurement equation.

8.3.1 DOE-Based State Equation

Define Chaser orbital element set as $\sigma = (a, \theta, i, q_1, q_2, \Omega)^T$ where $\theta = \omega + f$, $q_1 = e\cos\omega$ and $q_2 = e\sin\omega$. θ and f are the argument of latitude and the true anomaly, respectively. This set is sued because it can avoid the singularity problem of the classical orbital elements. The differential orbital elements and the relative motional state are $\delta\sigma = \sigma_T - \sigma_C$ and $X = (x, y, z, \dot{x}, \dot{y}, \dot{z})^T$, the components of which are along radial, in-track, and cross-track, respectively.

As in [7], the transition matrix of DOE and the geometric relationship between $X(t)$ and $\delta\sigma(t)$ can be described by

$$\delta\sigma(t) = \varphi(t, t_0)\delta\sigma(t_0) \tag{8.1}$$

$$X(t) = \phi(t)\delta\sigma(t) \tag{8.2}$$

where t_0 refers to the initial time. Note n as mean angular rate of the Chaser orbit and r as the distance of the Chaser to the Earth center. Without considering natural perturbations, $\varphi(t, t_0)$ and $\phi(t)$ can be be expressed as follows:

$$\varphi(t, t_0) = \begin{bmatrix} 1 & 0 & 0 & 0 & 0 & 0 \\ -\frac{3n(t_0)}{2a(t_0)}\frac{t-t_0}{G_\theta(t)} & \frac{G_\theta(t_0)}{G_\theta(t)} & 0 & \frac{G_{q_1}(t)-G_{q_1}(t_0)}{G_\theta(t)} & \frac{G_{q_2}(t)-G_{q_2}(t_0)}{G_\theta(t)} & 0 \\ 0 & 0 & 1 & 0 & 0 & 0 \\ & O_{3\times 3} & & & I_{3\times 3} & \end{bmatrix}.$$

in which

$$G_\theta = \frac{r\eta}{a(1+f_1)}, \eta = \sqrt{1-q_1^2-q_2^2}, f_1 = q_1\cos\theta + q_2\sin\theta, f_2 = q_1\sin\theta - q_2\cos\theta$$

$$G_{q_1} = \frac{q_2}{\eta(1+\eta)} + \frac{q_1 f_2}{\eta(1+f_1)} - \frac{r(a+r)}{\eta a^2}(q_2 + \sin\theta)$$

$$G_{q_2} = -\frac{q_1}{\eta(1+\eta)} + \frac{q_2 f_2}{\eta(1+f_1)} + \frac{r(a+r)}{\eta a^2}(q_1 + \cos\theta)$$

$$\phi(t) = \begin{bmatrix} \phi_{11}(t) & \phi_{12}(t) \\ \phi_{21}(t) & \phi_{22}(t) \end{bmatrix}.$$

$$\phi_{11} = \begin{bmatrix} r/a & rf_2/(1+f_1) & 0 \\ 0 & r & 0 \\ 0 & 0 & r\sin\theta \end{bmatrix},$$

$$\phi_{21} = \frac{n}{\eta}\begin{bmatrix} -f_2/2 & a - r/\eta^2 & 0 \\ -3(1+f_1)/2 & -af_2 & 0 \\ 0 & 0 & a(q_1+\cos\theta) \end{bmatrix}$$

$$\phi_{12} = \frac{1}{\eta^2}\begin{bmatrix} -2raq_1 - r^2\cos\theta/a & -2raq_2 - r^2\sin\theta/a & 0 \\ 0 & 0 & \eta^2 r\cos i \\ 0 & 0 & -\eta^2 r\sin i\cos\theta \end{bmatrix}$$

$$\phi_{22} = \frac{n}{\eta^3}\begin{bmatrix} aq_1f_2 + r\sin\theta(1+f_1) & aq_1f_2 + r\sin\theta(1+f_1) & 0 \\ (1+f_1)(3aq_1 + 2r\cos\theta) & (1+f_1)(3aq_2 + 2r\sin\theta) & \eta^2 af_2\cos i \\ 0 & 0 & \eta^2 a(q_2+\sin\theta)\sin i \end{bmatrix}$$

As pointed in [7], X is in a curvilinear coordinate system as shown in Fig. 8.1. Actually, it is equivalent to the Cartesian system when the relative range is negligible compared with the radius of the Earth. So, the Cartesian state transfer function is

$$X(t_2) = \Phi(t_2, t_1)X(t_1) \tag{8.3}$$

$$\Phi(t_2, t_1) = \phi(t_2)\varphi(t_2, t_0)\varphi^{-1}(t_1, t_0)\phi^{-1}(t_1) \tag{8.4}$$

The equation above presents the kinematic state equation. It can be used as the dynamic relative motion model from t_1 to t_2 without requirement of knowing the Target maneuver.

8.3.2 UCM-Based Measurement Equation

Considering the nonlinearity of the measurement equation, we adopt the UCM technique provided in [18] to transform measurements by multiplying a conversion matrix and make them unbiased.

The usual radar measurements consist of relative range, azimuth, and elevation, noted with ρ, α, and β, respectively. The real measurement is $Z = (\rho, \alpha, \beta)^T$ with the actual $\bar{Z} = (\bar{\rho}, \bar{\alpha}, \bar{\beta})^T$ and covariance matrix $Q_Z = \mathrm{diag}(\sigma_\rho^2, \sigma_\alpha^2, \sigma_\beta^2)$. Then, the UCM-based measurement equation has the form as

$$Y(t) = X(t) + \zeta(t) \tag{8.5}$$

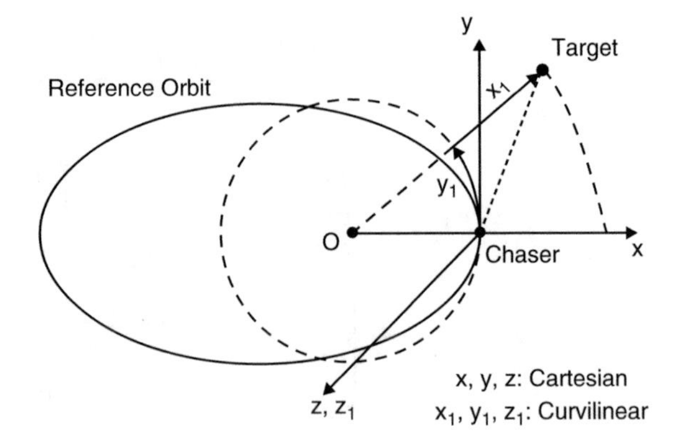

Fig. 8.1 The curvilinear and Cartesian coordinate systems

where

$$Y(t) = \begin{bmatrix} \rho(t)\cos\alpha(t)\cos\beta(t) \\ \rho(t)\cos\alpha(t)\sin\beta(t) \\ \rho(t)\sin\alpha(t) \end{bmatrix}, \quad X(t) = \begin{bmatrix} \bar{\rho}(t)\cos\bar{\alpha}(t)\cos\bar{\beta}(t) \\ \bar{\rho}(t)\cos\bar{\alpha}(t)\sin\bar{\beta}(t) \\ \bar{\rho}(t)\sin\bar{\alpha}(t) \end{bmatrix}.$$

and $\zeta(t)$ is the covariance matrix of the transformed measurement noise with the expression [4] as

$$Q(t) = \begin{bmatrix} Q_{11}(t) & Q_{12}(t) & Q_{13}(t) \\ Q_{12}(t) & Q_{22}(t) & Q_{23}(t) \\ Q_{13}(t) & Q_{23}(t) & Q_{33}(t) \end{bmatrix} \tag{8.6}$$

$$\begin{aligned}
Q_{11}(t) &= -\lambda_\alpha^2\lambda_\beta^2\rho^2(t)\cos^2\alpha(t)\cos^2\beta(t) \\
&\quad + \frac{1}{4}(\rho^2(t)+\sigma_\rho^2)(1+\lambda_\alpha^4\cos 2\alpha(t))(1+\lambda_\beta^4\cos 2\beta(t)) \\
Q_{22}(t) &= -\lambda_\alpha^2\lambda_\beta^2\rho^2(t)\cos^2\alpha(t)\sin^2\beta(t) \\
&\quad + \frac{1}{4}(\rho^2(t)+\sigma_\rho^2)(1+\lambda_\alpha^4\cos 2\alpha(t))(1-\lambda_\beta^4\cos 2\beta(t)) \\
Q_{33}(t) &= -\lambda_\alpha^2\rho^2(t)\sin^2\alpha(t) + \frac{1}{2}(\rho^2(t)+\sigma_\rho^2)(1-\lambda_\alpha^4\cos 2\alpha(t)) \\
Q_{12}(t) &= -\frac{1}{2}\lambda_\alpha^2\lambda_\beta^2\rho^2(t)\cos^2\alpha(t)\sin 2\beta(t) \\
&\quad + \frac{1}{4}(\rho^2(t)+\sigma_\rho^2)(1+\lambda_\alpha^4\cos 2\alpha(t))\lambda_\beta^4\sin 2\beta(t) \\
Q_{13}(t) &= -\frac{1}{2}(\rho^2(t)(1-\lambda_\alpha^2)-\sigma_\rho^2)\lambda_\alpha^2\lambda_\beta\sin 2\alpha(t)\cos\beta(t) \\
Q_{23}(t) &= -\frac{1}{2}(\rho^2(t)(1-\lambda_\alpha^2)-\sigma_\rho^2)\lambda_\alpha^2\lambda_\beta\sin 2\alpha(t)\sin\beta(t)
\end{aligned}$$

where $\lambda_\alpha = e^{-\sigma_\alpha^2/2}$, $\lambda_\beta = e^{-\sigma_\beta^2/2}$. The $\zeta(t)$ can be guaranteed zero-mean by the conversion matrix as

$$J = \begin{bmatrix} \lambda_\alpha\lambda_\beta & 0 & 0 \\ 0 & \lambda_\alpha\lambda_\beta & 0 \\ 0 & 0 & \lambda_\alpha \end{bmatrix} \tag{8.7}$$

8.4 Relative Tracking Algorithm

For Eqs. (8.3) and (8.5), showing the mathematic description of tracking, we use a generalized form as

$$X(k) = f(k, X(k-1)) + w(k) \tag{8.8}$$

$$Y(k) = H(k)X(k) + v(k) \tag{8.9}$$

where $k \in N$, $X \in \mathbb{R}^n$ and $Y \in \mathbb{R}^m$ is the measurement vector and $f(X)$ a nonlinear and differentiable function. The noises w and v are uncorrelated zero-mean white Gaussian with covariance matrices W and V.

8.4.1 Traditional Robust EKF

Due to the potential maneuver of an active target, the traditional Kalman filters, with optimal but rigid filtering gains, are hard to use under such disturbances from the Target. It becomes necessary to apply robust EKF methods such as in [16, 24] and [29]. Here we use an operator S to unify them and present a generalized form of robust EKF algorithms as the following steps.

Prediction One-step state prediction and covariance matrix

$$\hat{X}(k|k-1) = f(k, \hat{X}(k-1|k-1)) \tag{8.10}$$

$$P(k|k-1) = F(k-1)P(k-1|k-1)F^T(k-1) + W(k-1) \tag{8.11}$$

where $F = \partial f(X)/\partial X|_{X=\hat{X}}$.

Robustness Prediction error covariance matrix

$$\Sigma(k|k-1) = S(P(k|k-1)) \tag{8.12}$$

Innovation Measurements innovation and covariance matrix

$$\tilde{Y}(k) = Y(k) - H(k)\hat{X}(k|k-1) \tag{8.13}$$

$$P_Y(k) = H(k)\Sigma(k|k-1)H^T(k) + V(k) \tag{8.14}$$

Update Filtering gain and state estimates

$$K(k) = \Sigma(k|k-1)H^T(k)P_Y^{-1}(k) \tag{8.15}$$

$$\hat{X}(k|k) = \hat{X}(k|k-1) + K(k)\tilde{Y}(k) \tag{8.16}$$

$$P(k|k) = (\Sigma^{-1}(k|k-1) + H^T(k)V^{-1}(k)H(h))^{-1} \tag{8.17}$$

The robust EKF is an extended version of the EKF, equivalent to it if S is an identity operator. Through modifying the prediction error covariance matrix with S, the method generates robustness in price of sacrifice of filtering optimality. So, S reflects the filter robustness.

A convenient way is to use a diagonal operator [16] as

$$\Sigma(k|k-1) = \text{diag}(P(k|k-1)) \tag{8.18}$$

The operator resets the correlation part of prediction error variance at each iterative cycle so that the coupling terms of the measurement covariance matrix can be eliminated. The method is called uncoupled EKF (UEKF), showing anti-disturbance performance better than EKF. Nevertheless, it is not real robust because there is no robust index considered during design process.

A typical robust type of S given in [24], using an H_∞ index to guarantee the filtering stability when the linearization error reaches its upper bound, is as

$$\Sigma(k|k-1) = (P^{-1}(k|k-1) - \gamma^{-2}L^T(k)L(k))^{-1} \tag{8.19}$$

where $L(k) \in \mathbb{R}^n$ is a tuning matrix. γ is an attenuation factor larger than the ratio of the estimation error to the sum of all the errors (measurement noise, perturbation, target maneuver, modeling error), balancing the filtering robustness and optimality.

For given γ, the filter will work at constant robust status with sustained loss of filtering optimality, leading to excessive decrease of estimation precision. To avoid that, AREKF is developed in [29] by designing S as the following switching structure:

$$\Sigma(k|k-1) = \begin{cases} (P^{-1}(k|k-1) - \gamma^{-2}L^T(k)L(k))^{-1}, & \bar{P}_Y(k) > \alpha P_Y(k) \\ P(k|k-1), & \text{otherwise.} \end{cases} \tag{8.20}$$

where α is a redundancy factor, and $\bar{P}_Y(k)$ has the form as

$$\bar{P}_Y(k) = \begin{cases} \tilde{Y}(k)\tilde{Y}^T(k), & k = 0 \\ \frac{\rho P_Y(k-1) + \tilde{Y}(k)\tilde{Y}^T(k)}{\rho+1}, & k > 0. \end{cases} \tag{8.21}$$

in which ρ is a forgetting factor.

Because of the limitation of filtering stability, γ belongs to a bounded range. Denote estimation error as $\tilde{X}(k)$. The real prediction error and covariance matrix can be written as

$$\hat{X}(k|k-1) = \beta(k)F(k-1)\hat{X}(k-1) + w(k) \tag{8.22}$$

$$\hat{\Sigma}(k|k-1) = \beta(k)F(k-1)P(k-1)F^T(k-1)\beta(k) + W(k) \tag{8.23}$$

where $\beta(k)$ is a diagonal nonlinearity factor evaluating modeling error and external disturbances. γ should satisfy

$$\{\max eig[P(k|k-1)]\}^{1/2} < \gamma < \{\max eig[P^{-1}(k|k-1) - \bar{\Sigma}^{-1}(k|k-1)]\}^{-1/2} \tag{8.24}$$

such that $\Sigma(k|k-1)$ is positive definite, satisfying

$$\Sigma(k|k-1) \geq \bar{\Sigma}(k|k-1) \tag{8.25}$$

In (8.24), max $eig(\cdot)$ refers to the maximal eigenvalue of a matrix. Equation (8.24) clarifies the upper bound of γ to keep the estimation error divergent, and Eq. (8.25) gives a sufficient stability condition [29] for the AREKF. The condition can be transformed into a more applicable form by taking $L(k)$ as

$$L(k) = \gamma(P^{-1}(k|k-1) - \bar{\epsilon}_{\max}^{-2} I)^{1/2} \tag{8.26}$$

where $\bar{\epsilon}_{\max}^{-2} I$ is the compensation function satisfying

$$\bar{\epsilon}_{\max}^{-2} I \leq \bar{\Sigma}^{-1}(k|k-1) \tag{8.27}$$

8.4.2 RAREKF and Stability Condition

Practically, it is hard to obtain the actual state equation. Equation (8.10) has to be replaced with

$$\hat{X}(k|k-1) = F(k-1)\hat{X}(k-1|k-1) \tag{8.28}$$

so that the linearization error is generated. That results in the switching condition $\bar{P}_Y(k) > \alpha P_Y(k)$ always formed for the AREKF where $\alpha = 1$ and (8.20) is actually equivalent back to (8.19). To activate switching structure under nonlinear errors, we need to introduce some error-redundancy which can make part of the errors in an acceptable range, that is, let $\alpha > 1$.

To verify the stability of the AREKF when $\alpha \neq 1$, we give a redundant stability criterion based on Lemma 1 [23]. Let $\|\cdot\|$ denote two-norm of real vectors and spectral norm of real matrices.

Lemma 8.1 *There are stochastic processes $\zeta(k)$ and $\Gamma(\zeta(k))$ and positive real $\tau_{\min}$, $\tau_{\max}$, μ and $0 < \lambda \leq 1$ satisfying*

$$\tau_{\min}\|\zeta(k)\| \leq \|\Gamma(\zeta(k))\| \leq \tau_{\max}\|\zeta(k)\| \tag{8.29}$$

and

$$E\{\Gamma(\zeta(k))|\zeta(k-1)\} - \Gamma(\zeta(k-1)) \leq \mu - \lambda\Gamma(\zeta(k-1)) \tag{8.30}$$

formed with probability one. Then, $\zeta(k)$ is bounded in mean square, that is

$$E\{\|\zeta(k)\|^2\} \leq \frac{\tau_{\max}}{\tau_{\min}} E\{\|\zeta(k_0)\|^2\}(1-\lambda)^k + \frac{\mu}{\tau_{\min}} \sum_{i=1}^{k-1}(1-\lambda)^i \tag{8.31}$$

In other words, Eq. (8.30) is to keep the conditional expectation bounded. Because if

$$E\{\Gamma(\zeta(k))|\zeta(k-1)\} \leq \epsilon_{\max} \tag{8.32}$$

there must exist $\eta \in (0, 1]$ such that

$$E\{\Gamma(\zeta(k))|\zeta(k-1))\} - \Gamma(\zeta(k-1)) \leq \eta[\epsilon_{\max} - \Gamma(\zeta(k-1))] \tag{8.33}$$

which is exactly (8.30) if $\lambda = \eta$ and $\mu = \eta\epsilon_{\max}$. For filtering stability, Lemma 8.1 indicates that $\zeta(k)$ is gradually convergent if the Lyapunov function $\Gamma(\zeta(k))$ and its one-step conditional expectation are both bounded.

Then, a redundant stability criterion can be presented by Theorem 8.1.

Theorem 8.1 *For a nonlinear stochastic system described by (8.8) and (8.9), and the robust EKF expressed by (8.10)–(8.11), (8.13)–(8.17), and (8.19), there exist real $f_{\min}$, $f_{\max}$, $w_{\min}$, $w_{\max}$, $h_{\min}$, $h_{\max}$, $v_{\min}$, $v_{\max}$, $\beta_{\min}$, $\beta_{\max}$, $p_{\min}$, $p_{\max}$ satisfying*

$$f_{\min}I \leq F(k)F^T(k) \leq f_{\max}I,\ w_{\min}I \leq W(k) \leq w_{\max}I \tag{8.34}$$

$$h_{\min}I \leq H(k)H^T(k) \leq h_{\max}I,\ v_{\min}I \leq V(k) \leq v_{\max}I \tag{8.35}$$

$$\beta_{\min}I \leq \beta(k)\beta^T(k) \leq \beta_{\max}I,\ p_{\min}I \leq P(k) \leq p_{\max}I \tag{8.36}$$

Then, let $\beta_m = \|I - F_k^T H_k^T V_k^{-1}(V_k - H_k P_k H_k^T)V_k^{-1} H_k F_k P_{k-1}\|^{-1/2}$ and if

$$\beta_{\max} \leq \beta_m \tag{8.37}$$

$$\Sigma(k|k-1) \geq \beta^{-2}(k)\bar{\Sigma}(k|k-1) \tag{8.38}$$

there exist real $\mu_{\max} > 0$ and $0 < \lambda_{\min} \leq 1$ satisfying

$$E\{\|\tilde{X}(k)\|^2\} \leq \frac{p_{\max}}{p_{\min}} E\{\|\tilde{X}(k_0)\|^2\}(1-\lambda_{\min})^k + \frac{\mu_{\max}}{p_{\min}} \sum_{i=1}^{k-1}(1-\lambda_{\min})^i \tag{8.39}$$

In Theorem 8.1, it is not hard to derive that $\beta_m \in (1, (1 - \frac{1}{4} f_{\max}^2 p_{\max} p_{\min}^{-1})^{-1/2}]$, $\mu_{\max}$ and $\lambda_{\min}$ can be determined by using (8.30). The proof is specified in Appendix.

Considering that (8.34)–(8.36) are formed naturally, Theorem 8.1 demonstrates that the estimation error of the robust EKF is bounded if (8.37)–(8.38) holds. Compared with (8.25), (8.38) enlarges the range of $\Sigma(k|k-1)$ with (8.37) as a constraint, and is sufficient to guarantee the filtering stability of the robust EKF. So, with introduction of the coefficient $\beta^{-2}(k)$, the lower bound of $\Sigma(k|k-1)$ decreases to the linear part, providing an error-redundancy with most possibility for filtering stability.

To realize the error-redundant filtering, RAREKF is developed, by designing the compensation function as

$$\bar{\epsilon}_{\max}^{-2} I = f_{\alpha}^{-1}(k) P^{-1}(k|k-1) \tag{8.40}$$

where

$$f_{\alpha}(k) = \alpha^{-1} \text{diag}\left[\frac{\bar{P}_Y(k)}{P_Y(k)}\right] \otimes I_2 \tag{8.41}$$

is called the compensation factor and α is the redundancy factor, defined as an m-dimensional diagonal matrix and satisfying

$$I_2 \otimes \alpha \le \beta^2(k) \tag{8.42}$$

where $\otimes$ represents Kronecker multiplication, expanding α to an n-dimensional matrix space for dimension matching. Clearly, a two-order unit matrix should be used for $Z(k)$ to match the dimension of $X(k)$. It is easy to know that $f_\alpha > I$ when the filter switches to robust filtering.

Hence, the RAREKF can be expressed as (8.20)–(8.21), (8.26), and (8.40)–(8.42). The stability of the RAREKF can be guaranteed by Theorem 8.2.

Theorem 8.2 *Consider the filter expressed by (8.10)–(8.11), (8.20), (8.26), (8.40)–(8.42), and (8.13)–(8.17), and the assumptions of (8.8), (8.9), (8.34)–(8.36) similar to Theorem 8.1 and let* $m = 3$, $n = 6$. *Then, there exist real* $\mu_{\max} > 0$ *and* $0 < \lambda_{\min} \le 1$ *such that (8.39) is formed.*

Proof Based on the proof of Theorem 8.1, we only need to verify (8.38) under (8.20) and (8.40) to (8.42).

According to (8.9) and (8.23), we have

$$\bar{P}_Y(k) = H(k)\bar{\Sigma}(k|k-1)H^T(k) + V(k) \tag{8.43}$$

Subtracting α times (8.14), Eq. (8.43) becomes

$$\bar{P}_Y(k) - \alpha P_Y(k) = H(k)[\bar{\Sigma}(k|k-1) - (I_2 \otimes \alpha \Sigma(k|k-1))]H^T(k) \tag{8.44}$$

Table 8.1 Comparison of AREKF and RAREKF

Filtering algorithm	RAREKF	AREKF
Sufficient conditions	$\sum(t\|t-1) \geq \beta^{-2}(t)\bar{\Sigma}(t\|t-1)$	$\sum(t\|t-1) \geq \bar{\Sigma}(t\|t-1)$
Compensation function	$\bar{\epsilon}_{\max}^{-2} I = f_{\alpha}^{-1}(t)P^{-1}(t\|t-1)$	$\bar{\epsilon}_{\max}^{-2} I \geq \bar{\Sigma}(t\|t-1)$
Redundancy factor	$\\|\alpha\\| \in [1, \beta_m^2]$	$\\|\alpha\\| = 1$
Nonlinearity factor	$\beta_{\min} \leq \\|\beta(t)\\| \leq \beta_{\max} \leq \beta_m$	$\beta_{\min} \leq \\|\beta(t)\\| \leq \beta_{\max}$
Working status	factor β_m^2 $\beta^2(t)$ $\\|\alpha\\|$ 1 Linear Part O t	factor β_m^2 $\beta^2(t)$ 1 $\\|\alpha\\|$ Linear Part O t

If $\bar{P}_Y(k) \leq \alpha P_Y(k)$, noticing that $I_2 \otimes \alpha \leq \beta^2(k)$ from (8.42), we may form (8.38) by

$$\Sigma(k|k-1)) \geq (I_2 \otimes \alpha)^{-1}\bar{\Sigma}(k|k-1) \geq \beta^{-2}(k)\bar{\Sigma}(k|k-1) \tag{8.45}$$

Otherwise, if $\bar{P}_Y(k) > \alpha P_Y(k)$, substituting (8.40) and (8.41) into (8.20) yields

$$\Sigma(k|k-1) - \beta^{-2}(k)\bar{\Sigma}(k|k-1) \geq [(I_2 \otimes \alpha)^{-1} - \beta^{-2}(k)]\bar{\Sigma}(k|k-1) \geq 0 \tag{8.46}$$

which can be formed under (8.42). Therefore, (8.38) holds.

Based on Theorem 8.1, Theorem 8.2 is then proven.

Compared with AREKF, the RAREKF shows several advantages, which are listed in Table 8.1. First, $\Sigma(k|k-1)$ is remained at the linear part of $\bar{\Sigma}(k|k-1)$. That achieves maximum use of the stability redundancy given by Theorem 8.1. Second, all the user-defined variables are eliminated so that the adaptability of filtering gains is improved. Third, the value range of α is extended from the constant unit matrix to (8.42). Fourth, $\|\alpha\|$ makes the robust filtering status flexible, creating a tuning space for (8.20) so that the switching is triggered only when necessary.

The advantages help achieve two objectives: reduced switching frequency and compensation magnitude. Due to the tunable error-redundancy, the RAREKF is able to pick out the more interested target maneuver, leading to less expense of optimality. That is expected for active satellite tracking.

The entire tracking algorithm based on the RAREKF can be stated as follows: For tracking system (8.3) and (8.5) with relative measurement, apply (8.28), (8.11)–(8.17), where (8.12) is composed of (8.20)–(8.21), (8.26), and (8.40)–(8.42).

8.5 Tracking Error Evaluation

For the problem of active satellite tracking, the quality of both relative motion model and filtering algorithm determines the tracking accuracy and precision, so an integrated index is required to evaluate the tracking error.

For Eq. (8.8), $w(k) = 0$ and the initial state $X(k_0) = X_0$. A typical index is designed in [11], aiming to evaluate orbital models using various types of coordinate systems, such as Cartesian, polar, and orbital elements, and is expressed as

$$\sigma_J(k, k_0) = \frac{\|\Phi(k, k_0) - \bar{\Phi}(k, k_0)\|}{\|\bar{\Phi}(k, k_0)\|} \tag{8.47}$$

where $\bar{\Phi}(k, k_0)$ is the true state transfer matrix with true initial conditions $\bar{X}(k_0)$, and $\Phi(k, k_0)$ represents the real values considering noises. Through using 3σ and Monte-Carlo methods, the index verifies that models based on orbital elements have the smallest propagation error to noises. To evaluate modeling error, another index is designed in [1] with the form as

$$\sigma_A(k) = \max_{X(k_0)} \left| \frac{y^T(k)|_{X(k_0)} y(k)|_{X(k_0)}}{\bar{y}^T(k)|_{X(k_0)} \bar{y}(k)|_{X(k_0)}} - 1 \right| \tag{8.48}$$

where $y(k)$ is a nondimensional vector of $X(k)$ by multiplying a weight matrix M, as

$$y(k)|_{X(k_0)} = MX(k)|_{X(k_0)} \tag{8.49}$$

The index (8.48) takes the maximum function value of various initial conditions spreading throughout the orbit. Based on the concept of H_∞ norm, the index shows the worst case of the modeling error.

Clearly, (8.47) and (8.48) evaluate the model from a single viewpoint, from either noises or initial states, without considering filtering quality.

To evaluate the tracking error, we have to consider both model and filter in the evaluation index, so we design an integrated index function as

$$\sigma(k) = \frac{1}{N} \sum_{i=1}^{N} \sigma(k, k_0)|_{X_i(k_0)} \tag{8.50a}$$

$$\sigma(k, k_0)|_{X(k_0)} = \frac{\|\hat{y}(k, k_0)|_{X(k_0)} - \bar{y}(k, k_0)|_{X(k_0)}\|}{\|\bar{y}(k, k_0)|_{X(k_0)}\|} \tag{8.50b}$$

$$\hat{y}(k, k_0)|_{X(k_0)} = M\hat{X}(k, k_0)|_{X(k_0)} \tag{8.50c}$$

From the index function we can see, it is defined as a distance in an ℓ_2 inner-product space, reflecting both magnitude and direction of error vectors. Also, the index uses real estimation results to evaluate the tracking error. It makes the filtering quality, besides stochastic noises and initial state errors, taken into account. It is noticeable that we use expectation to replace maximization in (8.48) to avoid the probable failure under the uncertain maneuvers during active satellite tracking.

8.6 Simulations

We use two simulation cases to verify the tracking approach. The first assumes that the Target is flying freely. The second assumes that the Target has a potential maneuver generated by continuous finite thrust. The two satellites follow in-track formation [13] under which the relative orbit is always along in-track of the Chaser.

The initial orbital elements of the two satellites are given in Table 8.2. The initial relative position and velocity have the covariance $(1\ \text{m})^2$ and $(0.1\ \text{m/s})^2$, respectively. For the measurement angles and range, their covariance takes $(0.1\ \text{rad})^2$ and $(1\ \text{m})^2$. The simulations last 16,290 s, about 3 orbital periods. We take the usual Hill–Clohessy–Wiltshire (HCW) model and the DOE-based model as the tracking model; and EKF, UEKF, AREKF, and RAREKF as the filtering algorithm, respectively.

Figures 8.2 and 8.3 show the comparison result of the RAREKF to AREKF and other typical filters for the first simulation case.

The compensation functions of the two methods are the same as (8.40)–(8.41). Take $\alpha = I$ for the AREKF and $\alpha = \text{diag}(3.5, 2.5, 10)$ for the RAREKF. Figure 8.2b indicates that the RAREKF has a relatively smaller compensation factor. That helps reduce the excessive loss of the filtering optimality and increase the tracking precision. For the AREKF, Fig. 8.2a shows that the estimation precision is much lower.

Compared with other typical filters, the RAREKF achieves the best robustness and precision. From Fig. 8.3, we can see nonlinear modeling error leads to the inaccuracy of the non-robust filters. The filtering accuracy of the UEKF is lower than that of EKF, increasing linearly with time.

Now, we let the Target maneuver 200 s from the time 10,300 s, along radial and in-track directions both with an acceleration of $0.02\ \text{m/s}^2$. Figures 8.4 and 8.5 show the compared tracking results. Clearly, EKF-based tracking method is divergent and the UEKF has a remarkable wave during target maneuvering. The result of

Table 8.2 Initial orbital elements

OE	a (km)	e	i (deg)	Ω (deg)	ω (deg)	f (deg)
Target	6678.137	0.05	28.5	45	0	315
Chaser	6678.137	0.05	28.5	45	0.5	315

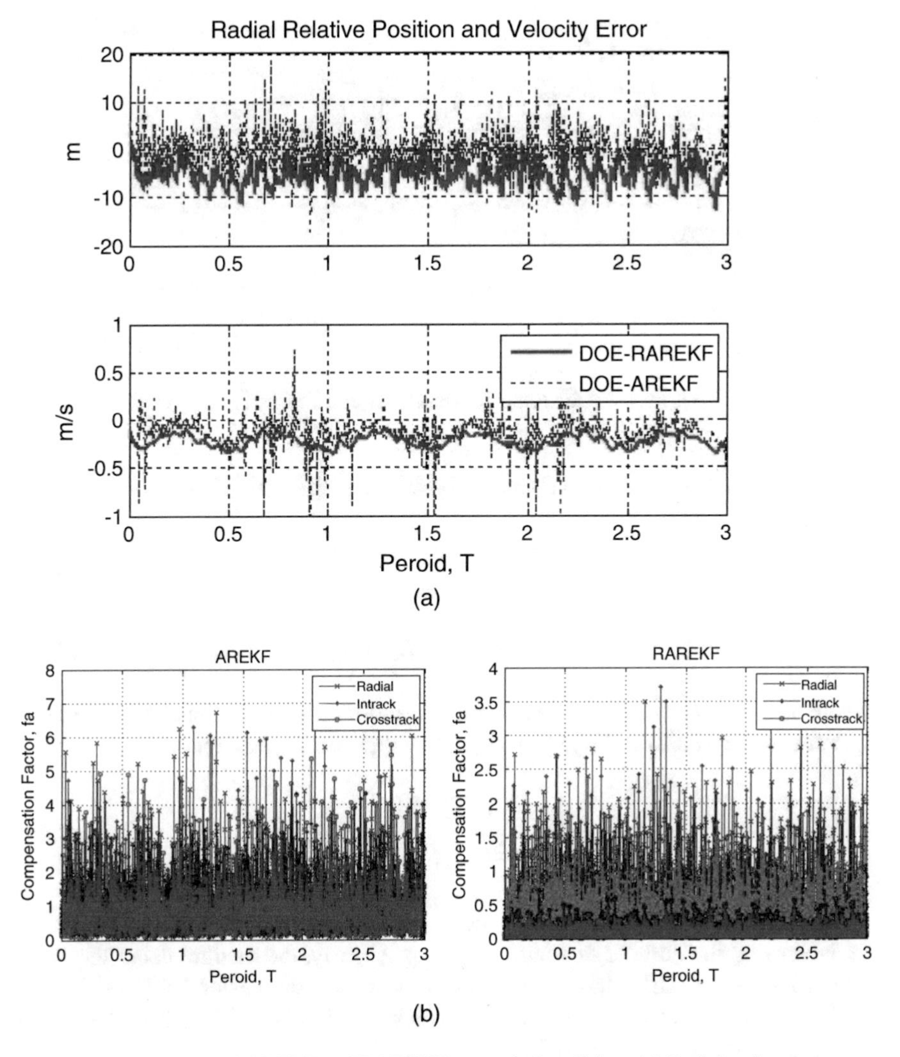

Fig. 8.2 Comparisons of AREKF and RAREKF, target is free-flying. (**a**) Radial estimation error. (**b**) Compensation factor

the RAREKF is desired, presenting a higher precision than that of the AREKF. We may learn that the error-redundancy improves the filtering robustness, which betters the fast tracking capability and tracking precision.

To evaluate the tracking error, we use the presented evaluation method. When $\delta a = \delta e = \delta i = \delta \Omega = \delta f = 0$, the in-track formation can be formed by letting

$$\rho(t) = a(1 - e \cos E(t))\delta\omega \in [a(1 - e)\delta\omega, a(1 + e)\delta\omega] \tag{8.51}$$

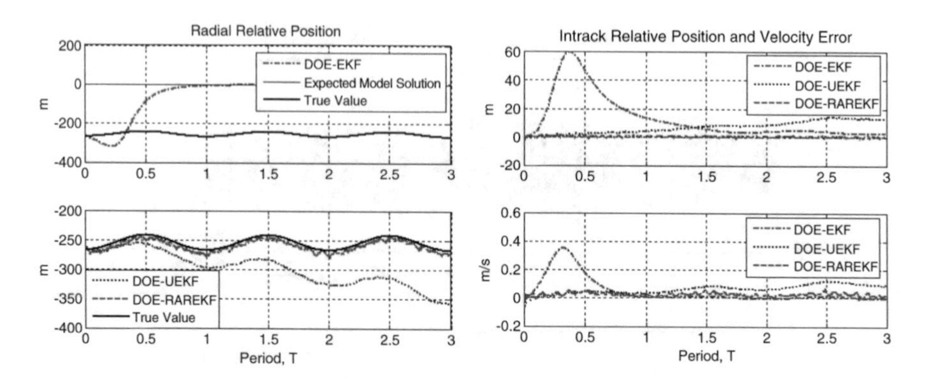

Fig. 8.3 Comparisons of EKF, UEKF, and RAREKF, target is free-flying

Select $N = 2L$ points along the relative orbit as $2L$ different initial states. Let

$$E(l) = \begin{cases} \cos^{-1}(-1 + 2\frac{l}{L}), & l = 0, \ldots, L-1 \\ \cos^{-1}(-1 + 2(\frac{l}{L} - 1)) - \pi, & l = K, \ldots, 2L-1. \end{cases} \tag{8.52}$$

$$M = \text{diag}(1/R_e, 1/R_e, 1/R_e, 1/n_e R_e, 1/n_e R_e, 1/n_e R_e) \tag{8.53}$$

where $n_e = \sqrt{\mu_e / R_e}$. R_e is the Earth radius, and μ_e the gravitational coefficient.

Substituting (8.52) and (8.53) into (8.50) expresses the index as

$$\sigma(k) = \frac{1}{2L} \sum_{l=1}^{2L} \sigma(k, k_0)|_{E(l)} \tag{8.54}$$

By taking $L = 6$, Fig. 8.6 shows the evaluation results, which can be smoothed by increasing the number of sampling points. Clearly, the method using the DOE model and the RAREKF has the lowest error level, whereas using HCW model and EKF gets the highest. The sequence from low to high is DOE with UEKF, DOE with EKF, and HCW with RAREKF.

Figure 8.7 compares the indexes of $t = 2T$ varying with parameters of the relative orbit. From the figures we can see, the indexes are insensitive to a due to the small ratio a/R_e. For variant $\delta\omega$, the errors of the DOE-based methods are much lower than that of the HCW-based methods and less sensitive to the Chaser eccentricity. Figure 8.7d indicates that when the Target maneuvers, all the methods using the RAREKF have much better evaluation results.

Therefore, we may conclude that the method based on the DOE model and the RAREKF has the best robustness and accuracy among all the compared methods.

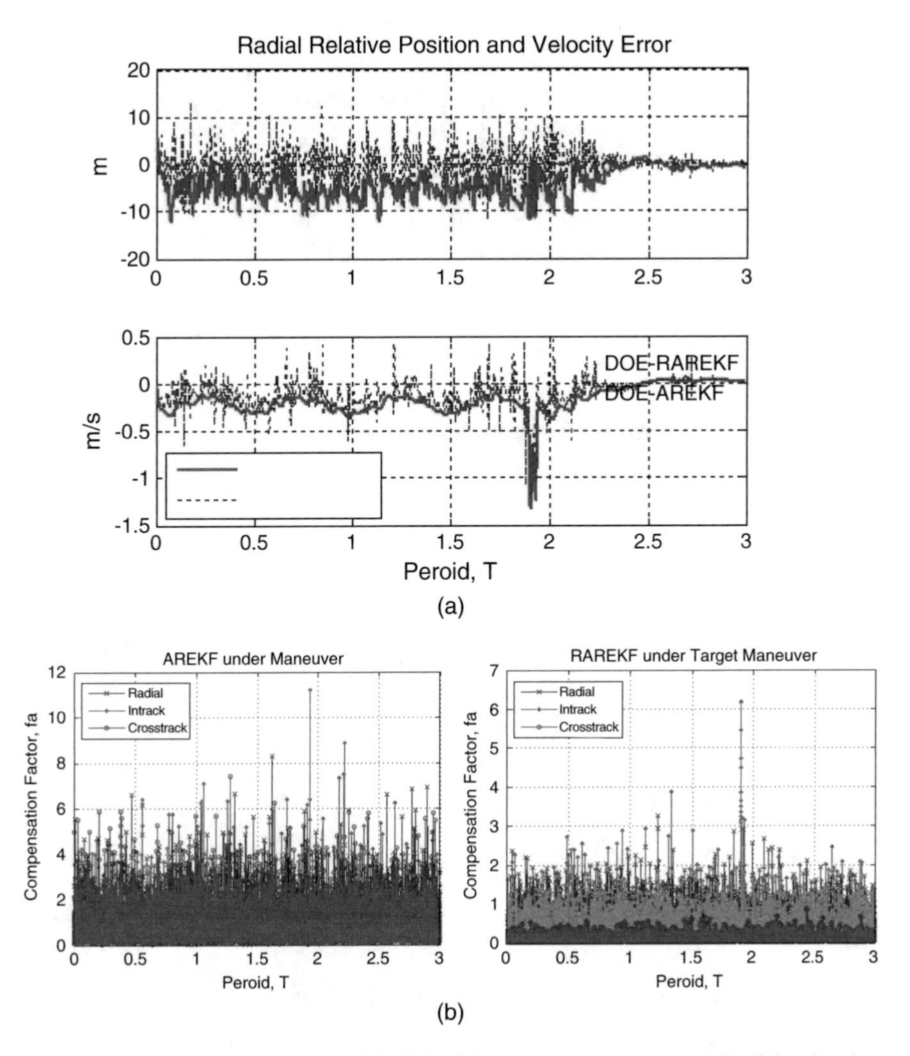

Fig. 8.4 Comparisons of AREKF and RAREKF, target has a maneuver. (**a**) Radial estimation error. (**b**) Compensation factor

8.7 Conclusion

A robust tracking approach is presented and applied to active satellite tracking. The approach is established with a DOE-based state model, a UCM-based measurement model, and an error-tolerant RAREKF algorithm. The RAREKF is an improved version of robust EKF, with the introduction of a redundancy factor to control the switching between robust and optimal filtering modes. Under external disturbances and inner modeling errors, the RAREKF-based tracking approach can obtain more accurate tracking results.

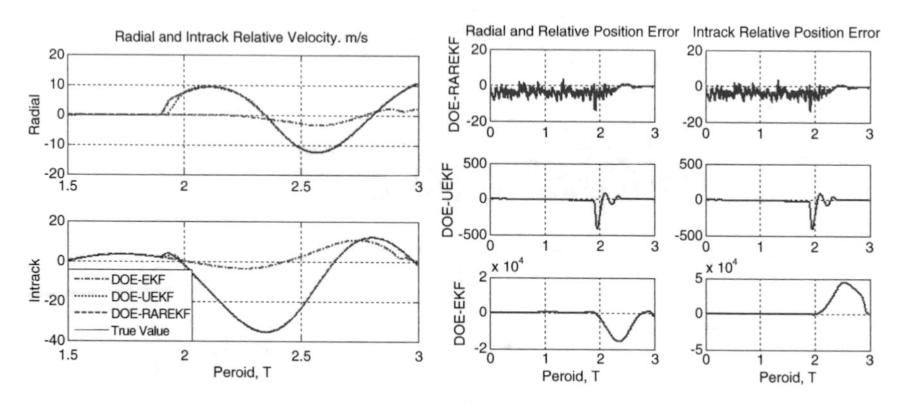

Fig. 8.5 Comparisons of AREKF and RAREKF, target has a maneuver

Appendix

Proof of Theorem 8.1

Proof In order for convenience, the time variable is denoted as subscript. Design Lyapunov function as

$$\Gamma(\tilde{X}_k) = \tilde{X}_k^T P_{k|k}^{-1} \tilde{X}_k \tag{8.55}$$

It is bounded if considering (8.36), as

$$p_{\max}^{-1} \|\tilde{X}_k\|^2 \leq \Gamma(\tilde{X}_k) \leq p_{\min}^{-1} \|\tilde{X}_k\|^2 \tag{8.56}$$

According to (8.9), (8.16), and (8.22), the estimation error of the robust EKF can be written as

$$\tilde{X}_k = (I - K_k H_k)\beta_k F_{k-1}\tilde{X}_{k-1} + (I - K_k H_k)w_k - K_k v_k \tag{8.57}$$

Substituting (8.57) into (8.55) and taking conditional expectation, we may have

$$E\{\Gamma(\tilde{X}_k)|\tilde{X}_{k-1}\} = \tau_k + \mu_k \tag{8.58}$$

where

$$\tau_k = \tilde{X}_k^T F_{k-1}^T \beta_k [\Sigma_{k|k-1}^{-1} - H_k^T (V_k^{-1} - V_k^{-1} H_k P_k H_k^T V_k^{-1}) H_k] \beta_k F_{k-1} \tilde{X}_{k-1} \tag{8.59}$$

$$\mu_k = E\{W_k^{-1}(I - K_k H_k)^T P_k^{-1}(I - K_k H_k) W_k + v_k^T K_k^T P_k^{-1} K_k V_k\} \tag{8.60}$$

Based on (8.23) and (8.38), it has

$$\Sigma_{k|k-1} \geq \beta_k^{-2}(\beta_k F_{k-1} P_{k-1} F_{k-1}^T \beta_k + W_k) \geq F_{k-1} P_{k-1} F_{k-1}^T \tag{8.61}$$

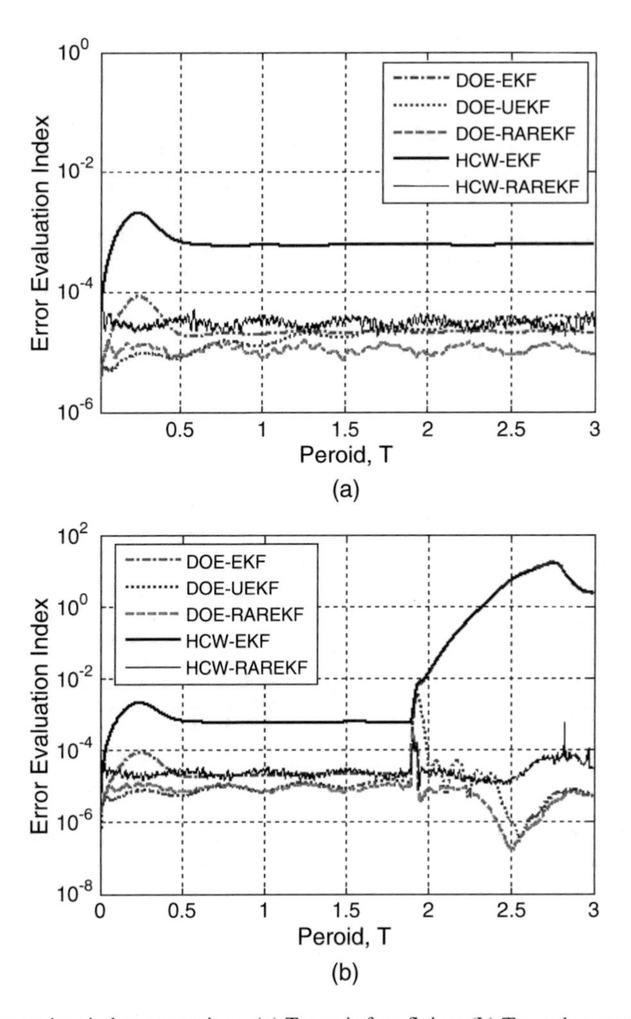

Fig. 8.6 Time-variant index comparison. (**a**) Target is free-flying. (**b**) Target has maneuvers

According to matrix theory, there is

$$V_k^{-1} - V_k^{-1} H_k P_k H_k^T V_k^{-1} = (H_k \Sigma_{k|k-1} H_k^T + V_k^{-1})^{-1} \tag{8.62}$$

Substituting (8.61) and (8.62) into (8.59) yields

$$E\{\Gamma(\tilde{X}_k)|\tilde{X}_{k-1}\} \leq \tilde{X}_{k-1}^T \beta_k [P_{k-1}^{-1} - F_{k-1}^T H_k^T (H_k \Sigma_{k|k-1} H_k^T + V_k^{-1})^{-1} H_k F_{k-1}]$$
$$\beta_k \tilde{X}_{k-1} + \mu_k \tag{8.63}$$

which can be reshaped into a form similar to (8.30), expressed as

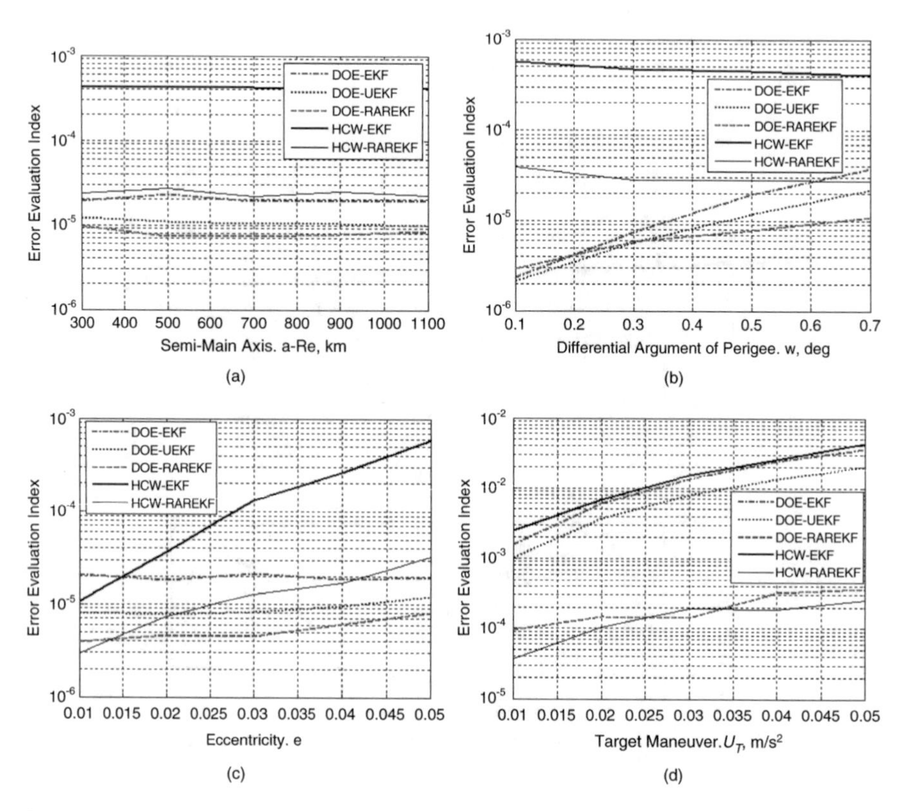

Fig. 8.7 Index comparisons at $t = 2T$ varying with parameters. (**a**) Semi-main axis a. (**b**) Differential argument of perigee $\delta\omega$. (**c**) Eccentricity e. (**d**) Target maneuver U_T

$$E\{\Gamma(\tilde{X}_k)|\tilde{X}_{k-1}\} - \Gamma(\tilde{X}_{k-1}) \leq \mu_k - \lambda_k\Gamma(\tilde{X}_{k-1}) \tag{8.64}$$

where

$$\lambda_k = 1-\tilde{X}_{k-1}^T\beta_k[P_{k-1}^{-1} - F_{k-1}^T H_k^T (H_k\Sigma_{k|k-1}H_k^T + V_k^{-1})^{-1} H_k F_{k-1}]\beta_k\tilde{X}_{k-1}\Gamma^{-1}(\tilde{X}_{k-1}) \tag{8.65}$$

According to lemma 8.1, the rest objective is to find the required $\mu > 0$ and $0 < \lambda \leq 1$. Firstly, it is obvious from (8.60) that $\mu_k > 0$ and

$$\mu_k \leq tr[(P_k^{-1} + H_k^T V_k^{-1} H_k P_k^{-1} H_k^T V_k^{-1} H_k)W_k] + tr(K_k^T P_k^{-1} K_k V_k) \tag{8.66}$$

Define the upper bound as

$$\mu_{\max} = (p_{\min}^{-1} + h_{\max}^4 v_{\min}^{-2} p_{\max})q_{\max}n + h_{\max}^2 p_{\max} v_{\min}^{-1} m \tag{8.67}$$

Secondly, it is clear that

$$\begin{aligned}\lambda_k &= 1 - \tilde{X}_{k-1}^T \beta_k [P_{k-1}^{-1} - F_{k-1}^T H_k^T (H_k \Sigma_{k|k-1} H_k^T + V_k^{-1})^{-1} \\ &\quad H_k F_{k-1}] \beta_k \tilde{X}_{k-1} \Gamma^{-1}(\tilde{X}_{k-1}) \\ &\le 1 - \tilde{X}_{k-1}^T \beta_k [P_{k-1}^{-1} - F_{k-1}^T H_k^T (H_k P_{k|k-1} H_k^T + V_k^{-1})^{-1} \\ &\quad H_k F_{k-1}] \beta_k \tilde{X}_{k-1} \Gamma^{-1}(\tilde{X}_{k-1}) \\ &= 1 - \tilde{X}_{k-1}^T \beta_k [P_{k-1} - P_{k-1} F_{k-1}^T H_k^T (H_k W_k H_k^T + V_k)^{-1} \\ &\quad H_k F_{k-1}] \beta_k \tilde{X}_{k-1} \Gamma^{-1}(\tilde{X}_{k-1}) \le 1\end{aligned} \tag{8.68}$$

Thirdly, we can know from (8.65) that $\lambda_k > 0$ if only

$$\beta_k < [I - F_{k-1}^T H_k^T (H_k \Sigma_{k|k-1} H_k^T + V_k^{-1})^{-1} H_k F_{k-1} P_{k-1}]^{-1/2} \tag{8.69}$$

Taking (8.62) into account, it becomes

$$\beta_k < [I - F_{k-1}^T H_k^T V_k^{-1} (V_k - H_k P_k H_k^T) V_k^{-1} H_k F_{k-1} P_{k-1}]^{-1/2} \tag{8.70}$$

Take spectral norm to both sides to get

$$\|\beta_k\| < \|I - F_{k-1}^T H_k^T V_k^{-1} (V_k - H_k P_k H_k^T) V_k^{-1} H_k F_{k-1} P_{k-1}\|^{-1/2} = \beta_m \tag{8.71}$$

Considering (8.37), we may know $\beta_{\max} \le \beta_m$. Then, (8.69)–(8.71) are formed, that is, $\lambda_k > 0$.

Therefore, $\lambda_k \in (0, 1]$ and there must exist a minimum $\lambda_{\min} \in (0, 1]$. Considering the $\mu_{\max}$ in (8.67) together and based on Lemma 8.1, we can conclude that (8.39) is formed. That means, $\tilde{X}_k$ is bounded.

Range Determination of β_m From (8.69) we may know, $\beta_m \ge 1$. Denoting $\delta_k = H_k^T V_k^{-1} H_k$, reform β_m as

$$\beta_m = \|I - F_{k-1}^T (\delta_k - \delta_k P_k \delta_k) F_{k-1} P_{k-1}\|^{-1/2} \tag{8.72}$$

Clearly, when $\delta_k = 0$ or $\delta_k = P_k^{-1}$, β_m reaches the minimum. Due to (8.17) and the bounded $\Sigma_{k|k-1}$, the inferior of β_m is open. If $\delta_k = \frac{1}{2} P_k^{-1}$, β_m reaches its maximum value, expressed as

$$\beta_m = \left\| I - \frac{1}{4} F_{k-1}^T P_k^{-1} F_{k-1} P_{k-1} \right\|^{-1/2} \tag{8.73}$$

Then, the superior of β_m is $(1 - \frac{1}{4} f_{\max}^2 p_{\max} p_{\min}^{-1})^{-1/2}$ if satisfying $f_{\max}^2 < 4 p_{\min} p_{\max}^{-1}$. Thus, $\beta_m \in (1, (1 - \frac{1}{4} f_{\max}^2 p_{\max} p_{\min}^{-1})^{-1/2}]$.

Acknowledgements The work was supported in part by China Natural Science Foundation (No. 60775022 and No. 60674107).

References

1. Alfriend K, Yan H (2005) Evaluation and comparison of relative motion theories. J Guid Control Dyn 28(2):254–261
2. Busse F, How J (2002) Real-time experimental demonstration of precise decentralized relative navigation for formation flying spacecraft. In: AIAA guidance, navigation, and control conference and exhibit, p 5003
3. Culp R (1984) Satellite-to-satellite orbit determination using minimum, discrete range and range-rate data only. In: Astrodynamics conference, p 2030
4. Duan Z, Han C, Li XR (2004) Comments on" unbiased converted measurements for tracking". IEEE Trans Aerosp Electron Syst 40(4):1374
5. Dunham JB, Long AC, Sielski HM, Preiss KA (1983) Onboard orbit estimation with tracking and data relay satellite system data. J Guid Control Dyn 6(4):292–301
6. Fang BT (1979) Satellite-to-satellite tracking orbit determination. J Guid Control Dyn 2(1):57–64
7. Gim DW, Alfriend KT (2003) State transition matrix of relative motion for the perturbed noncircular reference orbit. J Guid Control Dyn 26(6):956–971
8. Gurfil P, Mishne D (2007) Cyclic spacecraft formations: relative motion control using line-of-sight measurements only. J Guid Control Dyn 30(1):214–226
9. Hablani H (2003) Autonomous navigation, guidance, attitude determination and control for spacecraft rendezvous in a circular orbit. In: AIAA guidance, navigation, and control conference and exhibit, p 5355
10. Hablani HB (2009) Autonomous inertial relative navigation with sight-line-stabilized sensors for spacecraft rendezvous. J Guid Control Dyn 32(1):172–183
11. Junkins JL, Akella MR, Alfrined KT (1996) Non-gaussian error propagation in orbital mechanics. Adv Astronaut Sci 92, 283–298
12. Kawase S (1990) Intersatellite tracking methods for clustered geostationary satellites. IEEE Trans Aerosp Electron Syst 26(3):469–474
13. Lane CM, Axelrad P (2006) Formation design in eccentric orbits using linearized equations of relative motion. J Guid Control Dyn 29(1):146–160
14. Lee S, Schutz BE, PAM Abusali (2004) Hybrid precise orbit determination strategy by global position system tracking. J Spacecr Rocket 41(6):997–1009
15. Li Y (2010) Autonomous follow-up tracking and control of noncooperative space target. Ph.D. Dissertation, Shanghai Jiao Tong University
16. Li YK, Jing ZL, Hu SQ (2009) Transient relative model based kinematical parameter estimation for orbital maneuvering target. Control Decis 24(7):1059–1064
17. Li Y, Jing Z, Hu S (2010) Redundant adaptive robust tracking of active satellite and error evaluation. IET Control Theory Appl 4(11):2539–2553
18. Longbin M, Xiaoquan S, Yiyu Z, Kang SZ, Bar-Shalom Y (1998) Unbiased converted measurements for tracking. IEEE Trans Aerosp Electron Syst 34(3):1023–1027
19. Markley F (1984) Autonomous navigation using landmark and intersatellite data. In: Astrodynamics conference, p 1987
20. Mohiuddin S, Psiaki ML (2007) High-altitude satellite relative navigation using carrier-phase differential global positioning system techniques. J Guid Control Dynam 30(5):1427–1436
21. Psiaki ML (1999) Autonomous orbit determination for two spacecraft from relative position measurements. J Guid Control Dynam 22(2):305–312
22. Psiaki ML, Mohiuddin S (2007) Modeling, analysis, and simulation of gps carrier phase for spacecraft relative navigation. J Guid Control Dynam 30(6):1628–1639

23. Reif K, Gunther S, Yaz E, Unbehauen R (1999) Stochastic stability of the discrete-time extended Kalman filter. IEEE Trans Autom Control 44(4):714–728
24. Seo J, Yu MJ, Park CG, Lee JG (2006) An extended robust h_∞ filter for nonlinear constrained uncertain systems. IEEE Trans Signal Process 54(11):4471–4475
25. Subbarao K, McDonald J (2005) Multi-sensor fusion based relative navigation for synchronization and capture of free floating spacecraft. In: AIAA guidance, navigation, and control conference and exhibit, p 5858
26. Sun D, Zhou F, Jun Z (2004) Relative navigation based on ukf for multiple spacecraft formation flying. In: AIAA guidance, navigation, and control conference and exhibit, p 5137
27. Woffinden DC, Geller DK (2007) Relative angles-only navigation and pose estimation for autonomous orbital rendezvous. J Guid Control Dynam 30(5):1455–1469
28. Wu SC, Yunck TP, Thornton CL (1991) Reduced-dynamic technique for precise orbit determination of low earth satellites. J Guid Control Dynam 14(1):24–30
29. Xiong K, Zhang H, Liu L (2008) Adaptive robust extended Kalman filter for nonlinear stochastic systems. IET Control Theory Appl 2(3):239–250
30. Yu KK, Watson N, Arrillaga J (2005) An adaptive Kalman filter for dynamic harmonic state estimation and harmonic injection tracking. IEEE Trans Power Delivery 20(2):1577–1584

Chapter 9
Optimal-Switched H_∞ Robust Tracking for Maneuvering Space Target

9.1 Introduction

For maneuvering target tracking problem, robust filtering is an effective way to gain fast and accurate target trajectory in real time. The H_∞ filter (H∞F) is a conservative solution with infinite-horizon robustness, leading to excessive cost of filtering optimality and reduction of estimation precision. In order to retrieve the filtering optimality sacrificed by conservativeness of the H∞F design, in this chapter, an optimal-switched filtering mechanism is developed and established on the standard HF to propose an optimal-switched H_∞ filter (OSH∞F).

The optimal-switched mechanism adopts a switched structure that switches filtering mode between optimal and H_∞ robust by setting a switching threshold, and introduces an optimality-robustness cost function (ORCF) to on-line optimize the threshold such that the switching structure can be optimized. In the ORCF, a non-dimensional weight factor (WF) is used to quantify the ratio of the filtering robustness and optimality. As the only tunable parameter in the filter, when the WF is given, the proposed OSH∞F can obtain the optimal state estimates with filtering optimality and robustness kept at the WF-determined ratio.

With the conservativeness of the H∞F optimized, the developed OSH∞F can be used as a generalized H∞F form. We use a numerical example of space target tracking to demonstrate the superior estimation performance of the OSH∞F compared with that of Kalman filter and other typical H_∞ filters. This chapter is primarily based on the work [4, 7].

9.2 Space Target Tracking Models

For a free-flying orbital chaser and a space target, the relative motion model can be described by [9]

Z. Jing et al., *Non-Cooperative Target Tracking, Fusion and Control*,
Information Fusion and Data Science, https://doi.org/10.1007/978-3-319-90716-1_9

$$\ddot{R} = -2\omega_C \times \dot{R} - \dot{\omega}_C \times R - \omega_T \times (\omega_T \times R) + \frac{\mu}{r_C^3} \times \left(\rho_C - \frac{r_C^3}{r_T^3} \rho_T \right) + D^R \tag{9.1}$$

where ρ denote the position in the Earth centered frame (ECF) and $R = [R_x, R_y, R_z]$ the relative position defined in the radial, in-track and cross-track (RIC) frame of the chaser orbit. μ, r, and ω are the geocentric gravitational constant, geocentric distance, and orbital angular rate, respectively, and the subscript C or T represents chaser or target. D^R describes the external disturbances that include the oblateness perturbation of the Earth and the target maneuver U_T and have bounded energy. With assumption that the relative range is far less than the geocentric distance of the chaser, i.e., $\|R\| \ll r_C$, Eq. (9.1) can be approximated by a discrete-time state equation as [10]

$$X_{k+1} = \phi_{k+1} \varphi_{k+1,k} \phi_k^{-1} X_k + D \tag{9.2}$$

in which $X = [R, \dot{R}]^T$ is the relative state and D denotes the bounded uncertainties resulted from D^R. φ and ϕ are defined by

$$X_k = \phi_k \delta\sigma_k \tag{9.3}$$

$$\delta\sigma_{k+1} = \varphi_{k+1,k} \delta\sigma_k \tag{9.4}$$

where $\delta\sigma$ is the difference of the orbital element sets of the target and the chaser, i.e., $\delta\sigma = \sigma_T - \sigma_C$. Define the orbital element set as $\sigma = [a, u + f, i, e\cos u, e\sin u, \Omega]^T$ where a, e, i, u, f, and Ω are the classical elements that represent semimajor axis, eccentricity, inclination, argument of periapsis, true anomaly, and longitude of the ascending node, respectively. The detailed form of φ and ϕ can be found in [3].

Denote radar measurement of the chaser as $Y = [\rho, \theta_a, \theta_e]^T$ with noise covariance $V^Y = [V_\rho^Y, V_a^Y, V_e^Y]^T$, the components of which represent relative range, azimuth, and elevation, respectively. The measurement equation can be expressed as

$$CY_k = R_k + v_k \tag{9.5}$$

where C is an unbiased conversion matrix which is derived based on the unbiased converted measurement technique [8]. v_k is the converted noise with covariance V. C and V are determined by Y and V^Y, and the detailed expressions can be found in [2]. The estimation model for space target tracking is then given by (9.2) and (9.5).

9.3 Optimal-Switched H_∞ Filter

In this section, a type of switched H_∞ filter (SH∞F) is presented first. The filter is built on the H∞F given above and the switched mechanism as in [5, 6] where switching logic is introduced to control the optimality cost of estimation results as required. Then, an optimization mechanism is developed to optimize the switching logic and to form the optimal-switched H_∞ filter (OSH∞F).

9.3.1 *Switched Filtering Mechanism*

Rewrite the tracking model (9.2) and (9.5) as

$$X_{k+1} = F_k X_k + G_k w_k + D \tag{9.6}$$

$$Z_k = H_k X_k + v_k \tag{9.7}$$

where F, G, and H are matrices independent with the system states and $F_k = \phi_{k+1}\varphi_{k+1,k}\phi_k^{-1}$. Z denotes the measurement vector and equals to CY_k. w and v are process and measurement noises assumed white with covariance matrix W and V, respectively.

As shown in Sect. 2.2.3, the standard H_∞ Kalman filter can be described as

$$\hat{X}_{k+1|k} = F_k \hat{X}_{k|k} \tag{9.8}$$

$$P_{k+1|k} = F_k P_{k|k} F_k^T + G_k W_k G_k^T \tag{9.9}$$

$$\Sigma_{k+1|k} = \left(P_{k+1|k}^{-1} - \gamma^{-2} I\right)^{-1} \tag{9.10}$$

$$\tilde{Z}_{k+1} = Z_{k+1} - H_{k+1}\hat{X}_{k+1|k} \tag{9.11}$$

$$P_{k+1}^Z = H_{k+1}\Sigma_{k+1|k}H_{k+1}^T + V_{k+1} \tag{9.12}$$

$$K_{k+1} = \Sigma_{k+1|k}H_{k+1}^T\left(P_{k+1}^Z\right)^{-1} \tag{9.13}$$

$$\hat{X}_{k+1|k+1} = \hat{X}_{k+1|k} + K_{k+1}\tilde{Z}_{k+1} \tag{9.14}$$

$$P_{k+1|k+1} = \left(\Sigma_{k+1|k}^{-1} + H_{k+1}^T V_{k+1}^{-1} H_{k+1}\right)^{-1} \tag{9.15}$$

Replace (9.10) with the following switching function:

$$\Sigma_{k+1|k} = \begin{cases} \left(P_{k+1|k}^{-1} - \gamma^{-2} I_n\right)^{-1} & \bar{P}_{k+1}^Z > \alpha P_{k+1}^Z \\ P_{k+1|k} & \bar{P}_{k+1}^Z \leq \alpha P_{k+1}^Z \end{cases} \tag{9.16}$$

where $\bar{P}^Z$ represents the real measurement covariance, which includes the deviations caused by all random uncertainties and has the form

$$\bar{P}_{k+1}^Z = \frac{\rho P_k^Z + \tilde{Z}_{k+1}\tilde{Z}_{k+1}^T}{\rho + 1} \tag{9.17}$$

where ρ is the forgotten factor and takes 0.98 empirically [1]. In (9.16), α is a scalar/diagonal-matrix controlling the switching judgment logic. If $\bar{P}^Z$ exceeds α times P^Z, the filter adopts the H_∞ prediction covariance as in (9.10) and becomes a robust H∞F, otherwise, reverts to an optimal KF.

On the other hand, considering the inconvenience of finding a proper γ for (9.16), we design γ by [5]

$$\gamma^{-2} I_n = P_{k+1|k}^{-1} - f_\alpha^{-1} P_{k+1|k}^{-1} \tag{9.18}$$

where f_α is a gain adjustment function, defined as

$$f_\alpha = \alpha^{-1} \text{diag} \frac{\bar{P}_{k+1}^Z}{P_{k+1}^Z} \otimes I_2 \tag{9.19}$$

The Kronecker product is to match the dimension of the measurements to that of the state vector. From (9.19) it is shown that, f_α is determined by ratio of the real measurement covariance to the switching threshold given in (9.16). That means, the level of filtering robustness varies with real system uncertainties, instead of a constant as in H∞F. Subscribing (9.19) with (9.18) into (9.16) completes the switching mechanism. Li et al. [5] proved the switching stability.

Use the switching mechanism to replace the gain adjustment step of the H∞F, a switched H_∞ filter is then established, which can be described by (9.8), (9.9), (9.16) and (9.11) to (9.15). Due to the existence of the switching logic, the SH∞F can switch between robust filtering of H∞F and optimal filtering of KF by judging whether the amplitude of system uncertainties exceeds the given threshold, and the threshold is tunable according to specified systems and determined by a switching parameter. Clearly, the SH∞F needs no priori value to the uncertainty bound γ, and the only required parameter is the switching parameter α. Compared with γ in the H∞F, α actually uses a relative ratio to parameterize uncertainties rather than an absolute value as γ, and is easier to give properly with clear physical interpretation.

9.3.2 Switching Parameter Optimization

In the SH∞F, the intrinsic parameter γ is transformed by the switching mechanism into α. By tuning α, the filter can yield estimation result with various weights of filtering optimality and robustness. However, the quantitative relationship between α and the weights is unclear. That means it is hard to tell precisely that, for a certain α the result of the filter contains how much optimality or robustness, and also for a certain weight of optimality or robustness what α is the best choice. As a result, it is difficult to evaluate α and get the best performance of the filter. In order to achieve filtering optimization, we need to identify the relationship between α and weight of filtering optimality or robustness, and in the sense of the weight optimize α in real time.

We utilize a weighted sum function to quantify the gross performances of optimality and robustness at current iteration, which we named the optimality-robustness cost function (ORCF), having the form as

$$\Gamma = \lambda E\left[(Z_k - \hat{Z}_k)(Z_k - \hat{Z}_k)^T\right] + P_k^Z \tag{9.20}$$

where $\hat{Z}_k = H_k \hat{X}_{k|k}$ and λ is the weight factor (WF) with scalar/diagonal-matrix form. The first term at right-hand side evaluates the filter robustness and the second term shows the optimality. The ratio of them can be tuned by λ. To optimize the filtering performance, the ORCF should be minimized.

Firstly, substitute (9.14) into $\hat{Z}_k$ and notice that (9.12) is formed. There is

$$Z_k - \hat{Z}_k = V_k \left(P_k^Z\right)^{-1} \hat{Z}_k \tag{9.21}$$

Take (9.21) to (9.20) by considering that $P_k^Z = E(\hat{Z}_k \hat{Z}_k^Z)$, and the ORCF can be rewritten as

$$\Gamma = \lambda \left[V_k \left(P_k^Z\right)^{-1}\right] P_k^Z \left[V_k \left(P_k^Z\right)^{-1}\right]^T + \left[V_k \left(P_k^Z\right)^{-1}\right]^{-1} V_k \tag{9.22}$$

To simplify description, structure the WF with $\lambda I_m = \lambda' V_k$. Taking the derivative of (9.22) to $V(P_k^Z)^{-1}$ and noticing the positive definite second derivative, the minimum of Γ can be achieved if satisfying

$$P_k^Z = \left(2\lambda' \bar{P}_k^Z\right)^{1/3} V_k \triangleq \left(P_k^Z\right)^* \tag{9.23}$$

The superscript $*$ denotes the optimal value of the variable.

Further, define proportional parameters μ_1 and μ_2 having diagonal-matrix form and satisfying

$$H_k \Sigma_{k|k-1} H_k^T = \mu_1 H_k P_{k|k-1} H_k^T \tag{9.24}$$

$$P_k^Z = \mu_2 \left(H_k P_{k|k-1} H_k^T + V_k \right) \tag{9.25}$$

which denote the amplification ratio of the prediction and measurement covariance, respectively. There is

$$\mu_1 = \left(I_m + \frac{v_k}{H_k P_{k|k-1} H_k^T} \right) \left(I_m - \frac{v_k}{P_k^Z} \right) \mu_2 \triangleq k_k^\mu \mu_2 \tag{9.26}$$

where k_k^μ is the transformation matrix. Assuming that P_k^Z takes $(P_k^Z)^*$ as in (9.23), (9.25) can be reformed as

$$\mu_2 = \left(P_k^Z \right)^* T_k^{-1} \tag{9.27}$$

where $T_k = H_k P_{k|k-1} H_k^T + V_k$.

To match the dimension between measurements and states, according to (9.18), expand (9.24) by letting

$$\Sigma_{k|k-1} = f_\alpha P_{k|k-1} \tag{9.28}$$

where $f_\alpha = \mu_1 \otimes I_2$. Then we have

$$(\alpha^*)^{-1} \text{diag} \frac{\bar{P}_k^Z}{P_k^Z} = \mu_1 \tag{9.29}$$

With (9.26), (9.27), and (9.23) substituted into (9.29), therefore, α^* can be formulated by

$$\alpha_k^* = T_k V_k^{-1} \left(2\lambda V_k^{-1} \bar{P}_k^Z \right)^{-1/3} \left(k_k^\mu \right)^{-1} \text{diag} \frac{\bar{P}_k^Z}{P_k^Z} \tag{9.30}$$

Equation (9.30) shows the α that is optimized by the ORCF with a certain α. Let the α in (9.16) and (9.19) take α^* as (9.30), the SH∞F is then transformed into the OSH∞F, which can be fully described by (9.8), (9.9), (9.16) to (9.30) and (9.11) to (9.15). Because α determines the switching threshold, the OSH∞F makes the threshold time-variant and optimal in the sense of λ, that means, with λ given, the filter can optimize filtering robustness and on-line achieve the best performance.

9.4 Simulations

Consider that a free-flying on-orbit chaser tracks a space object that maneuvers potentially. Assume that the target maneuvers with constant thrust and relative motion signals containing relative range, azimuth, and elevation angles are detectable in real time by the chaser.

The OSH∞F is used as target tracking algorithm. To demonstrate the superior tracking performance, KF, H∞F, and SH∞F are also adopted for comparison.

The initial orbits of the target and the chaser have the same orbital elements as follows: $a = 6578.137\,\text{km}$, $e = 0$, $i = 28.5°$, $f = 180°$, and $\Omega = 45°$, except u for which $u_T = 0.05°$ and $u_C = 0°$.

Measurements are generated by Keplerian equations with variances assumed $(1\,\text{m})^2$ for range and $(0.1\,\text{rad})^2$ for elevation and azimuth. The initial state variance takes $(1\,\text{m})^2$ for position and $(0.1\,\text{m/s})^2$ for velocity.

For (9.6) and (9.7), filters are adopted as tracking algorithm to obtain the estimates of the relative state X with the existence of the unknown target maneuver. Suppose that the target has a constant thrust of 2000 s from the time 1000 s along radial and in-track direction. The thrust acceleration of the first 1000 s is negative and that of the second is positive, both of which have the same amplitude $0.001\,\text{m/s}^2$. The true target maneuver and the yielded true trajectory of the target relative to the chaser are shown as in Fig. 9.1.

Figure 9.2 demonstrates the in-track estimation errors of the OSH∞F compared with that of the KF, H∞F, and the SH∞F. Clearly, the KF yields serious deviation due to the optimal filtering gain that has no robustness such that the filter needs quite long time to converge. The H∞F obtains better estimation accuracy than KF due to the H_∞ filter gain, but the infinite-horizon robustness results in excessive cost of filtering optimality that decreases the estimation precision. Introduced with

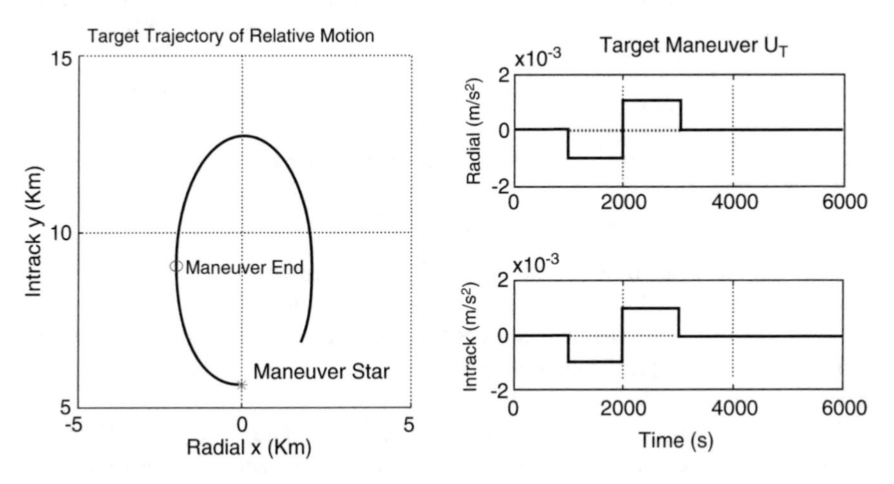

Fig. 9.1 The trajectory of the target maneuvering with constant thrust

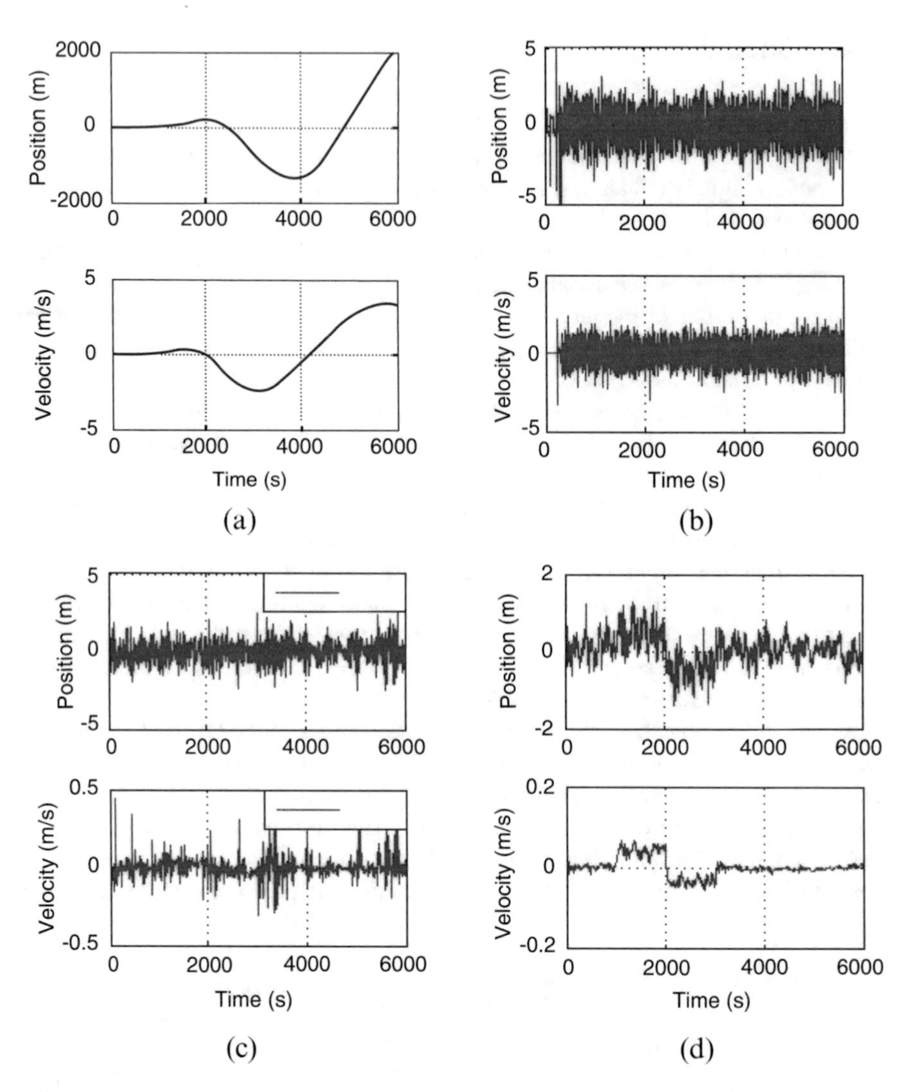

Fig. 9.2 Comparison of the tracking error for constant maneuvering target. (**a**) KF, (**b**) H∞F, (**c**) SH∞F, (**d**) OSH∞F

a switching mechanism of H_∞ and optimal filtering modes, the SH∞F saves part of the filtering optimality cost and improves the precision of the H∞F, as shown in Fig. 9.2c where $\alpha = 1$. The OSH∞F can optimally determine the α on-line. Hence, it achieves the best precision among all the compared filters. That can be drawn from Fig. 9.2d where we take $\lambda = 0.5$.

The tracking errors of OSH∞F varied with λ are presented in Fig. 9.3. The figure illustrates that robustness of the filter is enhanced with λ increasing. Obviously, the error of $\lambda = 0.05$ is smoother but larger than that of $\lambda = 5$, whereas $\lambda = 0.5$ achieves the most balanced accuracy and precision, i.e., filtering robustness and optimality, among the three.

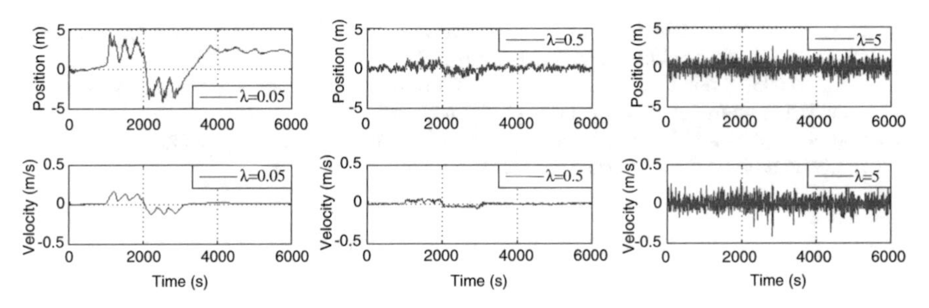

Fig. 9.3 The in-track estimation error of the OSH∞F with various λ

9.5 Conclusions

The OSH∞F is an optimal H_∞ filter developed to optimize the cost of filtering optimality caused by conservativeness of the H∞F. The filter is built on the H∞F and a switching structure, introduced with a designed ORCF to optimize the internal switching parameter in real time. With the only parameter WF showing the weight of filtering robustness and filtering optimality quantitatively, the filter can obtain optimal state estimates in the sense of WF, i.e., the conservativeness of the H∞F is optimized. Compared with KF, H∞F, and SH∞F, the presented OSH∞F can achieve superior estimation precision and accuracy and their weight can be controlled conveniently by setting the WF.

Acknowledgements This work was supported in part by the Project Sponsored by SRF for ROCS, SEM, and Fundamental Research Funds for the Central Universities (ZYGX-2014J098).

References

1. Bai M, Zhou DH, Schwarz H (1999) Identification of generalized friction for an experimental planar two-link flexible manipulator using strong tracking filter. IEEE Trans Robot Autom 15(2):362–369
2. Duan Z, Han C, Li XR (2004) Comments on "unbiased converted measurements for tracking". IEEE Trans Aerosp Electron Syst 40(4):1374
3. Gim DW, Alfriend KT (2003) State transition matrix of relative motion for the perturbed noncircular reference orbit. J Guid Control Dyn 26(6):956–971
4. Li Y (2010) Autonomous follow-up tracking and control of noncooperative space target. Ph.D. Dissertation, Shanghai Jiao Tong University
5. Li Y, Jing Z, Hu S (2010) Redundant adaptive robust tracking of active satellite and error evaluation. IET Control Theory Appl 4(11):2539–2553
6. Li Y, Jing Z, Liu G (2014) Maneuver-aided active satellite tracking using six-dof optimal dynamic inversion control. IEEE Trans Aerosp Electron Syst 50(1):704–719
7. Li Y, Zhang S, Ding L, Shi Z (2016) Optimal-switched h_∞ robust tracking for maneuvering space target. IFAC-PapersOnLine 49(17):415–419

8. Longbin M, Xiaoquan S, Yiyu Z, Kang SZ, Bar-Shalom Y (1998) Unbiased converted measurements for tracking. IEEE Trans Aerosp Electron Syst 34(3):1023–1027
9. Xu Y (2005) Sliding mode control and optimization for six dof satellite formation flying considering saturation. J Astronaut Sci 53(4):433–443
10. Yoon H, Agrawal BN (2009) Novel expressions of equations of relative motion and control in keplerian orbits. J Guid Control Dyn 32(2):664–669

Part III
Visual Tracking and Fusion

Chapter 10
Constrained Image Deblurring with Sparse Proximal Newton Splitting Method

10.1 Introduction

Image restoration has been widely studied and discussed, firstly explored in the 1960s [3, 22]. The main task of image restoration is to recover the latent image from a single or multiple degraded observations. It is a difficult problem depending on the imaging model and the occurred degradation. In the past years, a substantial amount of material was concerned on this subject. Some approaches have been developed in many applications, such as visual tracking, remote sensing, astronomical image processing, and medical image reconstruction. These restoration methods varied in prior models, regularization terms, and optimization methods. A linear imaging model is widely adopted, which consists of three components, i.e., an observation image z, latent image x, and additive independent noise η. This model is usually formulated as follows:

$$z = K \otimes x + \eta, \tag{10.1}$$

where K denotes the convolution operator or point spread function (PSF). More specifically, this operator is mainly assumed to be spatially invariant, which can be viewed as an approximation to the blurring process. The additive noise η varies under different imaging conditions, include Gaussian, salt-and-pepper noise, and so on. Based on the imaging process of the specific hardware platform, some researchers proposed some efficient methods to restore a clear scene, such as encoding first-order motion [29], gyroscopes and accelerometers [21].

Z. Jing et al., *Non-Cooperative Target Tracking, Fusion and Control*,
Information Fusion and Data Science, https://doi.org/10.1007/978-3-319-90716-1_10

10.1.1 Related Work

Recently, proximal splitting method [4, 5, 15, 19] is proved to be an efficient optimization scheme, which can be categorized as first-order method. The main idea of proximal splitting method is to split the objective function into sub-functions, which yield an easily implementable algorithm. Its key procedure is to solve the sub-problem, i.e., Moreau proximal minimization problem [15]. There are intensive research literatures on proximal splitting scheme [14]. Among these methods, forward-backward splitting method [5] and its variants are widely applied, including backward-backward algorithm [12], Beck-Teboulle proximal gradient method [5], Dykstra-like splitting method [17], and Douglas-Rachford splitting method [13]. Some optimization algorithms, extended from the proximal splitting scheme, have been developed to handle real-world's problems directly, such as image restoration [4, 5], compressive magnetic resonance (MR) image reconstruction [19], matrix completion [32], and machine learning [20, 33]. Among the above mentioned schemes, a class of accelerated proximal splitting algorithms were developed [4]. It can be noted that there is significantly less work to focus on the regularization parameter of the proximal term.

Another efficient way to handle this problem is variable splitting technology [1, 2], which is mainly combined with augmented Lagrangian method. This is achieved by converting the original problem into an equivalent constraint optimization problem. And this scheme has been applied to various application domains. For image restoration, an effective scheme [1, 2, 10, 31, 37] based on the alternating direction method of multipliers (ADMM) framework was proposed. A combined variable splitting and duality-based image restoration method for salt-and-pepper noise removal was proposed in [11]. For the multiple variables splitting case, Goldfarb [18] provided a novel solution. For TV/ℓ_1 and TV/ℓ_2 minimization problems and its variants, Min Tao [31] exploited the problems' structures and proposed some efficient minimization algorithms. This scheme also applied in some application domains, such as compressed sensing [9] (CS) and video restoration [10]. Although this scheme achieves a success in the real-world's application, it's convergence rate depends on the choice or update of the penalty parameter.

10.1.2 Motivation

The motivation of this chapter is to reduce the dimension limitations of proximal Newton splitting method in [6, 23]. It is well known that quasi-Newton methods suffer from the computational requirement and memory usage for obtaining the accurate representation of the true Hessian matrix as the dimension of the problem growing. In other words, when this framework applies to the large-scale problem, some issues come out, for instance the computation expense and the side effect of the approximation of the exact Hessian matrix. These shortcomings are a cue to develop an alternative approach, which is computationally cheap without destroying the convergence properties of the original framework. Then, sparse proximal Newton splitting scheme is proposed. As a result, an alternative solution to the resulting sub-problem is presented.

In this chapter, a fast image de-blurring algorithm, utilizing the sparse pattern of proximal Newton splitting framework, is proposed. To deal with the weighted Moreau proximal minimization problem, a sparse gradient projection method is proposed. Meanwhile, a modified symmetric rank-one (SR1) update is provided to improve the Hessian approximation. Then,we introduce an iteration-varying penalty or weighting matrix, which can control the progress of the estimated variable toward the optimal solution set dynamically. Specially, the proposed method maintains the sparse structure of the approximate Hessian matrix. The proposed method can be viewed as a unifying framework extended from the proximal splitting method. Furthermore, the theoretical and practical results of the proposed algorithm are discussed. The extensive experiments against other iterative shrinkage-thresholding algorithms indicate that the proposed method is more suitable for large-scale optimization problems. The presentation of this chapter is based on the work in [25, 26].

10.1.3 Chapter Outline

This chapter is organized as follows. In Sect. 10.3, total variation regularization model is introduced. In Sect. 10.4, a brief summary on related optimization methods is presented. In Sect. 10.5, the proposed method and its theoretical results are exhibited. Numerical experiments and conclusions are presented in Sects. 10.6 and 10.7 respectively.

10.2 Basic Notations

In this subsection, some notations are provided.

1. $\|x\|_F$ denotes the $Frobenius$ norm of a matrix $x \in R^{m \times n}$, where m, n denotes the size of the matrix.
2. Weighted $Frobenius$ norm is defined as $\|x\|_W$, where W is a symmetric positive matrix.
3. First-order gradient at the selected point x of the function $f(x) = \frac{1}{2}\|K \otimes x - z\|_F^2$ is defined with $\nabla f(x) = K^T(K \otimes x - y)$, where T denotes the transpose of the matrix.
4. $D(u, v)$ is the set of matrix-pairs (u, v), where $u \in R^{(m-1)\times n}$ and $v \in R^{m\times(n-1)}$ satisfy relationship

$$
\begin{aligned}
&u_{i,j}^2 + v_{i,j}^2 \leq 1, i = 1, \ldots, m-1, j = 1, \ldots, n-1 \\
&\left|u_{i,n}\right| \leq 1, i = 1, \ldots, m-1, \\
&\left|v_{m,j}\right| \leq 1, j = 1, \ldots, n-1.
\end{aligned} \tag{10.2}
$$

5. Linear operator $L : R^{(m-1)\times n} \times R^{m\times(n-1)} \to R^{m\times n}$ is defined as follows:

$$L(u, v)_{i,j} = u_{i,j} + v_{i,j} - u_{i-1,j} - v_{i,j-1},$$

where $u_{0,j} = u_{m,j} = v_{i,0} = v_{i,n}$, among every $i = 1, \ldots, m$ and $j = 1, \ldots, n$.
6. Adjoint operator $L^T : R^{m\times n} \to R^{(m-1)\times n} \times R^{m\times(n-1)}$ against on the operator L, which is defined as follows:

$$L^T(x) = (u, v), \tag{10.3}$$

where $u \in R^{(m-1)\times n}$ and $v \in R^{m\times(n-1)}$ are the matrices, defined by

$$
\begin{aligned}
u_{i,j} &= x_{i,j} - x_{i+1,j}, \quad i = 1, \ldots, m-1, j = 1, \ldots, n, \\
v_{i,j} &= x_{i,j} - x_{i,j+1}, \quad i = 1, \ldots, m, j = 1, \ldots, n-1.
\end{aligned} \tag{10.4}
$$

7. P_C is an orthogonal projection operator on the convex set C. C is defined by $B_{l,u}$, which can be explicitly given by

$$
P_{B_{l,u}}(x)_{ij} = \begin{cases} l & x_{ij} < l, \\ x_{ij} & l \leq x_{ij} \leq u, \\ u & x_{ij} > u. \end{cases} \tag{10.5}
$$

where $B_{l,u}$ is the bound condition, defined by $l < x < u$. And, l and u are constant lower and upper bounds.

10.3 Total Variation Regularization Model

In this section, we focus on constrained total variation-based deblurring problem. Total variation regularization model, proposed by Rudin-Osher and Fatemi [30] (ROF), can be exhibited as follows:

$$F(x) = f(x) + g(x) = \min_{x\in B_{l,u}} \frac{1}{2} \|K \otimes x - z\|_F^2 + \lambda \mathrm{TV}(x), \tag{10.6}$$

where $\|\cdot\|_F$ is a fit-to-data functional, z denotes the degrade image; $f(x)$ stands for a convex function $\frac{1}{2}\|K \otimes x - z\|^2$; $g(x)$ stands for the regularization term $\lambda\mathrm{TV}(x)$;

λ is a scalar positive parameter, which controls the regularization degree between $f(x)$ and $g(x)$. This parameter can be automatically selected by some methods or rules such as generalized cross-validation (GCV), discrepancy principle (DP), and L-curve [34]. $B_{l,u}$ is the bound constraint.

It should be noted that $F(x)$ is a non-smooth problem. ROF's regularization term has two forms, i.e., isotropic TV and ℓ_1-based isotropic TV. Isotropic total variation regularization model is given by

$$\begin{aligned} TV_I(x) = &\sum_{i=1}^{m-1} \left|x_{i,n} - x_{i+1,n}\right| + \sum_{j=1}^{n-1} \left|x_{m,j} - x_{m,j+1}\right| \\ &+ \sum_{i=1}^{m-1} \sum_{j=1}^{n-1} \sqrt{(x_{i,j} - x_{i+1,j})^2 + (x_{i,j} - x_{i,j+1})^2}\,. \end{aligned} \tag{10.7}$$

ℓ_1-based anisotropic TV regularization model is

$$\begin{aligned} TV_{l1}(x) = &\sum_{i=1}^{m-1} \sum_{j=1}^{n-1} \left|x_{i,j} - x_{i+1,j}\right| + \left|x_{i,j} - x_{i,j+1}\right| \\ &+ \sum_{i=1}^{m-1} \left|x_{i,n} - x_{i+1,n}\right| + \sum_{j=1}^{n-1} \left|x_{m,j} - x_{m,j+1}\right|. \end{aligned} \tag{10.8}$$

It is important to note that TV regularizer is non-smooth, which brings in some difficulties, such as non-differentiability, high storage requirement, and heavy computation demand. Recently, it has been applied successfully in different contexts, such as nonlocal means and graph-based total variation model [27]. Beyond these, several optimization algorithms were developed, such as projected gradient method [34], proximal splitting method [5], a class of iterative shrinkage/thresholding (IST) algorithms [4, 8], and sparse reconstruction by separable approximation (SpaRSA) [36]. Among these, IST algorithms were supposed to be efficient minimization methods due to these simple formulations, which vary in the different strategies on the step size and the iterate variable.

A general framework of first-order algorithms for the problem (10.6) is presented in Algorithm 1. The symbol $O_{\sigma,k}$ denotes a backward (implicit) step and involves a forward (explicit) step y_k. A common approach for solving $O_{\sigma,k}$ is proximal splitting method. t_k denotes the step size, which can be constant or iterate-varying. Some basic information about proximal splitting methods are outlined in Sect. 10.4.

Algorithm 1 General framework of first-order algorithms for constrained total variation-based deblurring problem

1: Input: x_0 and the iteration numbers N
2: **for** $k = 1$ to N **do**
3: %% y_k is a temporal variable
4: $y_k = x_{k-1} - t_k \nabla f(x_{k-1})$
5: $x_k = O_{\sigma,k}(y_k)$
6: **end for**

10.4 Proximal Splitting Method and Its Accelerations

In this subsection, some basic preliminaries about proximal splitting method and its acceleration are presented. Basic principles of these optimization methods are that the objective function can be decomposed into the direct sum of a generic differential function $f(x)$ with Lipschitz gradient and a non-differentiable function $g(x)$.

1. For proximal splitting method, $O_{\sigma,k}$, defined as an implicit procedure to be carried out, is expressed as follows:

$$x_k = O_{\sigma,k}(y_k) = prox_{\sigma}^{\mathrm{TV}}(y_k) = \arg\min_{x} \left\{\lambda \mathrm{TV}(x) + \frac{\sigma}{2}\|x - y_k\|_F^2\right\}, \tag{10.9}$$

   where λ is a positive regularization parameter, σ is a positive penalty parameter [7], determined at the beginning of the optimization method empirically [4, 5]. Its theoretical characteristics have been discussed and analyzed in [4]. Although some researchers have provided many nice solutions to this problem, the role of the penalty parameter σ has not been investigated fully.

   The main difference between proximal splitting method and ISTA-like algorithms is the treatment to the forward variable y_k and the mechanism to obtain the step size t_k. These options on the elements of the optimization method play an important role in the real application.
2. For IST, its solution is given by

$$x_k = O_{\sigma,k}(y_k) = (|y_k| - \alpha)\mathrm{sgn}(y_k), \tag{10.10}$$

   where α is a threshold value. $\mathrm{sgn}(\cdot)$ denotes Signum function. IST provides a convergence rate of $\mathcal{O}(1/k)$.
3. To alleviate the limitations of IST, Bioucasdias et al. [8] developed a faster method, named two-step iterative shrinkage/thresholding (TwIST), characterized by combining the previous two iterates' information x_{k-2} and x_{k-1}. Its primal idea is that

$$x_{k+1} = (1-\alpha)x_{k-2} + (\alpha - \beta)x_{k-1} + \beta prox_{\sigma}^{\mathrm{TV}}(y_k), \tag{10.11}$$

   where α, β are parameters. However, it should be noted that some approximating errors, introduced by the proximal term $prox_{\sigma}^{\mathrm{TV}}(y_k)$, may lead to the turbulence decrease of objective function value. To improve the stability of TwIST, they also proposed a monotone-version TwIST [8] (MTwIST). That is,

$$x_{k+1} = \begin{cases} y_k, & \text{if} \quad F(y_k) \le F(x_k), \\ prox_{\sigma}^{\mathrm{TV}}(y_k), & \text{if} \quad F(y_k) > F(x_k). \end{cases} \tag{10.12}$$

   This method can be viewed as a simple trust-region method.

To accelerate the convergence rate of proximal splitting method, fast iterative shrinkage/thresholding algorithm [4] (FISTA) was proposed. The main workflow is exhibited in Algorithm 2.

Algorithm 2 The framework of FISTA

1: Input: L : Lipschitz constant estimated from Δf, and $x_0, t_1 = 1$, step size μ and the maximum iteration numbers N.
2: **for** $k = 1$ to N **do**
3: $\quad y_k = x_k - \mu \nabla f(x_k)$
4: $\quad \hat{x}_{k+1} = prox_{\sigma}^{\mathrm{TV}}(y_k)$
5: $\quad t_{k+1} = \frac{1+\sqrt{1+4t_k^2}}{2}$
6: $\quad x_{k+1} = x_k + \frac{t_k - 1}{t_{k+1}}(\hat{x}_{k+1} - x_k)$
7: **end for**

Explicitly, the key idea of FISTA is the application of two-steps' extrapolation procedure. The convergence rate of FISTA is $\mathcal{O}(1/k^2)$.

Theorem 10.1 *Assuming the sequence x_k generated by FISTA, and it can be shown that [4]*

$$F(x_k) - F(x') \leq \frac{2\alpha \left\| x' - x_0 \right\|_2^2}{(k+1)^2} \quad k \geq 1\,, \tag{10.13}$$

where x' denotes the optimal solution, L denotes the Lipschitz constant of ∇f and α stands for the step size.

The power of FISTA is mainly derived from the procedures of line search and solving the proximal minimization problem $x_k = prox_{\sigma}^{\mathrm{TV}}(y_k)$. However, there are some limitations about FISTA. It is well-known that FISTA cannot solve the objective function with more than two regularization terms, for example,

$$x_{k+1} = \arg\min_{x \in C} \frac{1}{2} \left\| x - y_k \right\|_F^2 + \alpha \mathrm{TV}(x) + \beta |x|_1\,. \tag{10.14}$$

A general solution to these composite regularization terms is the alternating minimization technology [28]. However, some difficulties may arise, such as the choice of multi-regularization parameters.

10.5 Sparse Proximal Newton Splitting Method

In this section, the definition and the principles of sparse proximal Newton splitting method are summarized. Then, the scheme and the analysis to the weighted norms-based sub-problem is presented and analyzed. Finally, the specific iterative algorithm is outlined.

10.5.1 Definition and the Framework for Constrained Image Deblurring

The framework of sparse proximal Newton splitting method for constrained TV-based image deblurring problem is illustrated. Then, a modified symmetric rank (SR) 1 update is presented. At last, a close-form solution to the weighted primal-dual problem and sparse gradient projection method are proposed. At last, some Lemmas are presented.

Definition 10.1 Proximal Newton splitting method [6, 23] can be defined as follows:

$$\begin{aligned}\tilde{x} &= prox_W^{\mathrm{TV}}(y) \\ &= \arg\min_{x\in C}\left\{\lambda \mathrm{TV}(x) + \frac{1}{2}\|x-y\|_W^2\right\},\end{aligned} \tag{10.15}$$

where W denotes a symmetric and positive definite matrix, and y is the approximating point. A natural choice to W is Hessian matrix. This approximating process can be viewed as a semi-quadratic approximation to the selected point y.

The essence of proximal Newton splitting method can be expressed in detail. For the approximation point $y_k = x_k - W_k^{-1}\nabla f(x_k)$, the estimated point $\tilde{x}_{k+1}$ can be represented with

$$\begin{aligned}\tilde{x}_{k+1} &= prox_{W_k}^{\mathrm{TV}}(x_k - W_k^{-1}\nabla f(x_k)) \\ &= \arg\min_{x\in C}\left\{\frac{1}{2}\|(x-x_k) + W_k^{-1}\nabla f(x_k)\|_{W_k}^2 + \lambda \mathrm{TV}(y_k)\right\} \\ &= \arg\min_{x\in C}\left\{\langle\nabla f(x_k), x-x_k\rangle + \frac{1}{2}\|x-x_k\|_{W_k}^2 + \lambda \mathrm{TV}(y_k)\right\},\end{aligned} \tag{10.16}$$

where W_k^{-1} is the approximate inverse Hessian matrix at the iterate k.

Adding a constant term $f(x_k)$, the expression (10.16) can be rewritten as follows:

$$\tilde{x}_{k+1} = \arg\min_{x\in C} f(x_k) + \langle x-x_k, \nabla f(x_k)\rangle + \frac{1}{2}\|x-x_k\|_{W_k}^2 + \lambda \mathrm{TV}(y_k)\,. \tag{10.17}$$

The previous Eq. (10.17) can be reformulated as follows:

$$\tilde{x}_{k+1} = \arg\min_{x\in C}\frac{1}{2}\left\|x-(x_k - W_k^{-1}\nabla f(x_k))\right\|_{W_k}^2 + \lambda \mathrm{TV}(y_k)\,. \tag{10.18}$$

It can be seen that the connection to the proximal splitting method is a generalized procedure by converting the penalty parameter σ in Eq. (10.9), into a weighting matrix W_k. When W_k is chosen to be the scale identity matrix σI, Eq. (10.18) is reduced to the proximal splitting method. This general representation may provide an alternative approach to update the penalty parameter adaptively. Some basic properties of proximal Newton splitting method are concluded as follows:

1. Karush-Kuhn-Tucker (KKT) conditions for the first-order optimality condition,
2. $\tilde{x} = prox_W^{\mathrm{TV}}(y)$ exists and is unique for $x, y \in dom(f)$, where $dom(\cdot)$ denotes the efficient domain of the function f.
3. Non-expansive in W-norm.

For constrained total variation minimization-type problem, it should be noted that the original proximal Newton splitting method has some drawbacks that should be emphasized. First, the space-consuming introduced by the true or inverse Hessian matrix W_k^{-1} and W_k are huge, which particularly has an important influence on the large-scale optimization problem. Second, the computation of Hessian matrix is impracticable. Particularly, the calculation of approximate Hessian matrix W_{k-1}, based the inverse one W_k^{-1}, is assumed to be a computational-expense procedure as the dimension growing, such as classic Broyden-Fletcher-Goldfarb-Shanno [24] (BFGS) formula.

Based on above considerations, a modified version of the proximal Newton splitting scheme is proposed. This scheme is achieved by utilizing the sparse pattern of the approximate inverse Hessian matrix. The main framework for constrained TV-based image deblurring problem is presented in Algorithm 3.

Algorithm 3 Framework of sparse proximal Newton splitting method for constrained TV-based image deblurring problem

1: Input: $x_0 = z$, the step size t_k, $\nabla f(x)$ and the maximum iteration numbers N.
2: **for** $k = 1$ to N **do**
3: %% Update $\tilde{W}_k^{-1}$ with *Algorithm* 4.
4: $y_k = x_k - \tilde{W}_k^{-1}\nabla f(x_k)$
5: %% Update $\hat{x}_k$ by solving the sub-problem $prox_{\tilde{W}_k}^{TV}(\cdot)$ (see *Algorithm* 5).
6: $\hat{x}_{k+1} = prox_{\tilde{W}_k}^{TV}(y_k)$
7: $\Delta x_{k+1} = x_k - \hat{x}_{k+1}$
8: %% Backtracking line search by step size t_k
9: $x_{k+1} = x_k - t_k \Delta x_{k+1}$
10: **end for**

It can be noted that the two aspects of the proposed method have evolved very differently over the primal splitting method. First, the proposed framework avoids the inverse of Hessian matrix indirectly, which benefits from solving the corresponding sub-problem. Second, the proposed framework reduces huge memory requirement when obtaining the matrix $\tilde{W}_k^{-1}$. Essentially, the sparse pattern of inverse Hessian matrix $\tilde{W}_k^{-1}$ is exploited, which is demonstrated in Algorithm 4. An approximate approach, described in Algorithm 5, is proposed for solving the sub-problem. Then, many theoretical results are summarized in the Sect. 10.5.2.

The inverse Hessian matrix $\tilde{W}_k^{-1}$ defined in Eq. (10.15) is produced by Algorithm 4. The explicit solution to this sub-problem, i.e., Eq. (10.23), indicates that the calculation of the true Hessian matrix $\tilde{W}_k$ is not required directly. And, it just involved with the inverse Hessian matrix $\tilde{W}_k^{-1}$.

Algorithm 4 Modified symmetric rank (SR) one update for $\tilde{W}_k^{-1}$

1: Input: current iteration number k and variables $x_{k-1}, x_k, 0 < \gamma < 1, 0 < \omega_{\min} < \omega_{\max}$, β and termination criterion Tol.
2: %% the residual between the iterates x_{k-1} and x_k
3: $s_k = x_k - x_{k-1}$
4: $g_k = \nabla f(x_k) - \nabla f(x_{k-1})$
5: **if** $k == 1$ **then**
6: %%init the first inverse Hessian matrix.
7: $W_{tmp}^{-1} = \omega_{\min} I$ %% W_{tmp}^{-1} stands for the temporal matrix.
8: $e_k = 0$
9: **else**
10: $\omega_{BB} = \frac{\langle s_k, g_k \rangle}{\|g_k\|^2}$
11: %% range the ω_{BB} into $[\omega_{\min}, \omega_{\max}]$
12: $\omega_{BB} \mapsto [\omega_{\min}, \omega_{\max}]$
13: ${W_{tmp}}^{-1} = \gamma \omega_{BB} I$
14: **if** $\langle s_k - {W_{tmp}}^{-1} g_k, g_k \rangle \le tol \times \|g_k\|_F \left\| s_k - {W_{tmp}}^{-1} g_k \right\|_F$ **then**
15: %% drop the quasi-Newton procedure
16: $e_k = 0$
17: **else**
18: $e_k = (s_k - W_{tmp}^{-1} g_k) / \sqrt{\langle s_k - W_{tmp}^{-1} g_k, g_k \rangle}$
19: **end if**
20: **end if**
21: %% $diag(\cdot)$ stands for diagonalizing the input matrix
22: $W_k^{-1} = {W_{tmp}}^{-1} + diag(e_k {e_k}^T)$
23: $\tilde{W}_k^{-1} = diag(\min(\beta, \max(\frac{1}{\beta}, W_k^{-1})))$

For the most case, the value γ is chosen to be 0.3 empirically. The parameter β is configured to an acceptable penalty value. To preserve the positiveness of the inexact representation of inverse Hessian matrix, $\tilde{W}_{tmp}^{-1}$ is configured with $\gamma \omega_{BB} I$, where I stands for the identity matrix.

For solving the sub-problem efficiently, sparse inverse Hessian matrix $\tilde{W}^{-1}$ is reformulated as follows:

$$\tilde{W}_k^{-1} = \operatorname{diag}\left(\min\left(\beta, \max\left(\frac{1}{\beta}, W_k^{-1}\right)\right)\right). \tag{10.19}$$

Utilizing this sparse pattern of the matrix W_k^{-1}, this inexact representation of W^{-1} saves the storage greatly.

10.5.2 Theoretical Analysis

In this subsection, some theoretical aspects about the proposed method are presented. Then, an approximate solution is presented to avoid the intrinsic difficulties introduced by TV regularization model.

In order to get the dual formation of isotropic total variation regularization term, we need the following basic dual formula:

$$\begin{aligned} \sqrt{x^2+y^2} &= \max_{u_1,u_2\in D} \{u_1x+u_2y : u_1+u_2 \le 1\}, \\ |x|_1 &= \max_{u} \{ux : |u| \le 1\}. \end{aligned} \tag{10.20}$$

Based on Eq. (10.20), we get the dual formation of isotropic total variation regularization term corresponding to the sub-problem (10.15). Its formulation in detail, termed as weighted primal-dual problem, is written as

$$\min_{x\in C}\max_{(u,v)\in D}\left\{\frac{1}{2}\|x-y\|^2_{\tilde{W}} + \lambda Tr(L^T(u,v)x)\right\}, \tag{10.21}$$

The subscript k is omitted for convenience. And the expression $Tr(L(u,v)^T x)$ is expressed clearly as follows:

$$\begin{aligned} Tr(L(u,v)^T x) &= \sum_{i=1}^{m-1}\sum_{j=1}^{n-1}\left[u_{i,j}(x_{i,j}-x_{i+1,j}) + v_{i,j}(x_{i,j}-x_{i,j+1})\right] \\ &\quad + \sum_{i=1}^{m-1} u_{i,+n}(x_{i,n}-x_{i+1,n}) + \sum_{j=1}^{n-1} v_{m,j}(x_{m,j}-x_{m,j+1}). \end{aligned} \tag{10.22}$$

Proposition 10.1 *According to the primal-dual problem presented in (10.21), the optimal solution is*

$$x = P_C(y - \lambda\tilde{W}^{-1}L(u,v)). \tag{10.23}$$

Proof It is important to note that the characteristic of Eq. (10.21) is convex in C and concave in (u,v). Then the inner and the outer minimization procedures can be swap, which depend on the optimization theory in [7]. Equation (10.21) can be rewritten as

$$\begin{aligned} &\max_{(u,v)\in D}\min_{x\in C}\left\{\frac{1}{2}\|x-y\|^2_{\tilde{W}} + \lambda Tr(L(u,v)^T x)\right\} \\ &= \max_{(u,v)\in D}\min_{x\in C}\frac{1}{2}\|x-(y-\lambda\tilde{W}^{-1}L(u,v))\|^2_{\tilde{W}} - \|y-\lambda\tilde{W}^{-1}L(u,v)\|^2_{\tilde{W}} + \|y\|^2_{\tilde{W}}. \end{aligned} \tag{10.24}$$

The optimal solution to the inner minimization problem is

$$x = P_C(y - \lambda \tilde{W}^{-1} L(u, v)), \tag{10.25}$$

where $P_C(\cdot)$ is the orthogonal projection operator in the convex set C. Substituting the expression (10.25) into Eq. (10.23) and transforming into a minimization problem by multiplying with -1, a weighted *Frobenius* norm-based quadratic equation is obtained by

$$G(u, v) = \min_{(u,v)\in D} -\left\|\phi_D(y - \lambda \tilde{W}^{-1} L(u, v))\right\|_{\tilde{W}}^2 + \left\|y - \lambda \tilde{W}^{-1} L(u, v)\right\|_{\tilde{W}}^2, \tag{10.26}$$

where $\|\cdot\|_{\tilde{W}}^2$ denotes the weighted norm with a weighted matrix $\tilde{W}$. For simplicity, $\phi_D(x)$ is represented by

$$\phi_D(x) = x - P_D(x), x \in R^{m\times n}. \tag{10.27}$$

□

Lemma 10.1 *The gradient of the problem (10.26) can be obtained by*

$$\nabla G(u, v) = -2\lambda L^T P_D(y - \lambda \tilde{W}^{-1} L(u, v)). \tag{10.28}$$

Proof Redefine a function $\theta : R^{m\times n} \to R$

$$\theta(x) = \|\phi_D(x)\|_{\tilde{W}}^2, \tag{10.29}$$

And the function in Eq. (10.26) is reformatted as

$$G(u, v) = -\theta(y - \lambda \tilde{W}^{-1} L(u, v)) + \left\|y - \lambda \tilde{W}^{-1} L(u, v)\right\|_{\tilde{W}}^2. \tag{10.30}$$

Utilizing the quadratic formation of the function $\theta(x)$, the gradient $\nabla\theta(x)$ is

$$\nabla\theta(x) = 2\tilde{W}(x - P_C(x)). \tag{10.31}$$

Then, we get

$$\begin{aligned}\nabla G &= \nabla(-\theta(y - \lambda \tilde{W}^{-1} L(u, v)) + \|y - \lambda \tilde{W}^{-1} L(u, v)\|_{\tilde{W}}^2) \\ &= \lambda\nabla\theta(y - \lambda \tilde{W}^{-1} L(u, v)) - 2\lambda L^T(y - \lambda \tilde{W}^{-1} L(u, v)).\end{aligned} \tag{10.32}$$

substituting Eq. (10.31) into Eq. (10.32) and arranging, we have the gradient of G is given by

$$\nabla G(u, v) = -2\lambda L^T(y - \lambda \tilde{W}^{-1} L(u, v)). \tag{10.33}$$

□

10.5.3 Approximate Solution of Weighted Primal-Dual Problem

In this subsection, a sparse gradient projection method based on the results in Lemma 10.1 is presented. The proposed method within the framework of FISTA is demonstrated in Algorithm 5, which handles the corresponding primal-dual problem (10.21).

Algorithm 5 A sparse fast gradient projection method for the sub-problem (10.15)

1: Input: the blurred and noised image y, $\tilde{W}_k^{-1}$, $\lambda_{\max}$ from $\max(\tilde{W}_k^{-1})$, the regularization parameter λ, inner iterate number N.
2: Output: the optimal solution x^*
3: Initialization: $(r_1, s_1) = (u_0, v_0) = (0_{(m-1)\times n}, 0_{m\times(n-1)}), t_1 = 1.$
4: **for** $k = 1$ to N **do**
5: $\quad z_k = \frac{1}{8\lambda\lambda_{\max}} L^T(P_D(y - \lambda\tilde{W}_k^{-1}L(r_k, s_k)))$
6: $\quad (u_k, v_k) = P_p\left[(r_k, s_k) + z_k\right]$
7: $\quad t_k = \frac{1+\sqrt{1+4t_k^2}}{2}$
8: $\quad (r_{k+1}, s_{k+1}) = (u_k, v_k) + \frac{t_k - 1}{t_{k+1}}(u_k - u_{k-1}, v_k - v_{k-1})$
9: **end for**
10: $x^* = P_C[y - \lambda\tilde{W}^{-1}L(u_N, v_N)]$

10.5.4 Computational Complexity

The computational complexity of the proposed framework is $\mathcal{O}(n\log(n))$. The computational cost of the proposed approach can be summarized as follows:

1. The computation costs in the step 4 is $\mathcal{O}(n\log(n))$, which mainly involved with the operations with vectors adding, $\nabla f(\cdot)$ and $\tilde{W}_k^{-1}$ described in Algorithm 4. Its computation steps, in the steps of Algorithm 4 from step 3 to 23, involved with vectors adding, scale and the gradient $\nabla f(\cdot)$. Its costs are $\mathcal{O}(n)$, $\mathcal{O}(1)$ and $\mathcal{O}(n\log(n))$ independently. On account of the sparse pattern of $\tilde{W}^{-1}$, the $\tilde{W}^{-1}g_k$ comes into an element-wise product operation, whose cost is $\mathcal{O}(1)$ or $\mathcal{O}(n)$. Thus, the cost of step 4 is $\mathcal{O}(n\log(n))$.
2. As the introduction in [4], the sub-problem (10.15), i.e., the step 6 of Algorithm 3, can be solved in the computational complexity with $\mathcal{O}(n)$. The extra cost for solving this problem is the requirement on the storage of the weighting matrix, which is updated by Algorithm 4.
3. For the steps 7 and 9 of Algorithm 3, the costs are assumed to be $\mathcal{O}(n)$ or $\mathcal{O}(1)$.

In summary, the overall cost of the proposed method is $\mathcal{O}(n\log(n))$.

10.6 Computational Experiments and Discussion

To assess the computational efficiency of sparse proximal Newton (SPN) splitting method, a group of experiments on synthetic data is conducted to evaluate the performance with respect to blurring-levels, noise levels, and image dataset. The experimental settings of the referred methods and the synthetic examples are discussed in detail. Moreover, we compare the reconstruction quality with two other methods: IST [16] and FISTA [4]. The reconstructed quality of all the methods are demonstrated and discussed elaborately.

10.6.1 Experimental Settings

The numerical experiments are tested on a computer with a processor Core i5 2.6 GHz and 4G RAM. All algorithms are implemented in Matlab. Three types of blurring filters are considered, that is, Gaussian, uniform, and disk blurs, which are demonstrated in Table 10.1. All methods are applied to the boat, house, and girl images, which contain smooth or texture objects and complex structures. Gaussian noise with zero mean and a variance σ^2 is added to the corresponding blurred observations. To evaluate the performance for the estimated image, two criteria are used, that is, signal-to-noise (SNR) and structural similarity image measurement [35] (SSIM). The larger quality values of the SNR and SSIM mean the better quality of the reconstructed image. Based on these two criteria, some salient features can be analyzed. Our experiments focus on the strong blurring effect and low noise level. The point spread function in the experiments assumed to be normalized, that is, $\sum_{i,j}^{m,n} k(i, j) = 1$, where m, n denote the size of blurring kernel.

In the overall experiments, all the methods performed 100 iterates. All the images pixels' intensity values are rescaled between 0 and 1. And, the inner iteration number is 10. The source codes of IST[1] and FISTA[2] are downloaded from the authors' websites, respectively.

Table 10.1 shows the settings of five experiments in detail. The (1) and (2) experiments are performed for investigating the numerical behavior of the referred algorithms in different blurring operators with the same noise level. The regularization parameter λ of all the experiments is 7×10^{-4}. Some experiments are conducted for evaluating the impact of the regularization parameter λ, which varies from 10×10^{-4} to 10^{-4}. The sizes of the test images are 512×512, 256×256, and 256×256, respectively. The (4) and (5) experiments are taken for assessing the performance in different blurring-levels with noise level $\sigma^2 = 2 \times 10^{-3}$.

[1] http://www.lx.it.pt/~bioucas/code/TwIST_v2.zip.

[2] http://ie.technion.ac.il/~becka/papers/tv_fista.zip.

Table 10.1 Five experiments' settings

Experiment	Image	Blur kernel	Kernel size	Variance
(1)	Boat	Gaussian	9	4
(2)	Boat	Average	9	/
(3)	Boat	Disk	11	/
(4)	House	Average	9	4
(5)	Girl	Gaussian	11	4

Table 10.2 Criteria values for the first three experiments ($\lambda = 7 \times 10^{-4}$, $\sigma^2 = 1 \times 10^{-3}$)

Experiment	Algorithm	SNR(dB)	SSIM
(1)	IST	12.2489	0.7440
	FISTA	12.7636	0.7451
	SPN	**13.9288**	**0.7755**
(2)	IST	12.8570	0.7652
	FISTA	13.4728	0.7685
	SPN	**14.7286**	**0.7991**
(3)	IST	13.4528	0.7844
	FISTA	13.9685	0.7819
	SPN	**15.2045**	**0.8125**

10.6.2 First Three Experiments: Gaussian, Average, and Disk Blurs

In this subsection, the visual and numerical results of the first experiment on the boat image are demonstrated. For convenience, Table 10.2 lists the maximum values of the numerical results from the (1)–(3) experiments. The values of the SNR and SSIM for sparse proximal Newton splitting method are 13.9288 and 0.7755, separately. In term of the SSIM measures, the proposed method also shows a better performance, which indicates that the recovered latent image is closer than that of the other methods. Figure 10.1 shows the synthetic data and the reconstruction results of the three methods. In consideration of the contrast and the details of the reconstruct results, it can be noted that the result of the proposed method is slightly better than that of the other methods with regard to local regions' textures. For observing the reconstruction speed, Fig. 10.2 demonstrates the evolution of the SNR and SSIM against the iteration number. It can be noted that our method achieves a significant performance improvement.

As indicated by the bold number in the second and third rows of Table 10.2, the proposed method acts a better way than the other methods again. Although the diagonalized inverse Hessian matrix may lose some curvature or proper information, the proposed approximation method generates a sequence of penalty parameter and maintains the sparse pattern of the inverse Hessian matrix adaptively. On the account of the introduction and the update of the weighting matrix locally, our method can restore more visual details and lead to performance improvement. In other words, our method can complete the artifact suppression locally, and then reestablish the original edges and shapes efficiently. It can be concluded that the improvement of the visual effect arises from the thoughtfulness of the generalization process on the original proximal splitting method.

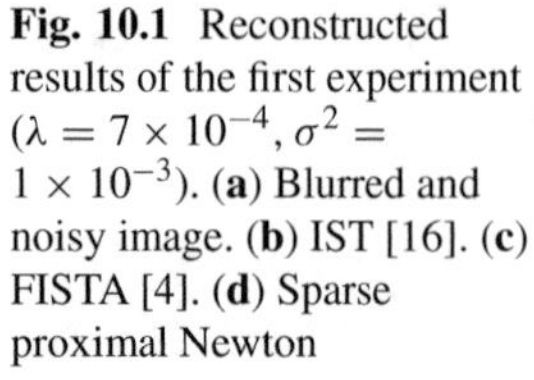

Fig. 10.1 Reconstructed results of the first experiment ($\lambda = 7 \times 10^{-4}, \sigma^2 = 1 \times 10^{-3}$). (a) Blurred and noisy image. (b) IST [16]. (c) FISTA [4]. (d) Sparse proximal Newton

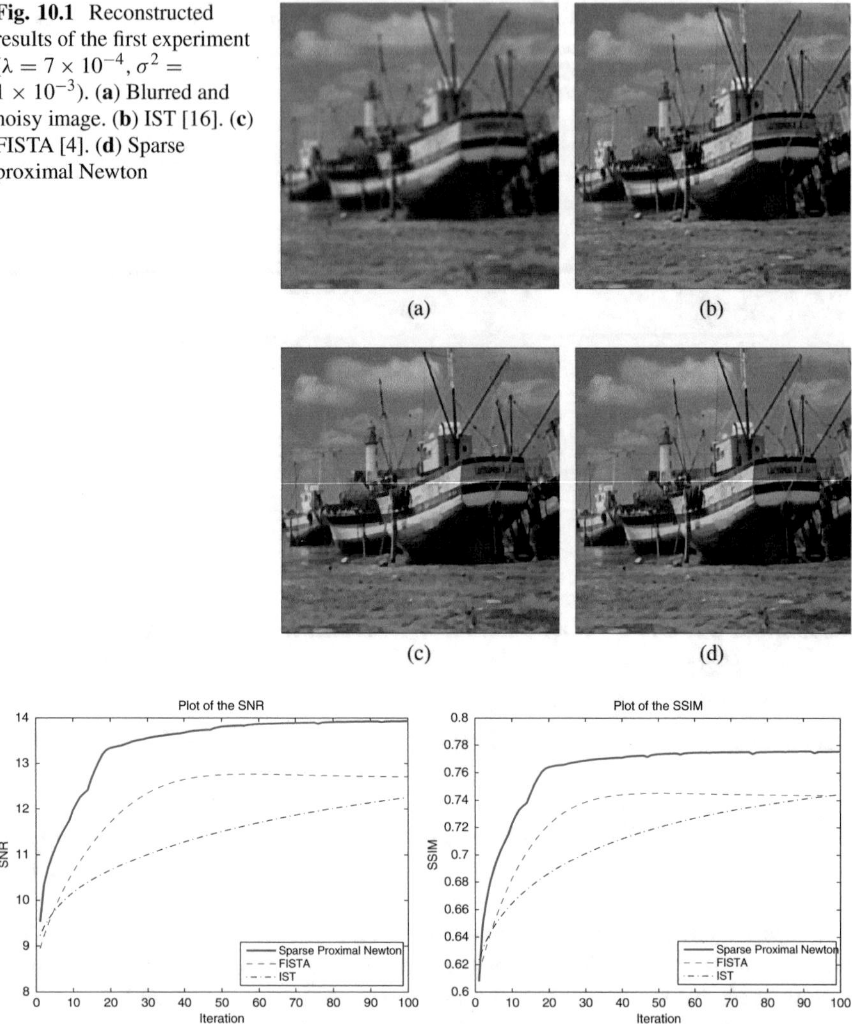

Fig. 10.2 Numerical performance of the first experiment (Gaussian, $\lambda = 7 \times 10^{-4}, \sigma^2 = 1 \times 10^{-3}$): (**a**) SNR vs. iterations, (**b**) SSIM vs. iterations

10.6.3 Fourth and Fifth Experiments

In these two experiments, the proposed method is tested in different blur strengths. In the fourth experiment, the noise level is $\sigma^2 = 2 \times 10^{-3}$. The two tests' regularization parameters are $\lambda = 7 \times 10^{-4}$.

The numerical results of the (4) and (5) experiments, depicted in Table 10.3, are presented. As indicated by the bold number, the qualitative measures of the proposed method achieve a better performance again. Figure 10.3 exhibits the two

Table 10.3 Criteria values of the (4) and (5) experiments ($\lambda = 7 \times 10^{-4}, \sigma^2 = 2 \times 10^{-3}$).

Experiment	Algorithm	SNR(dB)	SSIM
	IST	15.7048	0.8382
(4)	FISTA	17.0371	0.8512
	SPN	**17.8906**	**0.8613**
	IST	12.8484	0.8302
(5)	FISTA	13.2505	0.8258
	SPN	**14.1017**	**0.8406**

experiments' numerical results. Figure 10.3a, b demonstrate the outcomes from the house image. Figure 10.3c, d for the girl image. The illustrative results of these experiments are presented in Fig. 10.4. The visual perception of the proposed method is better than the other methods once more. Similarly, the outcomes of the proposed method produce better results in the presence of the different noise level and kernel size.

Based on the observed results, it can be concluded that the two experiments behave a similar way to the previous three experiments. The proposed method completes slightly better results than the other methods, as a result of the adaptive weighting matrix. The similar outcomes of the proposed method imply that the dynamic procedure for choosing the weighting matrix accomplishes the appealing feature against the other methods.

10.6.4 Sensitivity Analysis of Regularization Parameter

In this subsection, sensitivity analysis of the regularization parameter λ, presented in Fig. 10.5, is performed with a constant step 1×10^{-4}. Although, the practical meaning of λ is the balance between the regularization term and the data fidelity term, the behavior of all these numerical optimization methods has not been investigated fully. It can be noted that the reconstruction quality of these experiments is on the increase, when λ decreases. Due to the adjustment of the weighting matrix $\tilde{W}_k^{-1}$, the proposed method is better than other methods. The best λ is defined by the highest quality values of the referred methods, which can be expressed by $\max(V_i, V_j)$, where $i \neq j;\ i, j \in \{spn, fista, ist\}$. It is notable that $\lambda = 1 \times 10^{-4}$ is the best one among the (1)–(3) experiments separately. It has been annotated in Fig. 10.5. The curves of the IST and FISTA in Fig. 10.5b, d, and f imply that the advance of the FISTA depends on λ to some degree. It can be seen that the growth of the IST, relative to the amount of the SNR and SSIM, increases slowly. The main reason for this emergence by the IST is that the hard threshold on the iterate variable may diminish high-frequency details or image structures.

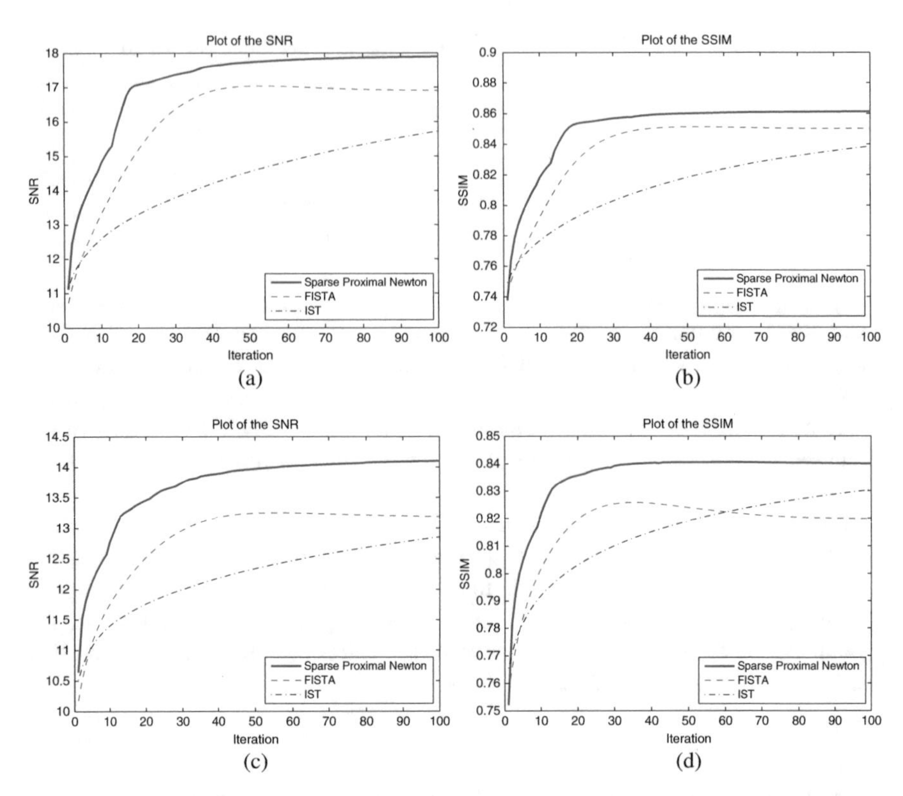

Fig. 10.3 Numerical performance of the (4) and (5) experiments ($\lambda = 7 \times 10^{-4}, \sigma^2 = 2 \times 10^{-3}$). (**a**) The curve of SNR vs. iterations by house. (**b**) The curve of SSIM vs. iterations by house. (**c**) The curve of SNR vs. iterations on girl. (**d**) The curve of SSIM vs. iterations on girl

10.6.5 Discussion

All the results of the five experiments indicate that the proposed method is suitable for the large-scale optimization problem. It can be noted that the proposed method leads to the performance improvement. Correspondingly, the presented method also can reconstruct more details. The robustness of the SPN is demonstrated by the fact that the similar parameters can be applied to the experiments on various images, noise level, and variant blurring effects. Moreover, the presence of the evaluation of the impact of λ on the experiments, presented in Fig. 10.5, also validates that the proposed method is much better than other methods. The visual quality, measured by the SSIM, indicates the proposed method shows better results. The difference may relate to the generalized representation of proximal splitting method and the introduction and maintenance of the sparse pattern of the iteration-varying inverse Hessian matrix. It is possible that the adaptive update of the weighting elements of the inverse Hessian matrix $\tilde{W}^{-1}$ produces better outcomes and is critical for

Fig. 10.4 Reconstructed results of the (4) and (5) experiments ($\lambda = 7 \times 10^{-4}$, $\sigma^2 = 2 \times 10^{-3}$). (**a** and **e**) Blurred and noisy image. (**b** and **f**) IST [16]. (**c** and **g**) FISTA [4]. (**d** and **h**) Sparse proximal Newton

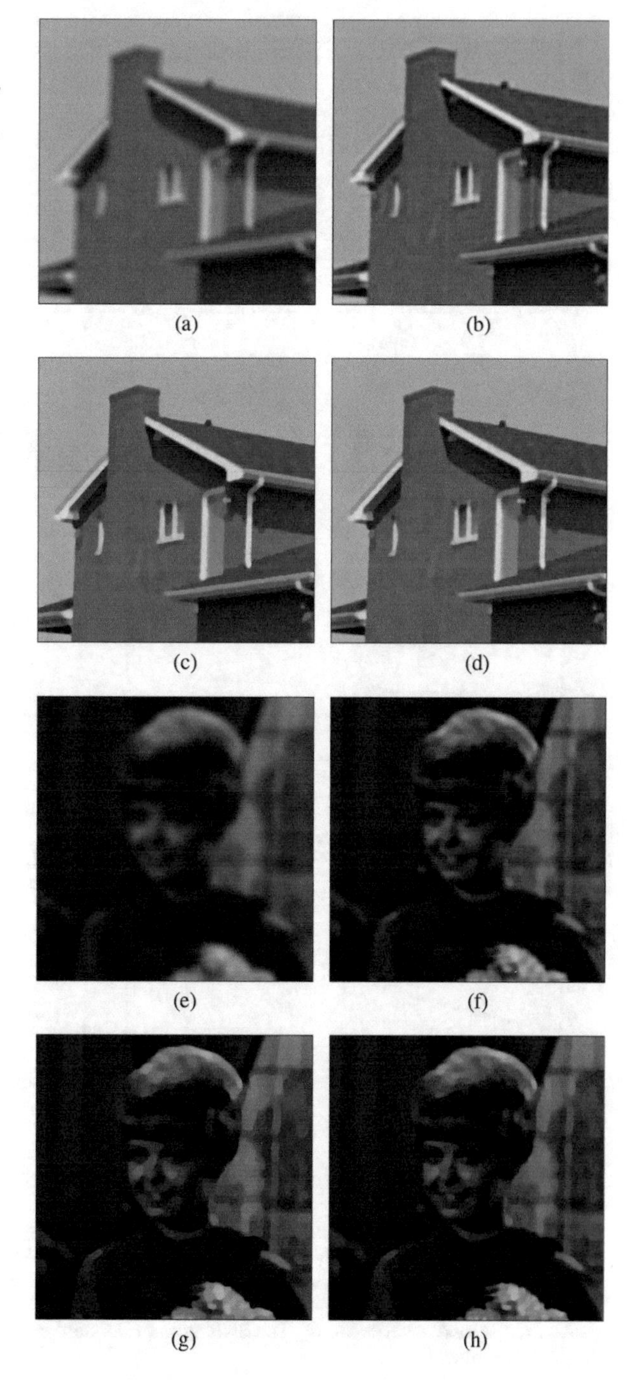

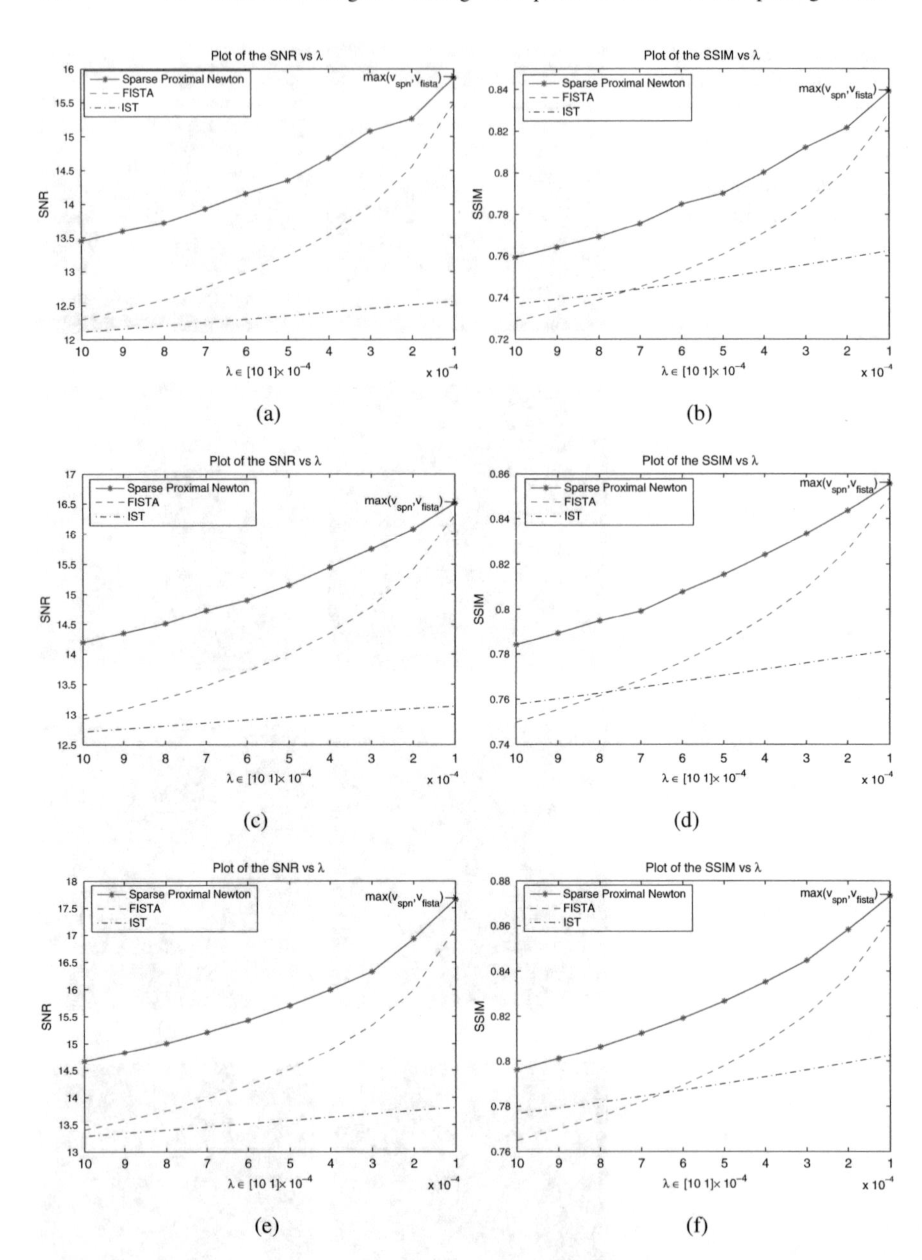

Fig. 10.5 Sensitivity analysis of regularization parameter λ by IST, FISTA, SPN in the (1)–(3) experiments ($\lambda = [10\ 1] \times 10^{-4}, \sigma^2 = 1 \times 10^{-3}$). (**a**) The SNR vs. λ of the first experiment. (**b**) The SSIM vs. λ of the first experiment. (**c**) The SNR vs. λ of the second experiment. (**d**) The SSIM vs. λ of the second experiment. (**e**) The SNR vs. λ of the third experiment. (**f**) The SSIM vs. λ of the third experiment

the success. Although our method makes use of the second derivative information partly, the proposed method potentially provides an alternative proximal splitting method for large-scale optimization problems.

The experiments also show some limitations of the proposed method. First, SPN suffers from the instability of the minimization process of the sub-problem slightly, which can be observed in Fig. 10.2. Second, two key procedures of the proposed method, which include maintaining the weighting matrix adaptively and solving the resulting primal-dual problem, may introduce some extra computational costs. Third, the sparse approximation of inverse Hessian matrix may bring in some estimate error to the recovery results.

Despite the constraint of the proposed method, it can be noted that SPN provides a variable penalty parameter proximal splitting method for constrained TV minimization problem and carries out better reconstruction results against IST and FISTA. Owing to the update of the entries of the weighting matrix locally, we argued that the proposed method can reduce the ring effects to some degree. And the experiments also confirm that our method is better than the other methods.

10.7 Conclusion

In this chapter, an efficient total variation minimization method for image deblurring is proposed. The key idea is the introduction of the penalty parameter matrix, which has a positive impact on the iterative process toward the optimal solution. As a result, a modified symmetric rank (SR) one update is provided. Meanwhile, an approximate solution to the weighted norm-based primal-dual problem is proposed. And a fast gradient projection method for the resulting sub-problem is presented. The superiority of the proposed method with respect to sequential approaches is faster and achieves a better performance improvement.

Acknowledgements This work is jointly supported by the National Natural Science Foundation of China (Grant Nos.61175028, 61603249 and 61271317) and the Ph.D. Programs Foundation of Ministry of Education of China (Grant No. 20090073110045). The authors also thank the excellent works of Dr. Jose M. Bioucas Dias and Dr. Amir by making their source codes public available.

References

1. Afonso MV, Bioucas-Dias JM, Figueiredo MA (2010) Fast image recovery using variable splitting and constrained optimization. IEEE Trans Image Process 19(9):2345–2356
2. Afonso MV, Bioucas-Dias JM, Figueiredo MA (2011) An augmented Lagrangian approach to the constrained optimization formulation of imaging inverse problems. IEEE Trans Image Process 20(3):681–695
3. Andrews HC, Hunt BR (1977) Digital image restoration. Prentice-Hall, Englewood Cliffs

4. Beck A, Teboulle M (2009) Fast gradient-based algorithms for constrained total variation image denoising and deblurring problems. IEEE Trans Image Process 18(11):2419–2434
5. Beck A, Teboulle M (2009) A fast iterative shrinkage-thresholding algorithm for linear inverse problems. SIAM J Imag Sci 2(1):183–202
6. Becker S, Fadili J (2012) A quasi-Newton proximal splitting method. In: Advances in neural information processing systems, pp 2618–2626
7. Bertsekas DP (2009) Convex optimization theory. Athena Scientific, Belmont
8. Bioucas-Dias JM, Figueiredo MAT (2007) A new twist: two-step iterative shrinkage/thresholding algorithms for image restoration. IEEE Trans Image Process 16(12):2992–3004
9. Candès EJ, Romberg J, Tao T (2006) Robust uncertainty principles: exact signal reconstruction from highly incomplete frequency information. IEEE Trans Inf Theory 52(2):489–509
10. Chan SH, Khoshabeh R, Gibson KB, Gill PE, Nguyen TQ (2011) An augmented Lagrangian method for total variation video restoration. IEEE Trans Image Process 20(11):3097
11. Clason C, Jin B, Kunisch K (2010) A duality-based splitting method for ℓ_1-tv image restoration with automatic regularization parameter choice. SIAM J Sci Comput 32(3):1484–1505
12. Combettes PL (2004) Solving monotone inclusions via compositions of nonexpansive averaged operators. Optimization 53(5–6):475–504
13. Combettes PL (2009) Iterative construction of the resolvent of a sum of maximal monotone operators. J Convex Anal 16(4):727–748
14. Combettes PL, Pesquet JC (2011) Proximal splitting methods in signal processing. In: Fixed-point algorithms for inverse problems in science and engineering. Springer, Berlin, pp 185–212
15. Combettes PL, Wajs VR (2005) Signal recovery by proximal forward-backward splitting. Multiscale Model Simul 4(4):1168–1200
16. Daubechies I, Defrise M, De Mol C (2004) An iterative thresholding algorithm for linear inverse problems with a sparsity constraint. Commun Pure Appl Math 57(11):1413–1457
17. Dykstra RL (1983) An algorithm for restricted least squares regression. J Am Stat Assoc 78(384):837–842
18. Goldfarb D, Ma S (2012) Fast multiple-splitting algorithms for convex optimization. SIAM J Optim 22(2):533–556
19. Huang J, Zhang S, Metaxas D (2010) Fast optimization for mixture prior models. In: European conference on computer vision (ECCV), pp 607–620
20. Ji S, Ye J (2009) An accelerated gradient method for trace norm minimization. In: Proceedings of the 26th annual international conference on machine learning. ACM, New York, pp 457–464
21. Joshi N, Kang SB, Zitnick CL, Szeliski R (2010) Image deblurring using inertial measurement sensors. In: ACM Transactions on Graphics (TOG), vol 29. ACM, New York, p 30
22. Kundur D, Hatzinakos D (1996) Blind image deconvolution. IEEE Signal Process Mag 13(3):43–64
23. Lee JD, Sun Y, Saunders MA (2012) Proximal Newton-type methods for minimizing convex objective functions in composite form. Adv Neural Inf Process Syst 25:827–835
24. Nocedal J, Wright SJ (2006) Numerical optimization, 2nd edn. Springer, New York
25. Pan H (2014) Image deblurring and dynamic fusion with sparse proximal operator. Ph.D. thesis, Shanghai Jiao Tong University
26. Pan H, Jing Z, Lei M, Liu R, Jin B, Zhang C (2013) A sparse proximal Newton splitting method for constrained image deblurring. Neurocomputing 122:245–257
27. Peyré G, Bougleux S, Cohen L (2008) Non-local regularization of inverse problems. In: European conference on computer vision (ECCV). Springer, Berlin, pp 57–68
28. Raguet H, Fadili J, Peyré G (2012) Generalized forward-backward splitting. SIAM J Imag Sci 6(3):1199–1226
29. Raskar R, Agrawal A, Tumblin J (2006) Coded exposure photography: motion deblurring using fluttered shutter. In: ACM transactions on graphics (TOG), vol 25. ACM, New York, pp 795–804
30. Rudin LI, Osher S, Fatemi E (1992) Nonlinear total variation based noise removal algorithms. Phys D Nonlinear Phenomena 60(1–4):259–268

31. Tao M, Yang J, He B (2009) Alternating direction algorithms for total variation deconvolution in image reconstruction. TR0918, Department of Mathematics, Nanjing University
32. Toh KC, Yun S (2010) An accelerated proximal gradient algorithm for nuclear norm regularized linear least squares problems. Pac J Optim 6(615–640):15
33. Tomioka R, Aihara K (2007) Classifying matrices with a spectral regularization. In: Proceedings of the 24th international conference on machine learning. ACM, New York, pp 895–902
34. Vogel CR (2002) Computational methods for inverse problems. SIAM, Philadelphia
35. Wang Z, Bovik AC, Sheikh HR, Simoncelli EP (2004) Image quality assessment: from error visibility to structural similarity. IEEE Trans Image Process 13(4):600–612
36. Wright SJ, Nowak RD, Figueiredo MA (2009) Sparse reconstruction by separable approximation. IEEE Trans Signal Process 57(7):2479–2493
37. Yang J, Zhang Y, Yin W (2010) A fast alternating direction method for tvl1-l2 signal reconstruction from partial fourier data. IEEE J Sel Top Sign Proces 4(2):288–297

Chapter 11
Simultaneous Visual Recognition and Tracking Based on Joint Decision and Estimation

11.1 Introduction

Visual target tracking and recognition are important topics in applications on computer vision, such as video surveillance, human–machine interfaces, medical imaging, and multimedia content analysis. The challenges in robust tracking and recognition are caused by complicated environments, such as the presence of noise, occlusion, background clutter, and illumination changes. One common solution is to cast tracking and recognition as two totally independent tasks without any combining work. For example, Lee et al. [15] presented a tracking and recognition algorithm modeled via a collection of sub-manifolds for human faces in video sequences. Although tracking and recognition are formulated as a maximum a posteriori estimation problem within one integrated framework, they are achieved separately in fact in the implementation procedure. The work in [41] embeds an appearance-adaptive model in a particle filter framework for visual tracking and recognition. The proposed adaptive-velocity model is derived based on the likelihood taken from a mixture of three appearance templates. However, this method still treats tracking and recognition as independent steps. Based on a pre-trained model, an online learning algorithm is presented [14] to construct an appearance manifold that is useful for recognition and tracking. Tzimiropoulos et al. [31] proposed a visual recognition and tracking framework based on the sparse representation of image gradient orientations. The problem of tracking and recognizing faces is addressed in [32] also via the incremental local sparse representation. In [19], an online learning framework is proposed with indefinite kernels which show superior performance for object tracking and recognition. Bhaskar [4] presented a framework that combines detection, identification, and tracking of human targets from a static camera. However, the above methods deal with tracking and recognition as two totally independent tasks and without any combining work.

Z. Jing et al., *Non-Cooperative Target Tracking, Fusion and Control*,
Information Fusion and Data Science, https://doi.org/10.1007/978-3-319-90716-1_11

Besides, some two-stage approaches solve this problem by tracking-before-recognition or recognition-before-tracking framework. An integrated detection, tracking, and recognition approach is presented [23] using a particle filter appearance model for infrared video-based vehicle classification. However, tracking result is in fact carried out before the classification procedure instead of joint achievement. Kim et al. [12] proposed an adaptive target appearance model in which the identity of the tracked subject is established by fusing pose and person discriminant features throughout the video sequence. However, recognition is formulated based on the tracking result so as to form a tracking-before-recognition system. Tang et al. [30] presented a novel method for hand tracking and pose recognition based on Kinect. Recognition is achieved based on tracking result which makes it a tracking-before-recognition system. Lu et al. [20] tackled the difficult task of detecting, tracking, and identifying multiple players in sports videos but fail to complete them with joint treatment. The problem of automatically tracking and identifying players in sports video is also solved in [34]. This is a problem of recognition-before-tracking since the player tracking is resolved based on the result of player number recognition. From the above, we can see that the major drawback of two-stage approaches is obvious. Take tracking-before-recognition methods for instance. The best tracking is made first disregarding recognition and then the recognition task is completed as if the tracking result were surely accurate. Tracking is made regardless of the recognition accuracy, and the possible tracking failures are completely ignored during the recognition process.

In the general case, tracking and recognition affect and facilitate each other, and thereby a simultaneous approach would be more promising than separate and two-stage ones [16]. Existing studies also show the effectiveness of simultaneous tracking and recognition. A combination approach of low-level object tracking and high-level recognition is proposed in [8]. The target category is actively recognized using tracking result, and the recognized result is then used to update tracking models so as to achieve better tracking performance. A particle filter-based algorithm is developed [9] using an initial generic appearance manifold so that target tracking and recognition can be achieved jointly in a seamless fashion. This method, however, is not implemented on real-world videos which may limit it from the visual tracking and recognition situations. In [22], tracking and target classification are combined together as a sparse minimization problem. However, the minimization framework in their work cannot ensure global optimization.

The optimal Bayes joint decision and estimation (JDE) [16] model is an integrated approach for simultaneous decision (recognition) and estimation (tracking) based on a novel Bayes risk. In the optimization theoretic parlance, this approach is able to guarantee global optimization, which inherently outperforms conventional two-stage or separate decision and estimation approaches, especially for problems where decision and estimation are highly correlated.

The sparse representation theory has played a fundamental role in many research areas by exploiting the compressibility of the true signal and using a lower-dimensional subspace to approximate it. Its robustness appears promising for visual tasks, but its low computational efficiency limits further application in practice.

This leads to the presentation of the structured sparse representation (SSR) theory which exploits the block structure on sparse representation so as to promote both the efficiency and recovery threshold significantly [7]. SSR achieves better performance than the original sparse representation because it includes the local information as well as the holistic information.

The goal of this chapter converts the problem of simultaneous visual recognition and tracking based on joint decision and estimation (JDE) model. The JDE framework is applied to solve the simultaneous visual tracking and recognition problem and its iteration learning procedure is used to achieve optimization. Besides, the SSR theory is used to represent the target appearance and is effective to improve the performance of the proposed method. Furthermore, the contribution of each test candidate is considered into the learning procedure with a kernel function. Then, the new joint weights of the kernel function provide flexibility with appearance changes and thus robustness to the dynamic scene. The presentation of this chapter is based on the work in [35].

The rest of this chapter is organized as follows. In Sect. 11.2, we discuss the general schemes of SSR, JDE, and the proposed method in detail. The experimental results are presented in Sect. 11.3. Section 11.4 concludes with a general discussion.

11.2 Simultaneous Visual Recognition and Tracking Based on Joint Decision and Estimation

11.2.1 Structured Sparse Representation

In this section, the structured sparse representation (SSR) theory is presented. In the SSR theory, the object appearance is modeled as a structured sparse linear combination of individual templates. Consider $\boldsymbol{y} \in \mathbb{R}^d$ is a high-dimensional signal and can be embedded into an extremely sparse signal $\boldsymbol{w} \in \mathbb{R}^L$ by a basis dictionary $T \in \mathbb{R}^{d \times L}$. This linear transformation is expressed as

$$\boldsymbol{y} = T\boldsymbol{w}. \tag{11.1}$$

T can be formulated by m blocks as $T = [T[1], T[2], \ldots, T[m]]$ where $T[j] \in \mathbb{R}^{d \times l_j}$, $j = 1, \ldots, m$ is the jth block and $L = \sum_1^m l_j$. As the sizes of the templates are normalized to the same, we consider the same $l_j = l$ and thus $L = ml$. The sparse coefficient vector $\boldsymbol{w}$ can be represented as $\boldsymbol{w} = [[\boldsymbol{w}[1]]^T, \ldots, [\boldsymbol{w}[m]]^T]^T$ where $\boldsymbol{w}[j] \in \mathbb{R}^{l_j}$, $j = 1, \ldots, m$ and here we also consider $l_j = l$.

SSR is used to find the minimal subspace which can reconstruct $\boldsymbol{y}$ with minimal error, that is to seek the sparsest nonzero entries in $\boldsymbol{w}$. This minimization task has been proved to be both numerically unstable and NP-Hard [5]. Normally, there are two approaches to approximate the solution: relaxation and greedy methods. Relaxation methods provide effective solution for the problem but their heavy computation and complex implementation make them unavailable for real-time visual tasks. Greedy methods, such as orthogonal matching pursuit (OMP) and block orthogonal matching pursuit (BOMP), are more efficient than relaxation methods and have comparable results [26]. In this chapter, in order to complete the minimization task, we introduce the BOMP algorithm to solve the ℓ_2-regularized problem

$$\boldsymbol{W} = \arg\min_{\boldsymbol{w}} \|\boldsymbol{y} - \boldsymbol{T}\boldsymbol{w}\|_2 \ \ s.t. \ \|\boldsymbol{w}\|_{2,0} < \varepsilon, \tag{11.2}$$

where $\|\boldsymbol{w}\|_{2,0} = \|(\|\boldsymbol{w}[1]\|_2, \ldots, \|\boldsymbol{w}[u]\|_2)\|_0$. The norm $\|\boldsymbol{w}\|_{2,0}$ is defined as counting the number of nonzero blocks $\boldsymbol{w}[q]$ $(q = 1, \ldots, u)$ in vector $\boldsymbol{w}$, where u denotes the total number of nonzero blocks in $\boldsymbol{w}$. $\|\cdot\|_0$ and $\|\cdot\|_2$ denote the ℓ_0 and ℓ_2 norms, respectively, and ε is a parameter to impose the sparsity prior.

11.2.2 Optimal Joint Decision and Estimation

Target tracking can be viewed as an estimation problem while recognition as a decision process. Estimation and decision are closely interrelated, and there is no clear boundary between them. However, the interdependence between estimation and decision often causes additional difficulties for treatment. Conventional solutions ignore this interdependence either completely (e.g., separate estimation and decision) or partially (e.g., decision-then-estimation or estimation-then-decision), which makes their performance suffer. The optimal Bayes joint decision and estimation (JDE) model is a general framework for jointly dealing with decision and estimation and has the potential of arriving at a global optimization.

The cornerstone of the optimal Bayes JDE framework is the following Bayes risk:

$$\bar{r} \triangleq \sum_i \sum_j \left\{\alpha_{ij} c_{ij} + \beta_{ij} E\left[C\left(x, \hat{x}\right) | R_i, H_j\right]\right\} p\left(R_i, H_j\right), \tag{11.3}$$

where c_{ij} stands for the cost of decision R_i when hypothesis H_j is true. $C(x, \hat{x})$ is the estimation cost function between state x and the estimation $\hat{x}$. $E[C(x, \hat{x})|R_i, H_j]$ denotes the expected cost conditioned on the fact that R_i is decided but H_j is true. α_{ij} and β_{ij} are nonnegative relative weights of decision and estimation costs, which allow a variety of special cases.

In this JDE framework, both estimation and decision are jointed with the Bayes risk in Eq. (11.3), and the decision cost, estimation error, and their couplings are all considered. The weight coefficients $\{\alpha_{ij}, \beta_{ij}\}$ provide flexibility to find the true algorithm [16].

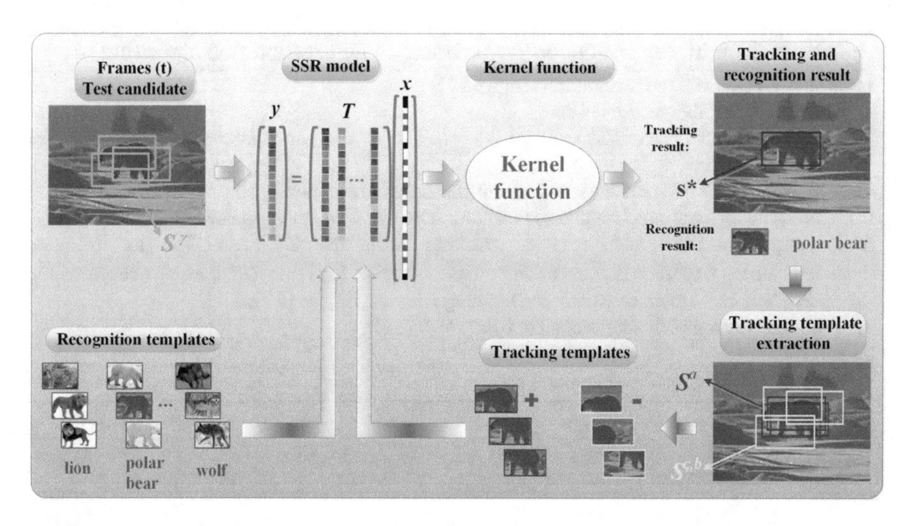

Fig. 11.1 General framework of the proposed method. The green thick arrows denote the operating direction of the proposed method. The blue thick arrows denote the updating direction of the templates. The green boxes denote the test candidates extracted by $S^{\gamma} = \{s \,|\, \|l_t(s) - l_{t-1}(s^*)\| < \gamma\}$. The blue box is the tracking and recognition result s^*. The pink boxes denote the recognition templates. The red and yellow boxes denote the positive and negative tracking templates extracted by $S^a = \{s \,|\, \|l_t(s) - l_t(s^*)\| < a\}$ and $S^{\varsigma,b} = \{s \,|\, \varsigma < \|l_t(s) - l_t(s^*)\| < b\}$, respectively. The symbols "+" and "–" in the tracking templates are the labels of positive and negative templates, respectively

11.2.3 Simultaneous Visual Recognition and Tracking

The general framework of the proposed method is described in Fig. 11.1. The tracking window in the first frame is located manually or by other detection approaches. Let $l_t(s) \in \mathbb{R}^2$ denote the location of sample s at the tth frame, and s^* denote the sample of the tracking and recognition result. At the tth frame, we select some patches $S^{\gamma} = \{s \,|\, \|l_t(s) - l_{t-1}(s^*)\| < \gamma\}$ surrounding the target location $l_{t-1}(s^*)$ in the $(t-1)$th frame and set them as the test candidates. Symbol y is one of the test candidates and is supposed to be sparsely represented by the template T and coefficient vector $\boldsymbol{w}$ based on the SSR model. Then, the kernel function is introduced to solve the simultaneous tracking and recognition problem and obtain the final tracking and recognition result. Next, we extract a set of positive tracking templates by randomly cropping some patches $S^a = \{s \,|\, \|l_t(s) - l_t(s^*)\| < a\}$ surrounding $l_t(s^*)$. Similarly, we crop some patches $S^{\varsigma,b} = \{s \,|\, \varsigma < \|l_t(s) - l_t(s^*)\| < b\}$ where $a < \varsigma < b$ and label them as negative tracking samples. We use these positive and negative tracking templates to constitute the template T of the $(t+1)$th frame with recognition samples. The recognition templates are collected from real-world images at different poses and illuminations, which can effectively prevent recognition failure when the target undergoes drastic pose and illumination changes. The main steps of the proposed tracking and recognition algorithm are summarized in Algorithm 1.

Algorithm 1 Main steps of the proposed tracking and recognition algorithm

1: **Input:** frames, and the recognition template T_R
2: **for** $t = 1$ to number of frames **do**
3: Sample a set of patches $S^{\gamma} = \{s \,|\|l_t(s) - l_{t-1}(s^*)\| < \gamma\}$ and make each of them as a test candidate y.
4: Formulate the SSR model $\boldsymbol{y} = T\boldsymbol{w}$ and calculate the solution $\boldsymbol{W}$.
5: Find the recognition class and the tracking state $\{\hat{R}, \hat{x}\}$ by the proposed method.
6: Sample positive and negative tracking templates T_T^+ and T_T^- by $S^a = \{s \,|\|l_t(s) - l_t(s^*)\| < a\}$ and $S^{\varsigma,b} = \{s \,|\varsigma < \|l_t(s) - l_t(s^*)\| < b\}$, respectively, and then make them constitute the tracking templates $T_T = [T_T^+, T_T^-]$.
7: Update the template dictionary by $T = [T_T, T_R, I]$.
8: **end for**
9: **Output:** tracking location and class

11.2.3.1 The Kernel Function

Let $i = 1, \ldots, N$ denote the ith test candidate among N candidates, and $j = 1, \ldots, m$ denote the jth class among m classes. $R = \{R_1, \ldots, R_i, \ldots, R_N\}$ stands for the recognition decisions on the class label (i.e., $1, \ldots, m$) of each test candidate. Specifically, the value of each R_i is a certain class label. $H = \{H_1, \ldots, H_j, \ldots, H_m\}$ represents the hypotheses of recognition classes. Specifically, the values of H_j are set as $H_j = j$ and are static throughout the tracking process. This theory of hypothesis testing method aims at matching one decision to all hypotheses and rejecting them except the optimal one. Then, the kernel risk is defined as

$$\bar{kr} \triangleq \sum_i Kernel\left(x_i, \hat{x}\right) \sum_j \left[\alpha_{ij} Cost_{ij} + \beta_{ij} mse\left(\hat{x}\,\middle|R_i, H_j\right)\right] p\left(H_j\,|y\right), \tag{11.4}$$

where the observation y is represented with the test candidate, and $x_i = (x_{i,x}, x_{i,y})$ is tracking state of the ith test candidate, where $x_{i,x}$ and $x_{i,y}$ denote the x-coordinate and y-coordinate location values, respectively. $\hat{x}$ is the estimated tracking state. $Cost_{ij}$ stands for the cost of recognition decision R_i when hypothesis H_j is true, and $p\left(H_j\,|y\right)$ is the posterior probability of hypothesis H_j. $Kernel\,(\bullet, \circ)$ denotes the kernel function. $mse\left(\hat{x}\,\middle|R_i, H_j\right)$ is the mean square error of $\hat{x}$ on condition of R_i and H_j [16]. Symbols α_{ij} and β_{ij} denote the joint weights.

Then, the recognition class is decided, and the tracking state $\{\hat{R}, \hat{x}\}$ is estimated by

$$\begin{aligned} \hat{R} &= \{R_i : C_i \leq C_h, \forall h = 1, \ldots, N\} \\ \hat{x} &= \sum_{i.j} x_i \bar{P}\left(R_i, H_j\,|y\right) \end{aligned}, \tag{11.5}$$

where

$$C_i = \sum_j \left[\alpha_{ij} Cost_{ij} + \beta_{ij} mse\left(\hat{x} \left| R_i, H_j \right.\right)\right] p\left(H_j \left| y \right.\right),$$

$$\bar{P}\left(R_i, H_j \left| y \right.\right) = \frac{1}{n_p} \beta_{ij} P\left(R_i, H_j \left| y \right.\right).$$

Symbol n_p stands for the normalization constant. In the next sections, we discuss Eq. (11.5) in detail.

While formulating $\bar{kr}$, if each candidate contributes equally in this learning procedure, the less important candidates may influence more on the final result. The so-called less important candidates are those samples that are farther from the location of estimated tracking state $\hat{x}$ than the others, which have relatively small possibilities of being the target. The proposed method integrates the candidate contributions into the efficient iteration learning procedure so that their appearances are much more discriminative. The Gaussian radial basis function [28] is used to represent the kernel function as

$$Kernel\left(x_i, \hat{x}\right) = \frac{1}{n_k} \exp\left[-\frac{\left\| l\left(x_i\right) - l\left(\hat{x}\right) \right\|^2}{2\sigma^2}\right], \tag{11.6}$$

where $l\left(\cdot\right) \in \mathbb{R}^2$ is the location function. Symbol n_k is the normalization constant, and σ denotes the variance. Equation (11.6) is a monotone decreasing function with respect to the distance $\left\| l\left(x_i\right) - l\left(\hat{x}\right) \right\|$ between the locations of the ith candidate and $\hat{x}$. Therefore, the candidates are weighted according to their locations, i.e., the ones near $\hat{x}$ contribute more to the joint tracking and recognition framework than those far from it.

In the original JDE framework, the determination of joint weights $(\alpha_{ij}, \beta_{ij})$ is fixed to $(1, 0)$, $(0, 1)$, $(0, c_{ij})$, or $(1, 1)$, which reduces the JDE problem to independent decision, independent estimation, decision-before-estimation, or simply the sum of traditional combination with equal and fixed importance. These weight options cannot cater for flexible circumstances if decision and estimation compete intensely with each other. α_{ij} and β_{ij} are computed as

$$\begin{cases} \alpha_{ij} = 1 - \beta_{ij} \\ \beta_{ij} = \dfrac{\exp\left(-\lambda_T\left[error\left(\boldsymbol{y}_i, \mathbf{W}_T^+\right)\right]^2\right)}{\exp\left(-\lambda_T\left[error\left(\boldsymbol{y}_i, \mathbf{W}_T^+\right)\right]^2\right) + \exp\left(-\lambda_T\left[error\left(\boldsymbol{y}_i, \mathbf{W}_T^-\right)\right]^2\right)} \end{cases}. \tag{11.7}$$

The symbol λ_T is a control parameter that controls the tracking importance. By doing this, joint weights α_{ij} and β_{ij} are adaptive with target appearance changes and thus robust to the dynamic scene.

11.2.3.2 Structure Sparse Representation-Based Error Function

The structured sparse representation (SSR) theory achieves better performance than the original sparse representation since it includes both partial and spatial information.

In the proposed method, we assume that each target candidate is sparsely represented in the space spanned by the target and trivial templates and apply SSR to solve this sparse approximation problem. Some visual tracking or recognition literatures [2, 21, 33] also use this assumption and present successful examples of SSR being suitable for visual tasks. Regular SSR solve visual tracking or recognition problems in two steps (i.e., appearance construction and reconstruction) which are not suitable for the problem of simultaneous visual tracking and recognition. Therefore, SSR is extended to the simultaneous visual tracking and recognition task by remodeling these two steps as discussed below.

Different from the regular SSR form for separate visual tracking and recognition, the basis dictionary T in the proposed framework includes both positive/negative tracking and recognition templates. The negative templates are added in order to improve the model's robustness against background clustering. In addition, both tracking and recognition templates are combined into the dictionary to jointly deal with recognition and tracking. Compared with the recognition template T_R which remains static all the time, the tracking template T_T updates dynamically during the tracking process. A trivial template is also added here for occlusion circumstances.

With the above consideration, the target candidate $\boldsymbol{y}$ is denoted as

$$\boldsymbol{y} = [T_T, T_R, I] \begin{bmatrix} \boldsymbol{w}_T \\ \boldsymbol{w}_R \\ \boldsymbol{e} \end{bmatrix} = T\boldsymbol{w}, \tag{11.8}$$

where $T = [T_T, T_R, I]$ is the template dictionary and $\boldsymbol{w} = [\boldsymbol{w}_T^T, \boldsymbol{w}_R^T, \boldsymbol{e}^T]^T$ is the coefficient vector. The tracking and recognition templates are $T_T = \left[T_T^+, T_T^-\right] \in \mathbb{R}^{d\times 2l}$ and $T_R = [T_R[1], \ldots, T_R[m]] \in \mathbb{R}^{d\times ml}$, where T_T^+ and T_T^- denote the positive and negative tracking templates, respectively. $\boldsymbol{w}_T = [[\boldsymbol{w}_T^+]^T, [\boldsymbol{w}_T^-]^T]^T \in \mathbb{R}^{2l}$ and $\boldsymbol{w}_R = [[\boldsymbol{w}_R[1]]^T, \ldots, [\boldsymbol{w}_R[m]]^T]^T \in \mathbb{R}^{ml}$ are the tracking and recognition coefficient vectors, where $\boldsymbol{w}_T^+$ and $\boldsymbol{w}_T^-$ denote the positive and negative tracking coefficient vectors, respectively. The trivial template $I \in \mathbb{R}^{d\times d}$ is set as an identity matrix to deal with the occlusion circumstances [33], and $\boldsymbol{e} \in \mathbb{R}^l$ is a trivial coefficient vector [22]. The graphical representation of the proposed SSR model in Eq. (11.8) is shown in Fig. 11.2. The candidate $\boldsymbol{y}$ is one of the test candidates and is sparsely represented by the tracking template T_T, recognition template T_R, and trivial template I. Symbols $T_T^{+(k)}$ and $T_T^{-(k)}, k = 1, \ldots, l$ denote positive and negative tracking templates extracted from the previous frame, respectively. $T_R[j] = [T_R^{(1)}[j], \ldots, T_R^{(l)}[j]], j = 1, \ldots, m$ are recognition templates of the jth class.

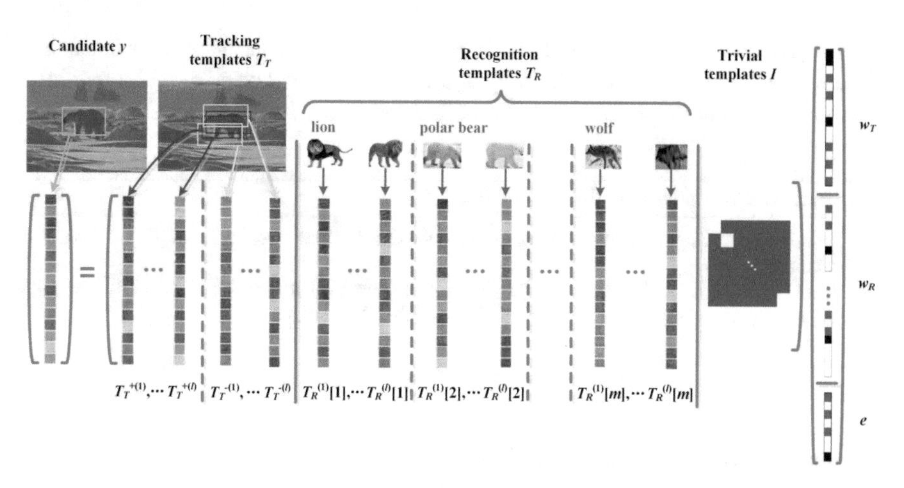

Fig. 11.2 Graphical representation of the proposed SSR model $\boldsymbol{y} = T\boldsymbol{w}$. The green box denotes one of the test candidates $\boldsymbol{y}$. The template dictionary T comprises the tracking template T_T, the recognition template T_R, and the trivial template I. The coefficient vector $\boldsymbol{w}$ comprises the tracking coefficient vector $\boldsymbol{w}_T$, recognition coefficient vector $\boldsymbol{w}_R$, and the trivial coefficient vector $\boldsymbol{e}$. The red and yellow boxes denote the positive and negative tracking templates $T_T^{+(k)}$ and $T_T^{-(k)}$, $k = 1, \ldots, l$, respectively. The recognition template $T_R = [T_R[1], \ldots, T_R[m]]$ comprises m classes, and $T_R[j] = [T_R^{(1)}[j], \ldots, T_R^{(l)}[j]]$, $j = 1, \ldots, m$ is template of the jth class

SSR is able to find the minimal subspace which can reconstruct $\boldsymbol{y}$ with minimal error. After solving the minimization problem using Eq. (11.2), we can obtain the optimal coefficient vector $\boldsymbol{W}$. The original reconstruction error $\|\boldsymbol{y} - T\boldsymbol{W}\|_2$ [17] in regular SSR cannot solve the simultaneous tracking and recognition problems [33], because the nonzero entries in $\boldsymbol{W}$ are expected to associate with the columns of T that are the recognition class and the tracking result. However, the JDE model needs the separate reconstruction errors, e.g., the error of tracking result or a certain recognition class, which calls for a separation method. Therefore, the characteristic function $\Delta(\circ)$ [33], a $L \times 1$ vector whose only nonzero entries are those in the vector $\circ$, is introduced to select the coefficients associated with the specified templates (i.e., the templates corresponding to $\circ$). $\Delta(\circ)$ has been used in face recognition [33] and image classification [29] algorithms to effectively select coefficients associated with different classes. Then, we define an operation symbol $\oplus$ as $\Delta(\circ \oplus *) = \Delta(\circ) + \Delta(*)$ to choose templates corresponding to $\circ$ and $*$ together. The proposed SSR-based error function is defined as $error(\bullet, \circ) = \|\bullet - T\Delta(\circ)\|_2$. By assigning the characteristic function to the ith and jth class templates, we can obtain the reconstruction error of recognition decision R_i when hypothesis H_j is true as $Cost_{ij} = error\left(\boldsymbol{y}_i, \boldsymbol{W}_R[R_i] \oplus \boldsymbol{W}_R[H_j]\right)$ to formulate the kernel risk $\bar{kr}$ in Eq. (11.4). And the likelihood between the test candidate and the jth recognition template is used to represent the posterior probability of hypothesis H_j as $p(H_j|y) = \exp\left(-\lambda_R\left[error\left(\boldsymbol{y}_i, \boldsymbol{W}_R[H_j]\right)\right]^2\right)$ where λ_R is a control parameter.

We use iteration to keep the learning procedure until the kernel risk $\bar{kr}$ is smaller than a threshold [18], so that a globally optimal solution will be achieved potentially. The main steps of the proposed method are summarized in Algorithm 2.

Algorithm 2 Main step of the kernel function

1: Initialization: Given an initial recognition decision $R^{(0)} = \{R_1^{(0)}, \ldots, R_N^{(0)}\}$ of each test candidate.

2: Tracking-Step: At each iteration k, for the given recognition $R^{(k)} = \{R_1^{(k)}, \ldots, R_N^{(k)}\}$ of each test candidate, compute the optimal state estimation of tracking

$$\hat{x} = \sum_{i.j} x_i \bar{P}\left(R_i, H_j \mid y\right),$$

where $\bar{P}\left(R_i, H_j \mid y\right) = \frac{1}{n_p} \beta_{ij} P\left(R_i, H_j \mid y\right)$.

3: Recognition-Step: Compute the posterior cost

$$C_{ij}^{(k+1)} \triangleq \left[\alpha_{ij} Cost_{ij} + \beta_{ij} mse\left(\hat{x} \mid R_i, H_j\right)\right] p\left(H_j \mid y\right),$$

in which the optimal recognition is given by $R^{(k+1)} = \{R_1^{(k+1)}, \ldots, R_N^{(k+1)}\}$ where $\hat{R}_i^{(k+1)} = \{j : C_{ij}^{(k+1)} \leq C_{ih}^{(k+1)}, \forall h = 1, \ldots, m\}$.

4: Update the joint weights $\{\alpha_{ij}, \beta_{ij}\}$ with

$$\begin{cases} \alpha_{ij} = 1 - \beta_{ij} \\ \beta_{ij} = \dfrac{\exp\left(-\lambda_T\left[error\left(\mathbf{y}_i, \mathbf{W}_T^{+}\right)\right]^2\right)}{\exp\left(-\lambda_T\left[error\left(\mathbf{y}_i, \mathbf{W}_T^{+}\right)\right]^2\right) + \exp\left(-\lambda_T\left[error\left(\mathbf{y}_i, \mathbf{W}_T^{-}\right)\right]^2\right)} \end{cases},$$

and the kernel function as

$$Kernel\left(x_i, \hat{x}\right) = \frac{1}{n_k} \exp\left[-\frac{\left\| l\left(x_i\right) - l\left(\hat{x}\right)\right\|^2}{2\sigma^2}\right].$$

5: Obtain the kernel risk by

$$\bar{kr} \triangleq \sum_i Kernel\left(x_i, \hat{x}\right) \sum_j \left[\alpha_{ij} Cost_{ij} + \beta_{ij} mse\left(\hat{x} \mid R_i, H_j\right)\right] p\left(H_j \mid y\right).$$

6: Repeat Step 2–5 until the reduction in the kernel risk $\bar{kr}$ is smaller than a threshold.

7: Output the final recognition decision and tracking estimation $\{\hat{R}, \hat{x}\}$ where $\hat{x}$ is computed in Step 2, and

$$\hat{R} = \{R_i : C_i \leq C_h, \forall h = 1, \ldots, N\},$$

where $C_i = \sum_j C_{ij}$.

11.3 Experiments and Evaluation

In this section, we test the proposed method on several challenging real-world sequences. This method is implemented using Visual Studio 2010 on an Intel Dual-Core 1.70 GHz CPU with 4 GB RAM. This procedure is operated 51 frames per second (FPS) on average.

The sample parameters are set as $a = 4$, $b = 30$, $\varsigma = 8$, and $\gamma = 20$ [39]. The control parameters λ_R and λ_T are set as 1.8. The total number of test candidates is set as $N = 300$ [22]. In [22], class number is set as $m = 4$ and the template number of each class is set as $l = 5$. We extend it to $m = 6$ and $l = 30$ for better applicability. The variance σ is set as 0.2 for a good performance. The initial values of decisions on recognition class are $R_1^{(0)} = \ldots = R_1^{(N)} = 1$ which means that they all belong to class 1 at the beginning of the iteration process.

Figures 11.3, 11.4, 11.5, 11.6, and 11.7 and Tables 11.1, 11.2, and 11.3 show the experimental results in nine challenging sequences named *David* [39], *Labman* (the lab man) [6], *MHuang* [27], *Fox*, *Skunk*, *Polarbear* (polar bear), *OE* (Orbital express) [25], *TMA* (Soyuz TMA-12 spaceship) [36], and *STS* (STS-126-Endeavour space shuttle) [36]. Sequences *Fox*, *Skunk*, and *Polarbear* are provided by the Animal world TV show, and Sequences *OE*, *TMA*, and *STS* are provided by the National Aeronautics and Space Administration (NASA). In the next section, we show both tracking and recognition results of the proposed method and the performance comparisons with separate trackers and recognizers as well as simultaneous tracking and recognition approaches.

11.3.1 Tracking Results

The tracking performance of the proposed method is compared with the state-of-the-art $\ell 1$ [22], MIL [1], MTT [37], VTD [13], FCT [39], ODFS [38], and JTR trackers. JTR is designed to observe the performance of the proposed method without the kernel function in Eq. (11.6). Two metrics, i.e., location error (pixel) [11] and success rate [10], are used to evaluate the tracking results of the proposed method quantitatively. The location error is computed as $error = \sqrt{(x_G - x_T)^2 + (y_G - y_T)^2}$, where (x_T, y_T) and (x_G, y_G) are the tracking bounding box center and the ground truth that is manually located. Before defining the success rate, we first introduce tracking overlapping which is defined as $area(ROI_G \cap ROI_T)/area(ROI_G \cup ROI_T)$. ROI_G and ROI_T denote the ground truth and tracking bounding box, respectively, and $area(\bullet)$ is rectangle area function. If overlapping is larger than 0.5, then the result is considered as a success in one frame. Besides, a smaller position error and a larger success rate indicate higher accuracy and robustness.

Figure 11.3 shows the tracking performances of all the eight methods in the nine challenge sequences. Due to the limitation of space, only four frames of

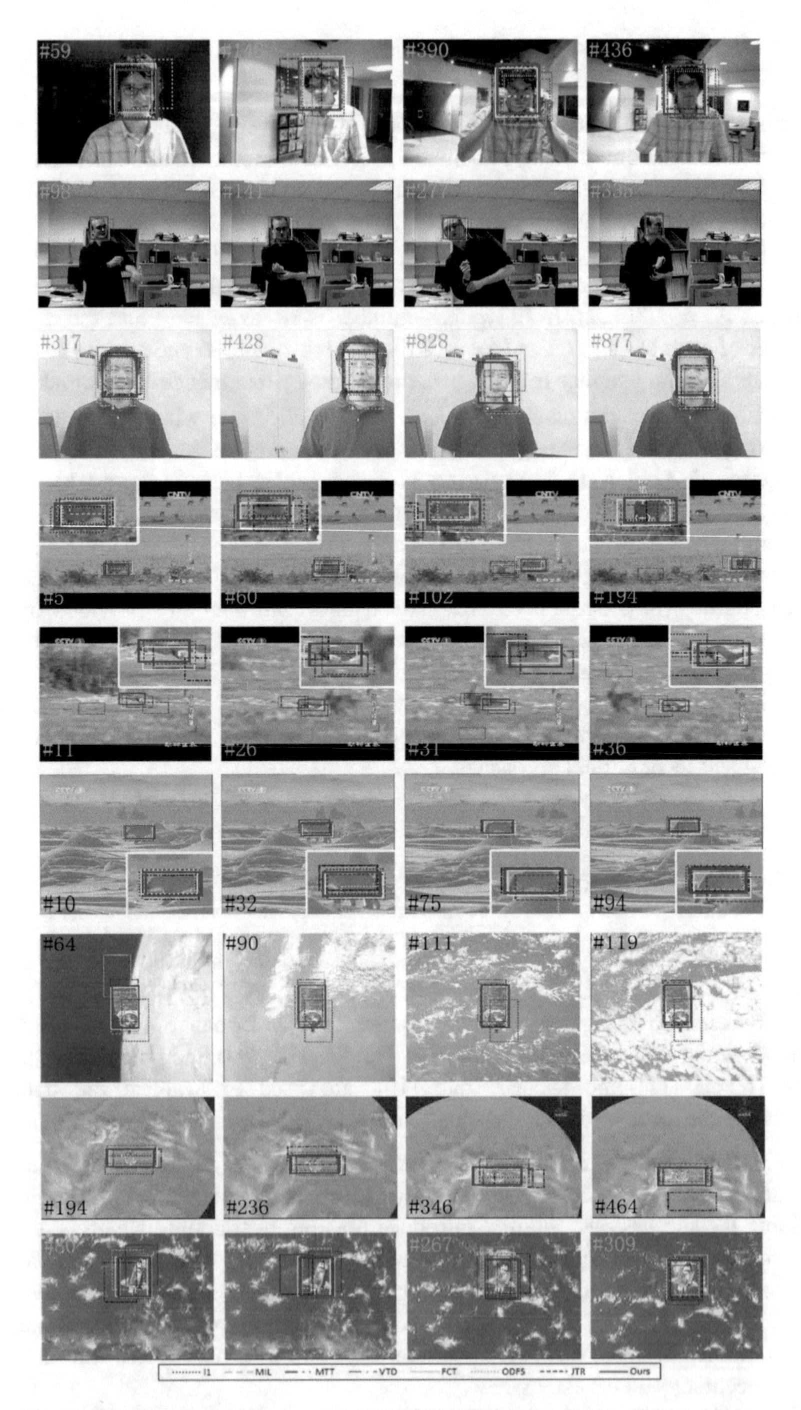

Fig. 11.3 Tracking performances by $\ell 1$, MIL, MTT, VTD, FCT, ODFS, JTR, and the proposed method in the challenging sequences with pose and illumination change (*David*, *Labman*, and *MHuang*), partial and full occlusion (*Fox*, *Skunk*, and *Polarbear*), and background clutter and abrupt change (*OE*, *TMA*, and *STS*) (from top to bottom)

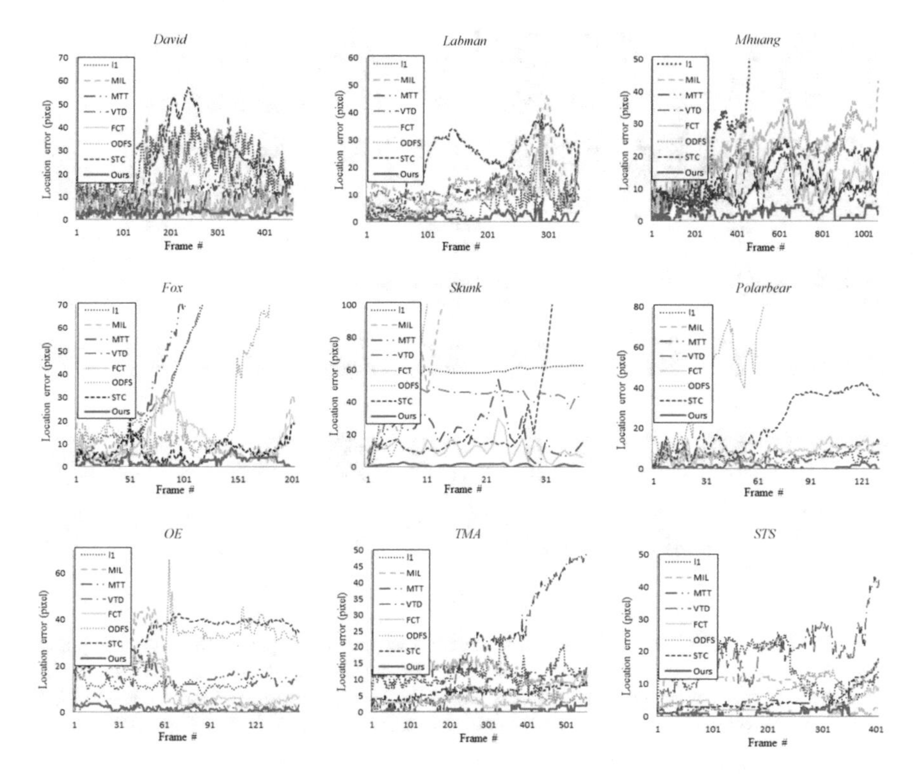

Fig. 11.4 Tracking location errors (pixel) of $\ell_1$1, MIL, MTT, VTD, FCT, ODFS, JTR, and the proposed method in Sequences *David*, *Labman*, *MHuang*, *Fox*, *Skunk*, *Polarbear*, *OE*, *TMA*, and *STS*. Vertical axis: location error (pixel). Horizontal axis: frame number

each sequence are displayed. Figure 11.4 shows the tracking location errors of these methods, and some of the curves are not shown entirely for presentation convenience. Table 11.1 shows the tracking results in term of success rate.

Pose and Illumination Change To demonstrate the efficiency of the proposed method, Sequence *David* (360 frames in total) is displayed with illumination and posture variations when the person walks out of the dark meeting room. The FCT and the proposed method perform well on this sequence. FCT uses the similar Haar-like features which have been shown to be robust to posture and illumination changes. The proposed method achieves better tracking result than FCT with smaller tracking error and higher success rate shown in Fig. 11.4 and Table 11.1, respectively. The target in Sequence *Labman* (360 frames in total) undergoes abrupt rotation and movement. At the tracking beginning, all these methods can almost track the target successfully. However, when the man begins to shake and turn around his head abruptly at around Frame #270, $\ell 1$, MIL, MTT, VTD, FCT, ODFS, and JTR fail to track the target accurately. The proposed method is able to overcome the abrupt appearance change and perform well on this video. The

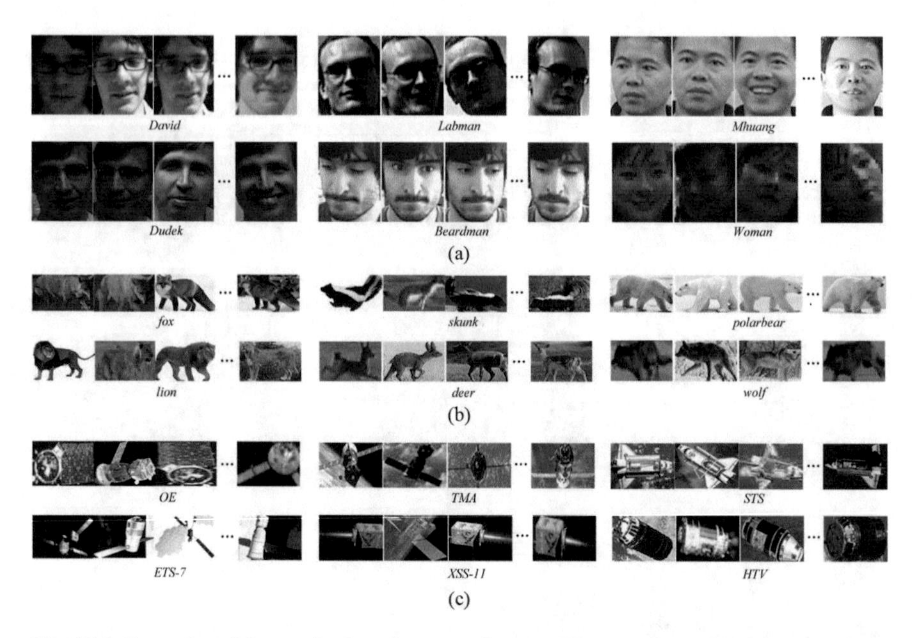

Fig. 11.5 Examples of face, animal, and spacecraft recognition templates: (**a**) Face recognition templates contain classes named David, Labman, MHuang, Dudek, Beardman, and Woman; (**b**) Animal recognition templates contain classes named Fox, Skunk, Polarbear, Lion, Deer, and Wolf; (**c**) Spacecraft recognition templates contain classes named OE, TMA, STS, ETS-7, XSS-11, and HTV

facial expression of the target man in Sequence *MHuang* (1071 frames in total) keeps changing throughout the tracking process. Once again, The proposed method performs the best among all these eight trackers. In Fig. 11.4, the tracking error of the ℓ_1 method grows larger and larger as this method uses holistic features which are less effective for large-scale appearance variations. The proposed method has the smallest position error and largest success rate for most of the frames.

Partial and Full Occlusion Sequences *Fox*, *Skunk*, and *Polarbear* are low-resolution recordings and present greater challenges in terms of occlusion. In the first part of Sequence *Fox* (203 frames in total), all the eight trackers can almost track the target successfully. However, when the target fox walks behind the bushes at around Frame #60, ℓ_1, MTT, VTD, FCT, and ODFS fail to track the target. And when the fox is occluded by the video subtitle at around Frame #190, MIL, FCT, and STC do not perform well. Figures 11.3 and 11.4 all indicate the tracking failure when occlusion occurs. The proposed method is able to overcome this partial occlusion and perform well on this video. Sequence *Skunk* (36 frames in total) contains an example of full occlusion in which the target skunk is occluded by another animal. The proposed method can handle the heavy occlusion because jointly combining tracking with recognition reduces outlier interference. In Sequence *Polarbear* (131 frames in total), tracking is challenged due to snow

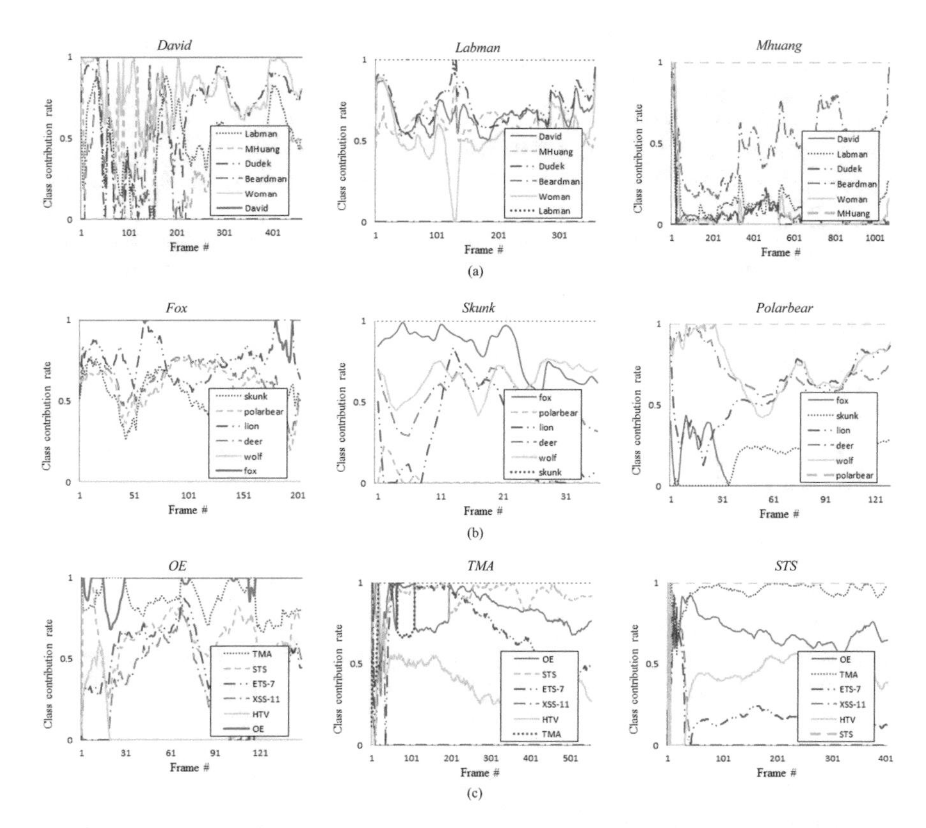

Fig. 11.6 Contribution rates (%) of each class for the (**a**) face recognition sequences (*David*, *Labman*, and *MHuang*), (**b**) animal recognition sequences (*Fox*, *Skunk*, and *Polarbear*), and (**c**) spacecraft recognition sequences (*OE*, *TMA*, and *STS*) by the proposed method. Vertical axis: class contribution rate. Horizontal axis: frame number

occlusion and similar background. As we see from Figs. 11.3 and 11.4, the proposed method performs the best due to the effectiveness and robustness of combining tracking and recognition into a generalized framework.

Background Clutter and Abrupt Change The space objects in Sequences *OE*, *TMA*, and *STS* undergo abrupt background changes. $\ell_1 1$, FCT, and the proposed method achieve favorable performances on Sequence *OE* (150 frames in total) in terms of both tracking error and success rate shown in Fig. 11.4 and Table 11.1, respectively. However, FCT drifts away from the target at around Frame #60. Sequences *TMA* and *STS* (554 and 405 frames in total, respectively) contain objects undergoing abrupt motion and background changes. The proposed method outperforms most of the other methods in most metrics (location accuracy and success rate).

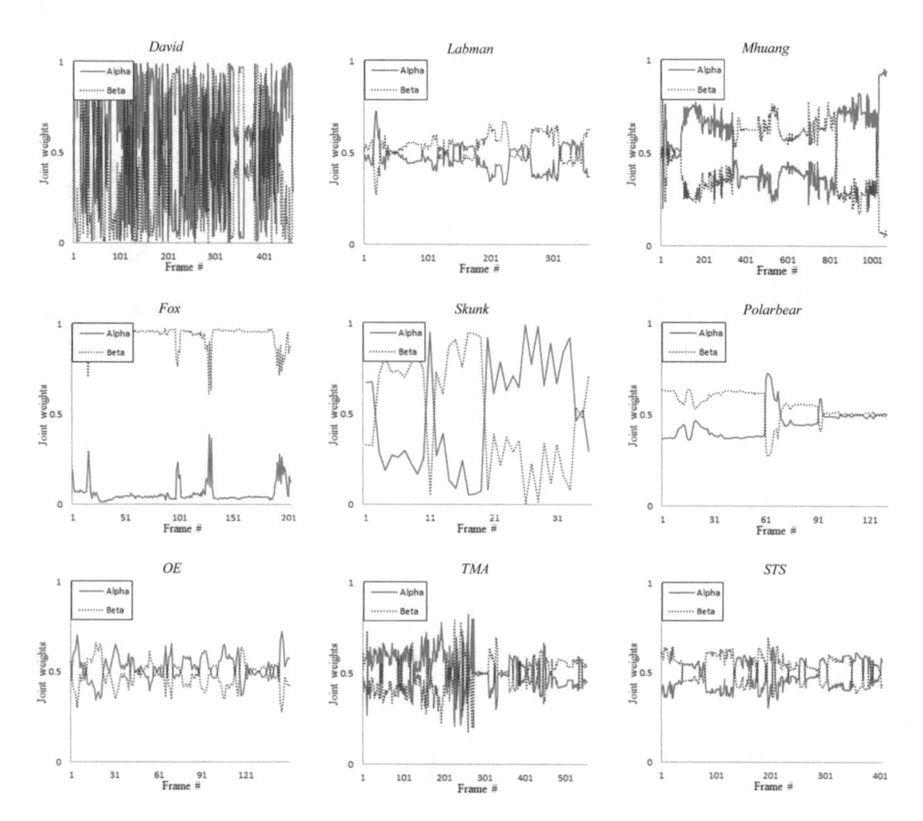

Fig. 11.7 Joint weights (α_{ij}, β_{ij}) of the proposed method in Sequences *David*, *Labman*, *MHuang*, *Fox*, *Skunk*, *Polarbear*, *OE*, *TMA*, and *STS*. Vertical axis: joint weights. Horizontal axis: frame number

11.3.2 Recognition Results

In this section, the proposed method is applied to the visual recognition task. The nine test sequences in Fig. 11.3 contain three kinds of recognition sequences, that is, the first three are face recognition sequences, the middle three are animal recognition sequences, and the last three ones are spacecraft recognition sequences. Each kind of recognition sequence shares the same recognition templates as shown in Fig. 11.5. In the recognition experiments, our work is to classify each frame into one of these m classes. The face recognition templates in Fig. 11.5a contain classes named David [39], Labman [6], MHuang [27], Dudek [31], Beardman [39], and Woman [40]. In the animal recognition templates, classes Fox, Skunk, and Polarbear are provided by the Animal world TV show, and classes lion, deer, and wolf are real-world Google images. In the spacecraft recognition templates, classes OE [25], TMA [36], and STS [36] are provided by NASA, and ETS-7, XSS-11, and HTV are also collected from Google. For each test sequence, we use the first half to do the experiments and the bottom half as some of the recognition templates.

Table 11.1 Tracking success rates (%) of $\ell 1$, MIL, MTT, VTD, FCT, ODFS, JTR, and the proposed method in Sequences *David*, *Labman*, *MHuang*, *Fox*, *Skunk*, *Polarbear*, *OE*, *TMA*, and *STS*

Sequences	*David*	*Labman*	*MHuang*	*Fox*	*Skunk*	*Polarbear*	*OE*	*TMA*	*STS*
$\ell 1$	54	77	12	33	38	**100**	77	96	87
MIL	85	35	25	97	39	99	66	75	98
MTT	99	65	89	42	83	98	53	97	99
VTD	38	45	746	40	53	93	97	68	63
FCT	98	71	81	82	87	92	71	**100**	**100**
ODFS	62	65	69	77	58	70	53	99	98
JTR	56	30	93	86	86	78	30	97	99
Ours	**100**	**99**	**99**	**100**	**100**	**100**	**99**	**100**	**100**

Bold fonts indicate the best performance

Table 11.2 Average recognition rates (%) of $\ell 1$, PCA, LDA, Eigen-face, Fisher-face, NN, JTR, and proposed method in the face recognition sequences (*David*, *Labman*, and *MHuang*), animal recognition sequences (*Fox*, *Skunk*, and *Polarbear*), and spacecraft recognition sequences (*OE*, *TMA*, and *STS*)

	Face			Animal			Spacecraft		
Sequences	*David*	*Labman*	*MHuang*	*Fox*	*Skunk*	*Polarbear*	*OE*	*TMA*	*STS*
$\ell 1$	92	82	89	87	96	89	90	88	93
PCA	86	92	86	90	95	86	83	89	97
LDA	94	83	91	90	86	92	88	90	**98**
Eigen-face	76	75	81	86	89	84	79	83	87
Fisher-face	77	91	85	86	93	**95**	82	84	84
NN	87	77	87	78	93	91	83	88	95
JTR	84	90	86	87	94	90	76	86	89
Ours	**97**	**95**	**96**	**94**	**98**	**95**	**91**	**93**	**98**

Bold fonts indicate the best performance

The metric contribution rate evaluates the normalized recognition score of each class at each frame as shown in Fig. 11.6. The calculation method is $contribution\ rate = \exp(-\lambda_R[error(\hat{x}, \boldsymbol{W}_R[j])]^2)/M$, $j = 1, \ldots, m$ where M is the normalized parameter and m is the total class number. As seen in Fig. 11.6, the recognition scores for the correct results are almost 1 throughout the process.

Since the target state is estimated frame by frame during tracking, we define the recognition rate as $recognition\ rate = N_S/N_F$, where N_S denotes the number of frames in which recognition success is achieved, and N_F denotes the total frame number. To demonstrate the recognition performance of the proposed method, it is compared with the state-of-the-art $\ell 1$, principal components analysis (PCA), linear discriminant analysis (LDA), Eigen-face [24], Fisher-face [3], nearest neighbor (NN) recognition methods, and JTR. The recognition rates for Sequences *David*, *Labman*, *MHuang*, *Fox*, *Skunk*, *Polarbear*, *OE*, *TMA*, and *STS* are shown in Table 11.2. Each method is trained with the same training templates in each

Table 11.3 Tracking success rates (%) and recognition rates (%) with respect to different joint weights $(\alpha_{ij}, \beta_{ij})$ of Sequences *David*, *Labman*, *MHuang*, *Fox*, *Skunk*, *Polarbear*, *OE*, *TMA*, and *STS*

$(\alpha_{ij}, \beta_{ij})$	Tracking success rate	Recognition rate	Tracking success rate	Recognition rate	Tracking success rate	Recognition rate
	David		*Labman*		*MHuang*	
(1, 0)	19	84	23	76	29	68
(0, 1)	53	31	31	35	56	17
$(0, c_{ij})$	45	31	77	54	27	33
(1, 1)	28	55	89	70	47	52
Ours	**100**	**99**	**99**	**97**	**99**	**98**
	Fox		*Skunk*		*Polarbear*	
(1, 0)	38	26	62	53	52	31
(0, 1)	86	79	40	79	60	34
$(0, c_{ij})$	65	53	24	50	90	92
(1, 1)	28	55	89	70	47	52
Ours	**100**	**94**	**100**	**100**	**100**	**95**
	OE		*TMA*		*STS*	
(1, 0)	77	59	30	51	33	81
(0, 1)	56	42	46	39	59	28
$(0, c_{ij})$	58	59	44	28	56	43
(1, 1)	63	50	69	66	70	51
Ours	**99**	**91**	**100**	**93**	**100**	**98**

Bold fonts indicate the best performance

class. The face recognition process is challenged with large variations in face pose, illumination, expression, and other conditions. The task of animal recognitions is difficult to achieve because of low resolution and heavy occlusion. Furthermore, the spacecraft recognition undergoes similar appearance between different classes and abrupt background changes. The proposed method significantly outperforms the other typical recognition methods and successfully achieves an average of 95% in recognition rate. This is because tracking performance can significantly impact the recognition result, and the feedback of recognition can improve tracking performance.

11.3.3 Performance with Different Joint Weights

The contribution comparisons on the joint weights $(\alpha_{ij}, \beta_{ij})$ of Eq. (11.7) are shown in Fig. 11.7. As we can see, the joint weights compete with each other frequently and fiercely for almost all these test sequences. In Sequence *Fox*, β_{ij} is always larger than α_{ij} because the recognition part in the kernel risk is more important than the tracking one. This can also be seen in Table 11.3 that the results of $(\alpha_{ij}, \beta_{ij}) = (0, 1)$ and $(\alpha_{ij}, \beta_{ij}) = (0, c_{ij})$ are much better than those of $(\alpha_{ij}, \beta_{ij}) = (1, 0)$ and $(\alpha_{ij}, \beta_{ij}) = (1, 1)$. Table 11.3 shows the performances with different joint weights $(\alpha_{ij}, \beta_{ij})$ in the nine test sequences. The tracking success rates and recognition rates are analyzed here to represent tracking and recognition performances. From Table 11.3, we see that if $(\alpha_{ij}, \beta_{ij}) = (1, 0)$, the JDE framework is concentrating on recognition whereas tracking is ignored. Thus, the recognition rates are almost much higher than the tracking ones. The performances of $(\alpha_{ij}, \beta_{ij}) = (0, 1)$ are the opposite. Then, if $(\alpha_{ij}, \beta_{ij}) = (0, c_{ij})$, the classification information is added into the estimation process, thereby resulting in a little better performance than $(\alpha_{ij}, \beta_{ij}) = (0, 1)$. Tracking and recognition are jointed together when $(\alpha_{ij}, \beta_{ij}) = (1, 1)$, whereas the weights are fixed and equal, and the results are not satisfying. However, the proposed method with the value $(\alpha_{ij}, \beta_{ij})$ set as Eq. (11.7) is able to cope with the dynamic tracking and recognition changes.

11.4 Conclusion

In this chapter, a simultaneous visual recognition and tracking method based on joint decision and estimation is presented. First of all, unlike existing algorithms, this chapter applies the optimal Bayes joint decision and estimation model for the visual tracking and recognition problem and uses its iteration procedure to achieve the optimization. Furthermore, structured sparse representation is adopted in this chapter for the appearance model and is effective to improve the performance of the kernel function. Then, the importance of each test candidate is considered into the iteration learning procedure with a kernel function. At last, the new joint weights of kernel function provide flexibility with target appearance changes and thus robustness to the dynamic scene. Numerous real-world video sequences were tested on the proposed method and other state-of-the-art algorithms, and here we only selected the representative ones. These experimental results demonstrated that the proposed method is highly accurate and robust.

Acknowledgements This work is supported by the National Natural Science Foundation of China (Grant Nos. 61175028, 61365009) and the Ph.D. Programs Foundation of Ministry of Education of China (Grant Nos. 20090073110045).

References

1. Babenko B, Yang MH, Belongie S (2011) Robust object tracking with online multiple instance learning. IEEE Trans Pattern Anal Mach Intell 33(8):1619–1632
2. Bai T, Li Y (2014) Robust visual tracking using flexible structured sparse representation. IEEE Trans Ind Inf 10(1):538–547

3. Belhumeur PN, Hespanha JP, Kriegman DJ (1997) Eigenfaces vs. fisherfaces: recognition using class specific linear projection. IEEE Trans Pattern Anal Mach Intell 19(7):711–720
4. Bhaskar H (2012) Integrated human target detection, identification and tracking for surveillance applications. In: 6th IEEE international conference intelligent systems (IS). IEEE, New York, pp 467–475
5. Chen SS, Donoho DL, Saunders MA (2001) Atomic decomposition by basis pursuit. SIAM Rev 43(1):129–159
6. Conaire CÓ, O'Connor NE, Smeaton A (2008) Thermo-visual feature fusion for object tracking using multiple spatiogram trackers. Mach Vis Appl 19(5–6):483–494
7. Eldar YC, Kuppinger P, Bolcskei H (2010) Block-sparse signals: uncertainty relations and efficient recovery. IEEE Trans Signal Process 58(6):3042–3054
8. Fan J, Shen X, Wu Y (2013) What are we tracking: a unified approach of tracking and recognition. IEEE Trans Image Process 22(2):549–560
9. Gong J, Fan G, Yu L, Havlicek JP, Chen D (2012) Joint view-identity manifold for target tracking and recognition. In: 19th IEEE international conference on image processing (ICIP). IEEE, New York, pp 1357–1360
10. Hare S, Golodetz S, Saffari A, Vineet V, Cheng MM, Hicks SL, Torr PH (2016) Struck: structured output tracking with kernels. IEEE Trans Pattern Anal Mach Intell 38(10):2096–2109
11. Jiang N, Liu W, Wu Y (2011) Learning adaptive metric for robust visual tracking. IEEE Trans Image Process 20(8):2288–2300
12. Kim M, Kumar S, Pavlovic V, Rowley H (2008) Face tracking and recognition with visual constraints in real-world videos. In: IEEE conference on computer vision and pattern recognition (CVPR). IEEE, New York, pp 1–8
13. Kwon J, Lee KM (2010) Visual tracking decomposition. In: IEEE conference on computer vision and pattern recognition (CVPR). IEEE, New York, pp 1269–1276
14. Lee KC, Kriegman D (2005) Online learning of probabilistic appearance manifolds for video-based recognition and tracking. In: IEEE Computer Society conference on computer vision and pattern recognition (CVPR), vol 1. IEEE, New York, pp 852–859
15. Lee KC, Ho J, Yang MH, Kriegman D (2005) Visual tracking and recognition using probabilistic appearance manifolds. Comput Vis Image Underst 99(3), 303–331
16. Li XR (2007) Optimal Bayes joint decision and estimation. In: 10th International conference on information fusion. IEEE, New York, pp 1–8
17. Li H, Shen C, Shi Q (2011) Real-time visual tracking using compressive sensing. In: IEEE conference on computer vision and pattern recognition (CVPR). IEEE, New York, pp 1305–1312
18. Liu Y, Li XR (2011) Recursive joint decision and estimation based on generalized Bayes risk. In: Proceedings of the 14th international conference on information fusion. IEEE, New York, pp 1–8
19. Liwicki S, Zafeiriou S, Tzimiropoulos G, Pantic M (2012) Efficient online subspace learning with an indefinite kernel for visual tracking and recognition. IEEE Trans Neural Netw Learn Syst 23(10):1624–1636
20. Lu WL, Ting JA, Little JJ, Murphy KP (2013) Learning to track and identify players from broadcast sports videos. IEEE Trans Pattern Anal Mach Intell 35(7):1704–1716
21. Luo M, Sun F, Liu H (2013) Hierarchical structured sparse representation for t–s fuzzy systems identification. IEEE Trans Fuzzy Syst 21(6):1032–1043
22. Mei X, Ling H (2011) Robust visual tracking and vehicle classification via sparse representation. IEEE Trans Pattern Anal Mach Intell 33(11):2259–2272
23. Mei X, Zhou SK, Wu H (2006) Integrated detection, tracking and recognition for ir video-based vehicle classification. In: IEEE international conference on acoustics, speech and signal processing, vol 5. IEEE, New York, pp 745–748
24. Pentland A, Moghaddam B, Starner T et al (1994) View-based and modular eigenspaces for face recognition. In: IEEE Computer Society conference on computer vision and pattern recognition (CVPR), vol 94, pp 84–91

25. Pinson R, Howard R, Heaton A (2008) Orbital express advanced video guidance sensor: ground testing, flight results and comparisons. In: AIAA guidance, navigation and control conference and exhibit, p 7318
26. Rauhut H (2007) Random sampling of sparse trigonometric polynomials. Appl Comput Harmon Anal 22(1):16–42
27. Ross DA, Lim J, Lin RS, Yang MH (2008) Incremental learning for robust visual tracking. Int J Comput Vis 77(1):125–141
28. Scholkopf B, Sung KK, Burges CJ, Girosi F, Niyogi P, Poggio T, Vapnik V (1997) Comparing support vector machines with gaussian kernels to radial basis function classifiers. IEEE Trans Signal Process 45(11):2758–2765
29. Sheng G, Yang W, Yu L, Sun H (2012) Cluster structured sparse representation for high resolution satellite image classification. In: IEEE 11th international conference on signal processing (ICSP), vol 1. IEEE, New York, pp 693–696
30. Tang C, Ou Y, Jiang G, Xie Q, Xu Y (2012) Hand tracking and pose recognition via depth and color information. In: IEEE international conference on robotics and biomimetics (ROBIO). IEEE, New York, pp 1104–1109
31. Tzimiropoulos G, Zafeiriou S, Pantic M (2011) Sparse representations of image gradient orientations for visual recognition and tracking. In: IEEE computer society conference on computer vision and pattern recognition workshops (CVPRW). IEEE, New York, pp 26–33
32. Wang C, Wang Y, Zhang Z, Wang Y (2013) Face tracking and recognition via incremental local sparse representation. In: Seventh international conference on image and graphics (ICIG). IEEE, New York, pp 493–498
33. Wright J, Yang AY, Ganesh A, Sastry SS, Ma Y (2009) Robust face recognition via sparse representation. IEEE Trans Pattern Anal Mach Intell 31(2):210–227
34. Yamamoto T, Kataoka H, Hayashi M, Aoki Y, Oshima K, Tanabiki M (2013) Multiple players tracking and identification using group detection and player number recognition in sports video. In: 39th Annual conference of the IEEE industrial electronics society (ECON). IEEE, New York, pp 2442–2446
35. Yun X, Zhongliang J (2016) Kernel joint visual tracking and recognition based on structured sparse representation. Neurocomputing 193:181–192
36. Zhang CL (2013) Research on visual tracking methods based on joint decision from multiple regions. Ph.D. thesis, Shanghai Jiao Tong University
37. Zhang T, Ghanem B, Liu S, Ahuja N (2012) Robust visual tracking via multi-task sparse learning. In: IEEE conference on computer vision and pattern recognition (CVPR). IEEE, New York, pp 2042–2049
38. Zhang K, Zhang L, Yang MH (2013) Real-time object tracking via online discriminative feature selection. IEEE Trans. Image Process 22(12):4664–4677
39. Zhang K, Zhang L, Yang MH (2014) Fast compressive tracking. IEEE Trans Pattern Anal Mach Intell 36(10):2002–2015
40. Zhang K, Zhang L, Liu Q, Zhang D, Yang MH (2014) Fast visual tracking via dense spatio-temporal context learning. In: European conference on computer vision (ECCV). Springer, New York, pp 127–141
41. Zhou SK, Chellappa R, Moghaddam B (2004) Visual tracking and recognition using appearance-adaptive models in particle filters. IEEE Trans Image Process 13(11):1491–1506

Chapter 12
Incremental Visual Tracking with ℓ_1 Norm Approximation and Grassmann Update

12.1 Introduction

Online learning of dynamic video objects is an important issue. Adaptive learning of dynamic object appearance and shape deformation is essential for reducing tracking drift and preventing tracking failure. In particular, online learning is desirable for video objects that experience large pose changes and other appearance changes. For adaptive object learning, many different methods have been proposed. They can be roughly categorized into two classes based on the appearance description, generative or discriminative. For generative model, the adaptive learning is based on linear/nonlinear subspace modeling and estimation [5, 9, 12, 14, 22, 24]. The tracking mechanism is realized by finding the most similar candidate object/object parts with minimal reconstruction error to the subspace, and then the tracked result is utilized as learning the new subspace. For discriminative model, classifiers are often trained to distinguish the object and the background/clutter. The classifier is usually updated online by the positive and negative samples (or bags) obtained from the current tracked results [3, 4, 11]. The tracker performance much depends on how discriminative the online-learned classifier is to the selected sets of positive and negative samples.

Since our proposed method belongs to the first type. We mainly focus on the first set of techniques. For subspace-based online learning, Ross et al. [22] proposed a method that performs incremental subspace learning in vector space with a sample mean update. However, significant nonplanar (or out-of-plane) pose changes from large video objects still cause tracking drift or tracking failure. This is probably due to the reason that video objects actually reside in nonlinear spaces or smoothly changing spaces, not in a linear vector space. In such scenarios, nonlinear manifolds are more desirable, and the manifold learning techniques may generate much more robust results than linear learning techniques.

Z. Jing et al., *Non-Cooperative Target Tracking, Fusion and Control*,
Information Fusion and Data Science, https://doi.org/10.1007/978-3-319-90716-1_12

There has been growing interest in manifold subspace tracking in signal processing and computer vision. An early work on manifold subspace tracking in [8] uses a conjugate gradient and Newton's method for subspace tracking on Grassmann and Stiefel manifolds with applications to orthogonal procrustes. Srivastava et al. [23] proposes piecewise geodesics on complex Grassmann manifolds using projection matrices for subspace tracking with simulations on synthetic signals from an array of sensors.These methods do not address visual objects in images. Balzano et al. [5] present online subspace tracking algorithm from highly incomplete observations. It uses a ℓ_2-norm cost function, which is problematic when facing data corruption or noise distributed other than Gaussian. Instead, He et al. [12] uses ℓ_1-norm cost function to estimate and track nonstationary subspaces when the streaming data vectors are corrupted with outliers. Except synthetic data, these two literatures are applied to separate the background and foreground in video.

Balzano et al. [5] views online face tracking as a subspace tracking problem from a Grassmann manifold with the subspace varying over time. It views the Grassmann manifold from the lie group. The movement direction of the subspace was updated using Kalman filter while a particle filter is used to search the object. Khan et al. [14] describes the object appearance by a basis matrix and its velocity. The object appearance is estimated or updated based on a nonlinear dynamic model. The Gaussian-distributed principal angle between the SVD decomposition-based observation basis matrix and the predicted manifold point is used as appearance likelihood computation. Except for particle filter used in appearance estimation, another particle filter is used for object tracking. [12] also studies the subspace tracking using particle filtering on Grassmann manifold.

Despite reasonably good results from these manifold-based tracking methods, some common challenges remain for tracking visual objects through complex scenes, e.g., when objects have significant pose changes or long-term partial occlusions. The reasons could be due to the lack of the robust manifold learning for the appearance, the lack of handling outliers or occlusions, and/or the lack of dynamic models on affine movements of object.

Multi-task learning is an important topic in machine learning. It aims to model the correlation among multiple tasks via jointing these tasks into single model and sharing information such as parameters or prior knowledge. It can improve the performance of these individual tasks through sharing information. Multi-task learning has very successful applications in image classification [17, 26], image annotations [21], or visual tracking [26]. In [26], authors formulated the visual tracking in particle filter framework as multi-task sparse learning problem. This work is a generalization of [19]. The latter considered each particle individually, instead the former mines the correlation among particles via constraining joint sparsity and thus improves the tracking performance. However, Mei et al. [19] has modeled the target appearance in original space, so the model library update strategy is heuristic and it cannot handle complex appearance variation. Therefore, we utilize traditional subspace to model the target appearance, and convert the appearance learning of multiple particles to multi-task sparsity learning in subspace, in order to overcome background variation or outlier. Visual tracking is performed in particle filtering framework on affine group *Aff*(2) which can provide effective search mechanism. The state is the 2-D affine transformation of the object and its dynamic can be described well by a first-order AR model. Based on [12], the subspace update is regarded as incremental gradient descent on the Grassmann manifold. The selection of step size is very important, and an adaptive step-size strategy is proposed. The dual variable is introduced to the likelihood definition, in order to account for the occlusion case.

Our method is inspired by, however significantly different from [12], where the subspace tracking is designated for matrix completion and the separation of background from foreground in video. Moreover, the multilevel adaptive step-size rule is not suitable for the subspace update from visual tracking application. In order to compensate the occlusion, new likelihood function is defined through the introduction of dual variables. The presentation of this chapter is based on the work in [18].

12.2 Preliminaries

12.2.1 Grassmann Manifold

Differential manifolds can be considered as topological spaces that consist of open sets of Euclidean space $\mathscr{R}^m$, glued together and are differentiable. This means that a manifold is locally similar to the Euclidean space, whereas globally not. For every point on a k-dimensional manifold, there exists a neighborhood that is topologically equivalent (homeomorphic) to an open set in $\mathscr{R}^m$.

This section briefly reviews the manifold theories, tangent spaces, and mapping functions for the sake of mathematical convenience in subsequent sections. A good introduction to the geometry of the Stiefel and Grassmann manifolds can be found in [1, 2, 8], which introduced gradient methods on these manifolds in the context of eigenvalue problems.

In Grassmann manifold, every element is a vector subspace, or numerically the column space of a full-rank matrix U. This manifold can be mathematically written as,

$$Gr(m,r) = \{\mathscr{U} = col(U) : U \in \mathscr{R}_*^{m\times r}\}. \tag{12.1}$$

The notation $\mathscr{R}_*^{m\times r}$ denotes the set of full-rank $m \times r$ matrices. For numerical reasons, we only use the orthonormal matrices $U \in St(m,r)$ to represent subspaces. The set $St(m,r)$ is the (compact) Stiefel manifold:

$$St(m,r) = \{U \in \mathscr{R}_*^{m\times r} : U^T U = I_r\}. \tag{12.2}$$

If $Gr(m,r)$ is endowed with a unique Riemannian metric such that $Gr(m,r)$ is a Riemannian quotient of $St(m,r)$. Let

$$O(r) = \{Q \in \mathscr{R}_*^{r\times r} : Q^T Q = I_r\} \tag{12.3}$$

denote the set of $r \times r$ orthogonal matrices. We have $Gr(m,r) = St(m,r)/O(r)$, or $[U] = \{UQ : Q \in O(r)\}$.

The Grassmann is a manifold, and as such we may define a tangent space to $Gr(m,r)$ at each point $\mathscr{U}$ as $T_{\mathscr{U}}Gr(m,r)$. For practical application, we represent $\mathscr{U}$ by the orthonormal matrix U, the tangent space to $Gr(m,r)$ at U is the set:

$$T_U Gr(m,r) = \{H \in \mathscr{R}^{m\times r} : U^T H = 0\}. \tag{12.4}$$

The orthogonal projector from $\mathscr{R}^{m\times r}$ onto the tangent space $T_U Gr(m,r)$ is given by

$$P_U : \mathscr{R}^{m\times r} \longmapsto : H \longmapsto P_U H = (I - UU^T)H. \tag{12.5}$$

Let $\mathscr{L}$ be differential mapping from $Gr(m,r)$ to $\mathscr{R}$, its Riemannian gradient $\nabla\mathscr{L}(U)$ at a point U is a vector in the tangent space $T_U Gr(m,r)$. The computation of Riemannian gradient on $Gr(m,r)$ can be relatively simple. The Riemannian gradient of $\mathscr{L}$ is the ordinary gradient on Euclidean space followed by linear projection P_U to tangent space $T_U Gr(m,r)$.

With the tangent space and Riemannian gradient, we can discuss the optimization on smooth manifolds. For iterative optimization algorithms, it is important to be able to move along the geodesic curve which goes through U in some prescribed direction specified by a tangent vector H: This is indeed the basic operation required by all descent methods. The geodesic is uniquely defined by two parameters, typically its initial position and derivative. For Grassmann manifold, using an initial condition $U(0) = U$ and a tangent vector $\dot{U}(0) = H$, the geodesic can be computed by the following formula:

$$U(t) = (UV \quad R)\begin{pmatrix} \cos \Sigma t \\ \sin \Sigma t \end{pmatrix}. \tag{12.6}$$

where $R\Sigma V$ is the compact SVD of H, and $\sin(\cdot)$ and $\cos(\cdot)$ act elements.

Generally, for many manifolds, the geodesic curve is numerically expensive or difficult to compute. Fortunately, cheaper ways of moving on a manifold, known as retractions, can be used instead of geodesics without compromising the convergence properties of many optimization algorithms.

12.2.2 Geometrical Particle Filter

We consider the 2-D affine motion of the object image in Lie group (affine group), not in vector space. The 2-D affine motion tracking is formulated as a filtering problem on affine group *Aff* (2) proposed in [16], in which visual tracking is realized via particle filtering on *Aff* (2) with the geometrically well-defined state equation on *Aff* (2) and other related geometric necessities. The 2-D affine transformation of object template coordinates is realized via multiplication in the homogeneous coordinates with a matrix $\begin{bmatrix} G & t \\ 0 & 1 \end{bmatrix}$. Next, we briefly review the visual tracking framework based on particle filtering on *Aff* (2).

In particle filtering framework, tracking problem is casted as an inference task in a Markov model with hidden state variables:

$$p(\mathbf{X}_k|\mathbf{V}_k) \propto p(v_k|\mathbf{X}_k) \int p(\mathbf{X}_k|\mathbf{X}_{k-1}) p(\mathbf{X}_{k-1}|\mathbf{V}_{k-1}) d\mathbf{X}_{k-1}, \tag{12.7}$$

where $\mathbf{X}_k$ is the state variable at time k and $\mathbf{V}_k = \{v_1, v_2, \ldots, v_k\}$ is the set of observations up to time k. The tracking process is governed by the observation model $p(v_k|\mathbf{X}_k)$ and dynamic model $p(\mathbf{X}_{k-1}|\mathbf{V}_{k-1})$. Here, $\mathbf{X}$ is a 2-D affine transformation matrix:

$$\mathbf{X} = \begin{bmatrix} G & b \\ 0 & 1 \end{bmatrix}, \tag{12.8}$$

where G is an invertible 2×2 real matrix, and b is a $\mathscr{R}^2$ translation vector. All affine transformations form the affine group.

Affine Group $\boldsymbol{G}$ is a differentiable manifold, and the tangent space to the identity element $\mathbf{I}$ of this group forms a Lie algebra $\boldsymbol{g}$ which has a matrix structure as follows:

$$\mathbf{x} = \begin{pmatrix} x_1 & x_3 & x_5 \\ x_2 & x_4 & x_6 \\ 0 & 0 & 0 \end{pmatrix}, \tag{12.9}$$

where $x_i \in \mathscr{R}$. Lie group $\boldsymbol{G}$ and its Lie algebra $\boldsymbol{G}$ can be related via the exponential map, $\exp : \mathbf{g} \to \mathbf{G}$. This Lie algebra is equivalent to a six-dimensional vector space.

State Equation on the Affine Group The geometrically well-defined state transition function is on *Aff* (2) is

$$\mathbf{X}_k = \mathbf{X}_{k-1} \cdot \exp(\mathbf{A}(\mathbf{X}, k)) \cdot \exp(d\Omega_k), \tag{12.10}$$

where a is the AR process parameter and $d\Omega_k = \Sigma_{i=1}^{6} \omega_{t,i} \mathbf{E}_i$ is a Gaussian noise on $\boldsymbol{g}$ with $\omega_k = \{\omega_{t,1}, \ldots, \omega_{k,6}\}$ sampled from a Gaussian distribution $\mathcal{N}(0, \Sigma_g)$. $\Sigma_g = \text{diag}\{\phi_s^2, \phi_\alpha^2, \phi_\varphi^2, \phi_\varpi^2, \phi_x^2, \phi_y^2\}$ and $\mathbf{E}_i$ is the basis element of $\boldsymbol{g}$ chosen as

$$\mathbf{E}_1 = \begin{pmatrix} 1 & 0 & 0 \\ 0 & 1 & 0 \\ 0 & 0 & 0 \end{pmatrix}, \mathbf{E}_2 = \begin{pmatrix} 1 & 0 & 0 \\ 0 & -1 & 0 \\ 0 & 0 & 0 \end{pmatrix}, \mathbf{E}_3 = \begin{pmatrix} 0 & -1 & 0 \\ 1 & 0 & 0 \\ 0 & 0 & 0 \end{pmatrix},$$
$$\mathbf{E}_4 = \begin{pmatrix} 0 & 1 & 0 \\ 1 & 0 & 0 \\ 0 & 0 & 0 \end{pmatrix}, \mathbf{E}_5 = \begin{pmatrix} 0 & 0 & 1 \\ 0 & 0 & 0 \\ 0 & 0 & 0 \end{pmatrix}, \mathbf{E}_6 = \begin{pmatrix} 0 & 0 & 0 \\ 0 & 0 & 1 \\ 0 & 0 & 0 \end{pmatrix}. \tag{12.11}$$

The main geometrical transformation modes corresponding to $\mathbf{E}_{1:6}$ are respectively scale, aspect ratio, rotation, skew angle, x translation and y translation.

In this chapter, the state transition equation is described by first-order AR process as follows:

$$\mathbf{X}_t = \mathbf{X}_{k-1} \cdot \exp(A_{k-1} + d\Omega_k), \tag{12.12}$$

$$\mathbf{A}_{k-1} = a \log(\mathbf{X}_{k-2}^{-1} \mathbf{X}_{k-1}). \tag{12.13}$$

12.3 Problem Formulation

Consider the following appearance representation:

$$z = Uv + e, \tag{12.14}$$

where U is a subspace which describes target appearance variation and satisfies $U^T U = I$, and e is a sparse vector which may have an arbitrarily large value and be used to account for occlusion.

In traditional subspace tracking proposed by Ross et al. [22], e in (12.14) is assumed to be Gaussian. It works well when the magnitude of noise e is small. In real applications, however, data is usually polluted by large noise or outlier. This appearance representation also is different from the ℓ_1 tracker presented by Mei et al. [19]. In [19], U is not orthogonal and is composed of target template library, so not only the error but also the representation coefficient is required to be sparse. In Eq. (12.14), only the error is constrained to be sparse. In [26] which is based on multi-task learning, U is the same as in [19] and is the set of target templates. He et al. [12, 13] utilized Eq. (12.14) to separate foreground and background, e models foreground and U reflects background variation. Instead, the proposed method adopts Eq. (12.14) to model the object appearance in visual tracking.

We make use of particle filter in affine group to estimate object state. At time t, multiple particles $\mathbf{X}_t^k$, $k = 1, \ldots, N$, are generated according to first-order AR dynamic model, and the corresponding target appearance feature vectors are denoted as z_t^k. Based on subspace U_{t-1} at time $t-1$, each observation can be represented by the subspace as follows:

$$z_t^k = U_{t-1} v_t^k + e_t^k. \tag{12.15}$$

Each object appearance representation (12.15) is a single task learning. In visual object tracking, collaboratively considering multiple tasks can improve tracking performance to a certain extent. Thus, (12.15) can be rewritten in matrix form as

$$Z_t = U_{t-1} V_t + E_t, \tag{12.16}$$

where $Z_t = \left[z_t^1, \ldots, z_t^N\right]$, $V_t = \left[v_t^1, \ldots, v_t^N\right]$, $E_t = \left[e_t^1, \ldots, e_t^N\right]$.

Mutual dependence or correlation among particles can be reflected in E_t; E_t is required to be jointly sparse. Thus, unknowns V_t and E_t could be evaluated by the following objective function:

$$\min\{1/2 \cdot \|Z_t - U_{t-1} V_t - E_t\|_F^2 + \lambda \|E_t\|_{1,q}\}, \tag{12.17}$$

where $\|\cdot\|_F$ is Frobenius norm and $\|E\|_{1,q} = \sum_{i=1}^N \|e^i\|_q$. (12.17) is equivalent to

$$\begin{aligned} &\min \|E_t\|_{1,q} \\ &s.t.\quad U_{t-1} V_t + E_t = Z_t. \end{aligned} \tag{12.18}$$

This kind of constrained minimization problem could be transformed into unconstrained augmented Lagrange form. Then, the augmented Lagrangian of (12.18) is

$$\begin{aligned} \mathscr{J}(V_t, E_t, C_t) = \|E_t\|_{1,q} &+ \langle C_t, U_{t-1} V_t + E_t - Z_t \rangle \\ &+ \frac{\rho}{2} \langle U_{t-1} V_t + E_t - Z_t, U_{t-1} V_t + E_t - Z_t \rangle, \end{aligned} \tag{12.19}$$

where C_t is the dual matrix, $\langle A, B \rangle = trace(A^T B)$. This optimization problem can be efficiently solved using ADMM [7]. That is, V_t, E_t, C_t are updated in an alternating fashion:

$$\begin{cases} V_{t,i+1} = \arg\min_V \mathscr{J}(V, E_{t,i}, C_{t,i}) \\ E_{t,i+1} = \arg\min_E \mathscr{J}(V_{t,i+1}, E, C_{t,i}) \\ C_{t,i+1} = C_{t,i} + \rho(U_{t-1} V_{t,i+1} + E_{t,i+1} - Z_t) \end{cases} \tag{12.20}$$

For first subproblem $V_{t,i+1} = \arg\min_V \mathscr{J}(V, E_{t,i}, C_{t,i})$ in (12.20), its objective function is quadratic; thus, its solution could be obtained by setting its derivative to zero:

$$V_{t,i+1} = \frac{1}{\rho} (U_{t-1} U_{t-1}^T)^{-1} U_{t-1} \left(\rho(Z_t - E_{t,i}) - C_{t,i}\right) \tag{12.21}$$

The second subproblem $E_{t,i+1} = \arg\min_E \mathscr{J}(V_{t,i+1}, E, C_{t,i})$ in (12.20) involves ℓ_q norm. When $q \geq 1$, this problem is a convex optimization. If $q = 1$, this subproblem is partitioned by column into N individual parts, which amounts to omitting the correlation among particles:

$$\min_e \left\{ \|e\|_1 + c_{t,i}^T (U_{t-1} v_{t,i+1}^k + e - z_t^k) + \frac{1}{\rho} \|U_{t-1} v_{t,i+1}^k + e - z_t^k\| \right\}, k = 1, \ldots, N. \tag{12.22}$$

After rearranging, (12.22) is equivalent to

$$\min_e \frac{1}{2} \left\| e - \left(z_t^k - U_{t-1} v_{t,i+1}^k - \frac{1}{\rho} c_{t,i}^k \right) \right\|_2^2 + \frac{1}{\rho} \|e\|_1, \tag{12.23}$$

and this equivalent has analytical solution, i.e., $e_{t,i+1}^k = \mathscr{S}(z_t^k - U_{t-1} v_{t,i+1}^k - \frac{1}{\rho} c_{t,i}^k)$, here $\mathscr{S}_{\frac{1}{\rho}}$ is the element-wise soft thresholding operator [6].

Next, the observation likelihood $p(v_t^k|\mathbf{X}_t^k)$ associated to $\mathbf{X}_t^k$ can be measured by the reconstruction error of each observed image patch:

$$p(v_t^k|\mathbf{X}_t^k) \propto \exp\{-\|U_{t-1} v_t^k - z_t^k\|_1\}. \tag{12.24}$$

However, (12.24) does not consider occlusion. Thus, we use a mask to factor out non-occluding and occluding parts:

$$p(v_t^k|\mathbf{X}_t^k) \propto \exp\{-(\|h_t^{kT} (U_{t-1} v_t^k - z_t^k)\|_1 + \sigma \|1 - h_t^k\|_1)\}, \tag{12.25}$$

where the dual vector indicates occlusion or not. The first part of the exponent accounts for the reconstruction error of non-occluded proportion of the target object, and the second term aims to penalize labeling any pixel as being occluded.

Visual tracking is based on particle filter framework in which importance function is chosen as motion transition model and object state is evaluated through the maximum a posteriori (MAP) criterion. Thus, the estimated state is the sample which has maximum likelihood.

Remark 12.1 This chapter mainly has considered the case $q = 1$. Equation (12.18) also can be reformulated as $\|U_{t-1} V_t - Z_t\|_{1,1}$; it uses norm ℓ_1 to quantify the subspace representation error, not the norm ℓ_2 in traditional representation. Norm $\|\cdot\|_1$ is used to measure the error or residual, which gives relatively large weights to small residual components and small weights to large residuals. It has robustness when multiple outliers exist in observations.

Remark 12.2 In terms of maximum likelihood estimation, for the norm $\|\cdot\|_1$ defined residuals, the prior is supposed that the noise density is the Laplacian distribution. For the norm $\|\cdot\|_2$ defined residuals, the noise is supposed to follow a Gaussian distribution. The Laplacian density has larger tails than the Gaussian, i.e., the probability of a very large noise component is far larger with a Laplacian than a Gaussian density. As a result, the associated maximum likelihood method expects to see greater numbers of large residuals.

12.4 Subspace Update on Grassmann Manifold

In visual tracking, tracked object usually undergoes various appearance changes, such as pose, illumination, scale, variation, or clutter, or occlusion. In other words, appearance subspace changes go with time; therefore, subspace update is very crucial. $Gr(m, r)$ is the set of all r-dimension subspace, and the subspace update can be considered as an optimization problem on Grassmann manifold. We need to construct an objective function on $Gr(m, r)$ as done in [12, 13]. We also adopt the augmented Lagrangian as the subspace loss function once we have estimated (e^*, v^*, c^*), from the previous U_{t-1}, and the tracked object's observation or feature z_t^*. The new loss function is stated as

$$\mathscr{J}(U) = \|e^*\|_1 + c_t^{*T}(Uv_t^* + e_t^* - z_t^*) + \frac{\rho}{2}\|Uv_t^* + e_t^* - z_t^*\|_2^2. \tag{12.26}$$

Since $\mathscr{J}(U) = \mathscr{J}(UQ)$, where $Q \in Q(r) = \{Q \in R_*^{r \times r} : Q^T Q = I_r\}$, i.e., function $\mathscr{J}(U)$ is rotation-invariant. When U is determined, the triple (e_t^*, v_t^*, c_t^*) is uniquely determined accordingly. For subspace UQ, the triple solution is $(e_t^*, Q^{-1}v_t^*, c_t^*)$. Therefore, function (12.26) is defined on manifold $Gr(m, r)$, and subspace update can be seen as the optimization problem on manifold $Gr(m, r)$.

12.4.1 *Grassmann Geodesic Gradient*

In order to take a gradient step along the geodesic of the Grassmann, according to [12, 13], it is firstly needed to derive the gradient formula $grad\,\mathscr{J}(U)$ of the real-valued loss function $\mathscr{J}$ at point U. The gradient computing on manifold could be finished in two steps. Firstly, the ordinary gradient $\frac{d\mathscr{J}}{dU}$ is computed, then it is orthogonally projected on the tangent space $T_U Gr(m, r)$ at point U. Thus, the geodesic gradient $grad\,\mathscr{J}(U)$ is

$$grad\,\mathscr{J}(U) = (I - UU^T)[c_t^* + \rho(Uv_t^* + e_t^* - z_t^*)]v_t^{*T}. \tag{12.27}$$

Denotes $\Gamma = (I - UU^T)[c_t^* + \rho(Uv_t^* + e_t^* - z_t^*)]$, then $grad\,\mathscr{J}(U) = \Gamma v_t^{*T}$. As pointed out in [12, 13], it is easily noted that rank of $grad\,\mathscr{J}(U)$ is one. Thus the singular value decomposition of $grad\,\mathscr{J}(U)$ is trivial to compute, and is used to compute the geodesic.

The sole nonzero singular value is $\theta = \|\Gamma\|\|v_t^*\|$, and the corresponding left and right singular vectors are $\frac{\Gamma}{\|\Gamma\|}$ and $\frac{v_t^*}{\|v_t^*\|}$ respectively. Then

$$grad\,\mathscr{J} = \left[\frac{\Gamma}{\|\Gamma\|}, x_2, \ldots, x_r\right] \times \mathrm{diag}(\theta, 0, \ldots, 0) \times \left[\frac{v_t^*}{\|v_t^*\|}, u_2, \ldots, u_r\right], \tag{12.28}$$

where $x_2, \ldots, x_r$ be an orthonormal set orthogonal to Γ and $u_2, \ldots, u_r$ be an orthonormal set orthogonal to v_t^*.

According to (12.6), initial value $U(0) = U_{t-1}$ and gradient vector $-grad\,\mathscr{J}$ can uniquely determine a geodesic, and after simple algebraic manipulations the geodesic is as follows:

$$U(\tau) = U_{t-1} + \left[U_{t-1}\frac{v_t^*}{\|v_t^*\|}(\cos(\theta\tau) - 1) - \sin(\theta\tau)\frac{\Gamma}{\|\Gamma\|}\right]\frac{v_t^{*T}}{\|v_t^*\|} \tag{12.29}$$

where τ is the step size and has great influence on the convergence of the algorithm.

He et al. [12, 13] pointed out, at each subspace update step, (12.29) does not remove outliers explicitly. In fact, the gradient of the augmented Lagrangian $\mathscr{J}(U)$ which exploits the dual vector y^* is used to leverage the outlier effect. This is the key to success.

12.4.2 Selection of Step Size

The question of how large a gradient step to take along the geodesic is an important issue. Balzano et al. [5] adopted a constant step-size rule. He et al. [12] proposed multilevel adaptive step size for subspace tracking. The step size satisfies the property $\theta_t \rightarrow 0 (t \rightarrow \infty)$. However, it is not sufficed for visual object tracking, since slow or abrupt appearance changes may happen due to the unpredictable intrinsic or extrinsic influence. The ideal step size should track the slow or abrupt appearance changes. The subspace update will be too conservative when θ_t is too small. The update will be overfitting when θ_t is too large.

Our proposed step-size strategy is simple. It is inspired by Klein et al. [15] as in [12]: If two consecutive gradients $grad\,\mathscr{J}_{t-1}$ and $grad\,\mathscr{J}_t$ are in the same direction, i.e., $\langle grad\,\mathscr{J}_{t-1}, grad\,\mathscr{J}_t\rangle > 0$, it intuitively means that the current estimated U_t is relatively far away from the true subspace. If this is the case, heuristically we should take a large step along $\mathscr{J}_t$. Otherwise, if two consecutive gradients $grad\,\mathscr{J}_{t-1}$ and $grad\,\mathscr{J}_t$ are not in the same direction, i.e., $\langle grad\,\mathscr{J}_{t-1}, grad\,\mathscr{J}_t\rangle < 0$, again intuitively it means that the current estimated U_t is relatively close to the true subspace and again heuristically we should take a small step along $\mathscr{J}_t$. The strategy presented as follows: if $\langle grad\,\mathscr{J}_{t-1}, grad\,\mathscr{J}_t\rangle > 0$, $\tau_t = c_1$. Otherwise, if $\langle grad\,\mathscr{J}_{t-1}, grad\,\mathscr{J}_t\rangle < 0$, $\tau_t = c_2$. The constraints posed on parameters c_1 and c_2 are $c_1, c_2 \in [0, 1]$ and $c_1 > c_2$.

12.5 Experiments

Nine image sequences are used to test the robustness and efficiency of our tracking algorithm. These test images captured by a moving/static camera contain challenging objects with pose changes or illumination changes, scale changes. Comparisons are performed with five state-of-the-art trackers denoted as L1T [19], MIL [4], TLD [27], CT [25], and IVT [22]. These trackers are implemented using publicly available source codes or binaries provided by authors. They were initialized using their default parameters.

12.5.1 Qualitative Comparison

The first sequence *pktest02* is obtained in footnote 1 below.[1] It is an infrared image sequence. The target-to-background contrast is very low and the sequence shows a vehicle which undergoes even but drastic illumination changes and mild occlusion. Some samples of the tracking results are presented in Fig. 12.1. Our tracker, IVT, MIL, TLD, and L1T track the target faithfully throughout the sequences. The CT tracker fails to track the target at the beginning probably due to the low resolution.

The second sequence *egtest01* is from the vivid dataset. The target-to-background contrast is very low and the car turns around and then goes ahead in straight line.

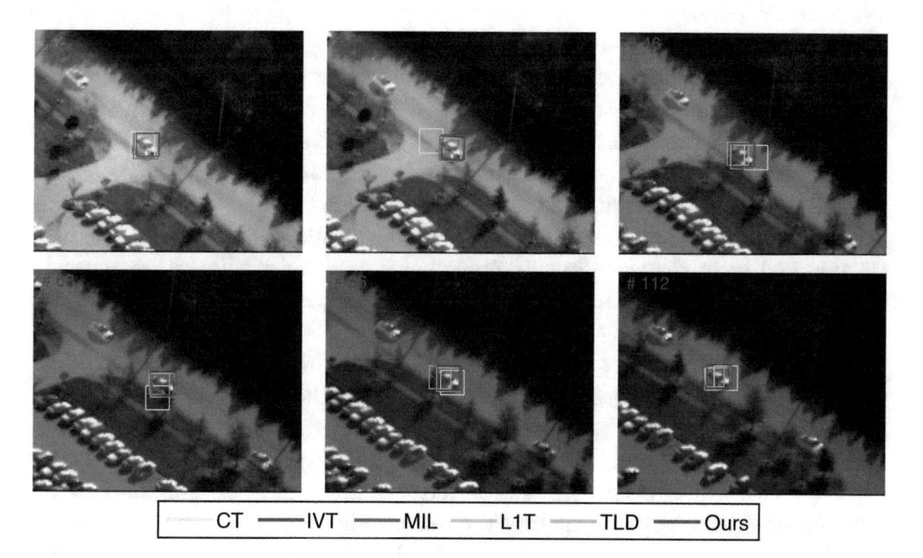

Fig. 12.1 *pktest02*

[1] http://www.ist.temple.edu/hbling/codedata.htm.

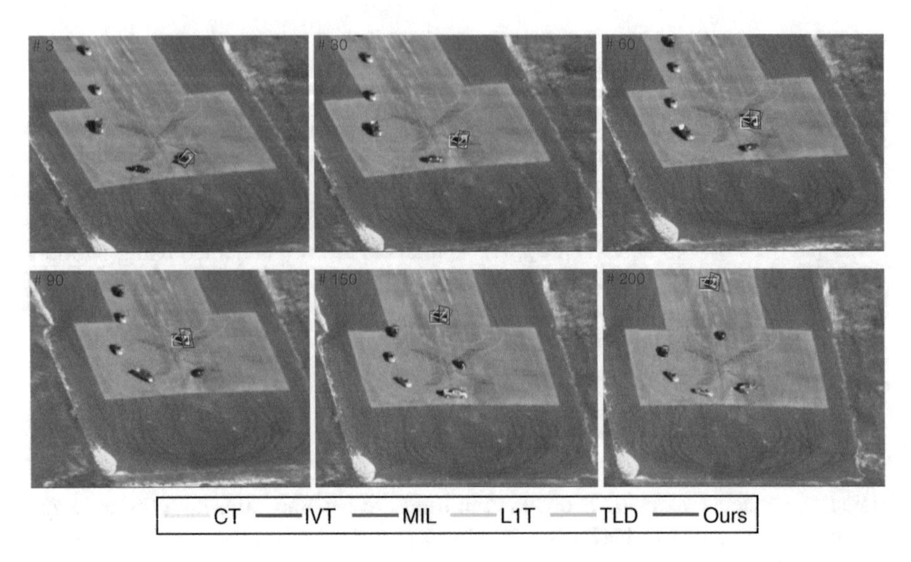

Fig. 12.2 *egtest01*

The car undergoes large in-plane rotation. Some representative tracking results are presented in Fig. 12.2. Our tracker follows the car successfully throughout the 200 frames. IVT, L1T fail to adapt to the pose variation, leading to location drift, or even lose the object. Although MIL, TLD, and CT also deviate to the true location due to fixed-scale window.

The object in the third sequence *coupon book* undergoes significant appearance change about the 60th frame, and then the other coupon book appears. Some representative tracking results are presented in Fig. 12.3. The TLD method is distracted to track the other coupon book while our tracker successfully locks on the correct one. IVT, L1T, MIL, CT drift away from the true location when significant appearance change happens about the 60th frame.

The fourth sequence *cube* is also tested. The cube undergoes large in-plan rotation and scale changes. Some representative tracking results are presented in Fig. 12.4. Our tracker faithfully tracks the cube throughout the whole sequence. All the compared methods fail to track the object.

The fifth image sequence *race1* is from [20], and the tracked racing car undergoes some pose and scale changes; tracking results are presented in Fig. 12.5. Our tracker shows a little drifting at frame 180 while the racing car changes directions. It is then recovered quickly and performs excellently. IVT estimates the position with highest precision but behaves poorly in target's size. Others do not adapt to the scale variation and deviate away from the true position.

The sixth sequence *davidin300* is to track a moving face, which has pose and illumination variance. Some representative tracking results are presented in Fig. 12.6. IVT and the proposed tracker track perform better than the other methods throughout the whole sequence. This can be attributed that appearance change of the object can be well approximated by a subspace at fixed pose.

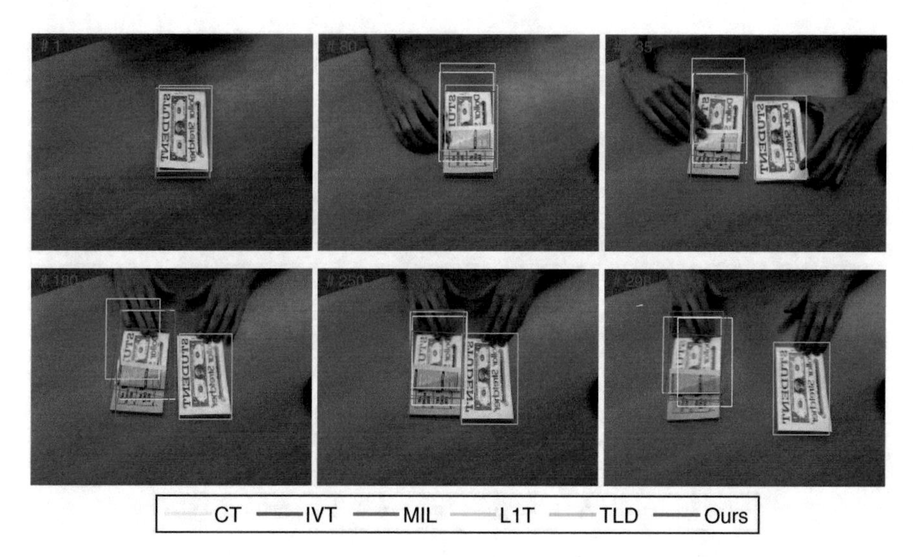

Fig. 12.3 *coupon book*

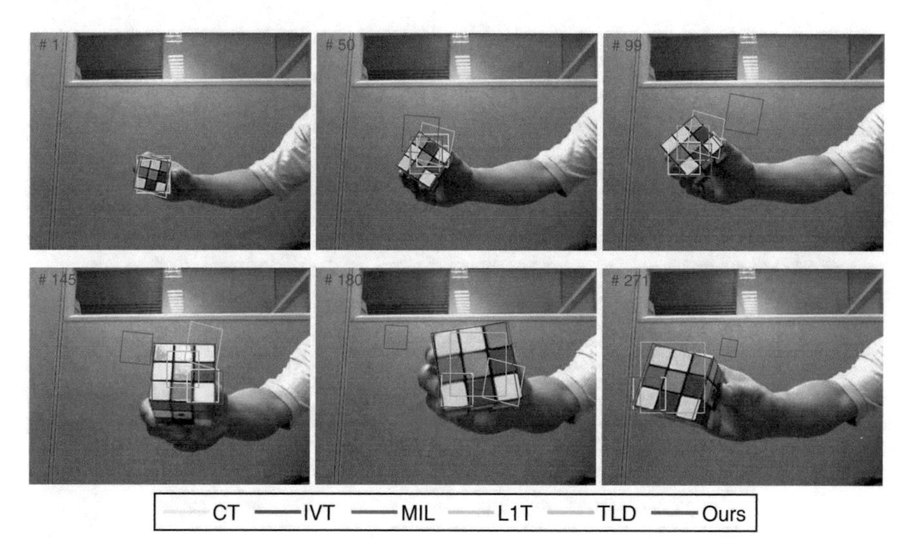

Fig. 12.4 *cube*

Some results of the seventh test sequence *faceocc* are shown in Fig. 12.7. In this sequence, we show the robustness of our algorithm in handling occlusion. Our tracker and L1T perform better, since these two methods take partial occlusion into account. The L1T tracker uses sparse representation with trivial templates, and our method models the occlusion as sparse outliers. IVT shows some drifting once occlusion occurs and recovers when occlusion disappears. Other methods lack the occlusion handling mechanism.

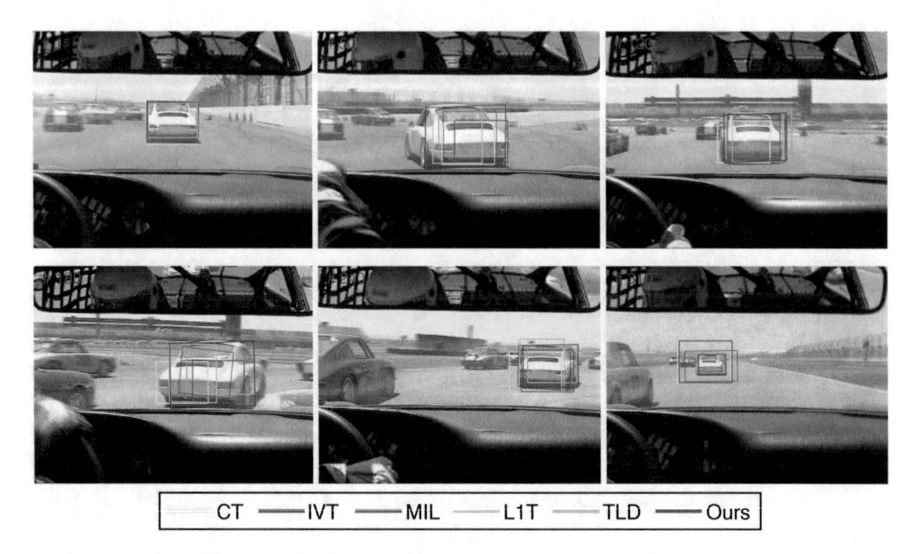

Fig. 12.5 *race1*

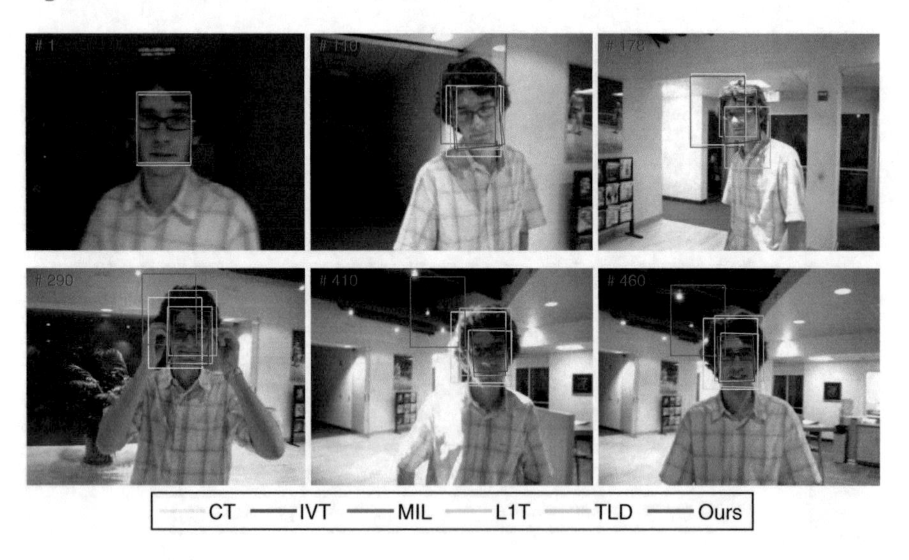

Fig. 12.6 *davidin300*

The eighth test image sequence *car4* is to track a car which undergoes drastic illumination changes and large-scale variation. Some samples of the final tracking results are demonstrated in Fig. 12.8. IVT and the proposed tracker perform better than the other methods throughout the whole sequence. The other trackers finally lose the car.

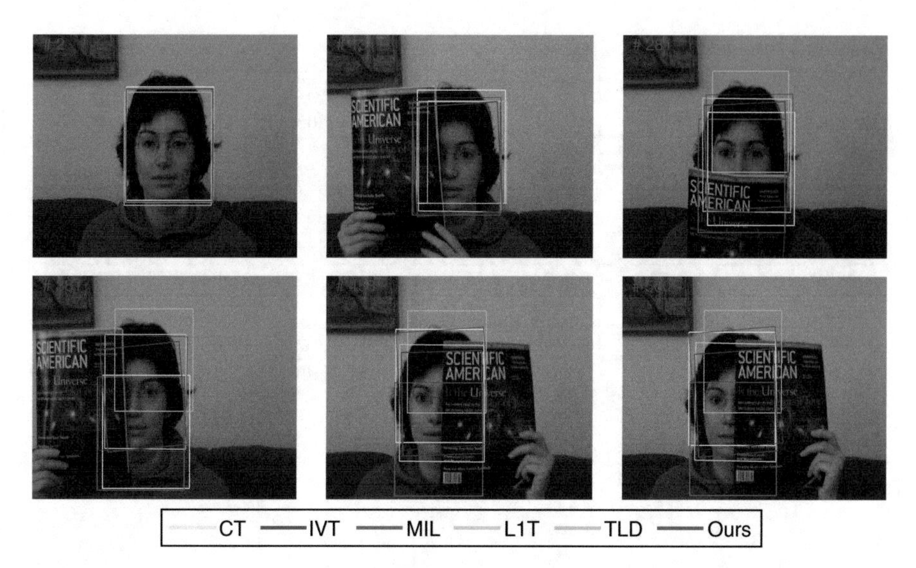

Fig. 12.7 *faceocc*

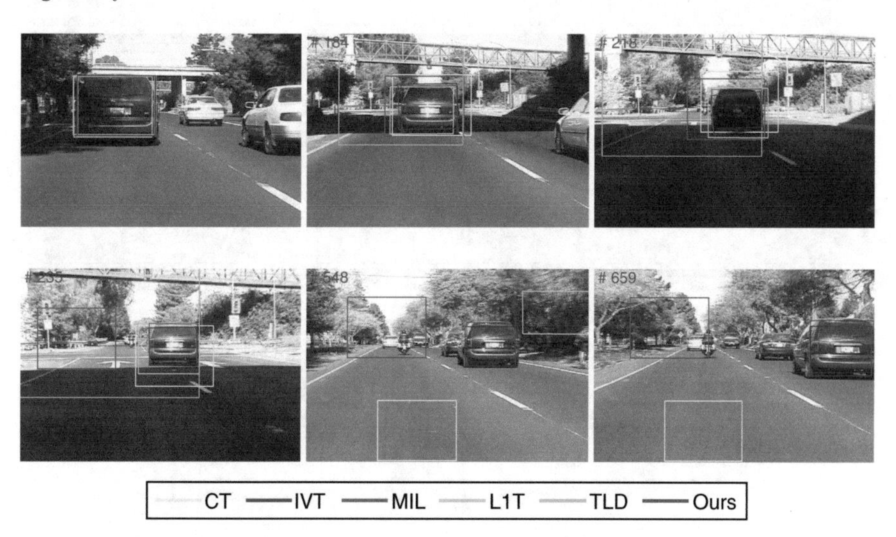

Fig. 12.8 *car4*

12.5.2 Quantitative Comparison

Performance evaluation is an important issue that requires sound criteria in order to fairly assess the strength of tracking algorithms. Here two quantitative evaluation criteria are used. One is to compute the difference between the tracked locations and the ground truth center locations, as well their average values. Table 12.1 summarizes the results in terms of average tracking errors. As indicated by the bold

Table 12.1 The average tracking errors

	CT	IVT	MIL	TLD	L1T	Ours
pktest02	0.1396	0.0145	0.0352	0.0303	0.0173	0.0292
car4	0.3233	0.0068	0.2656	NaN	0.1300	0.0138
coupon book	0.0578	0.0680	0.0646	0.2037	0.0902	0.0123
faceocc	0.0579	0.0143	0.0598	0.0661	0.0107	0.0099
race1	0.0410	0.0174	0.0407	0.0338	0.0290	0.0270
cube	0.0898	0.2424	0.1014	0.0183	0.0975	0.0085
egtest01	0.0492	0.0252	0.0413	—	0.0437	0.0190
davidin300	0.0373	0.0178	0.0798	0.0332	0.0316	0.0187
Ave.	1.4930	6.7410	0.3772	0.3240	0.1847	**0.0187**

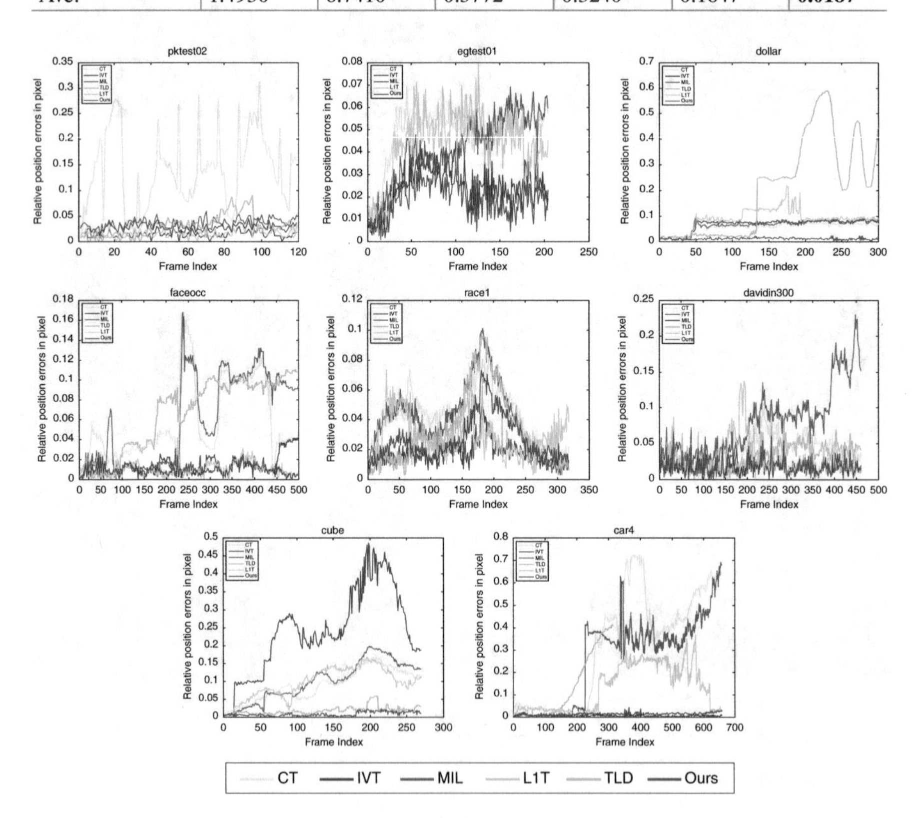

Fig. 12.9 Average center errors

number, our algorithm achieves lowest tracking errors in almost all the sequences, which is provided in Fig. 12.9. The second is to compute the tracking overlap rate which indicates stability of each algorithm since it takes the size and pose of the target object into account. Given the tracking result of each frame R_T and the corresponding ground truth R_G, the overlap rate is defined by the PASCAL VOC

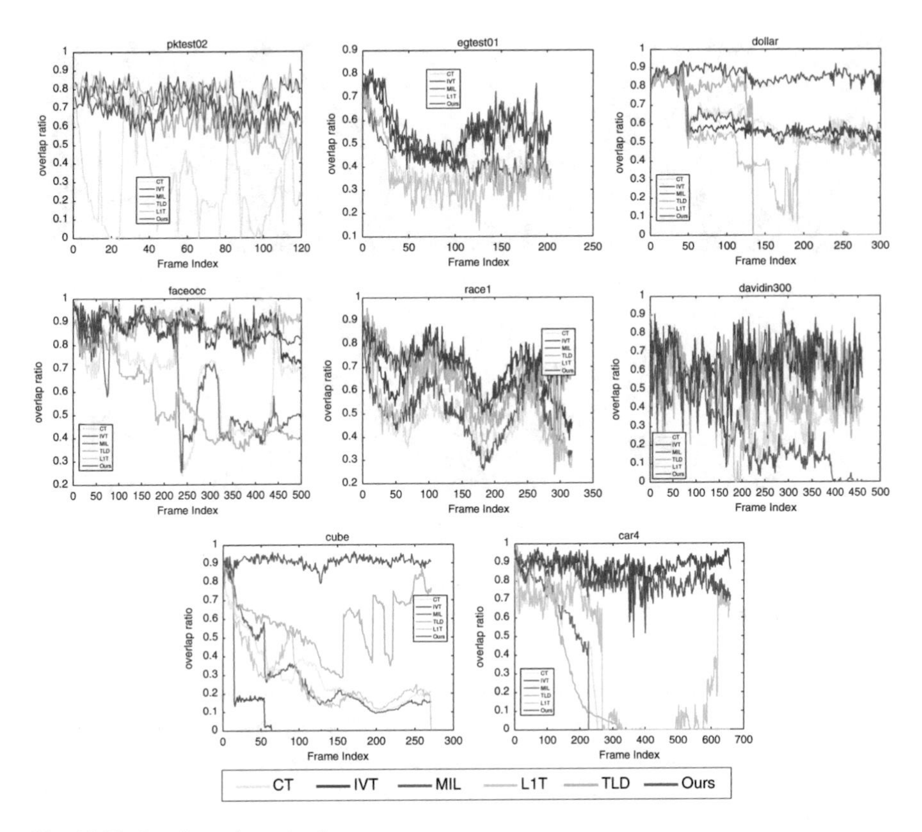

Fig. 12.10 Overlap ratio evaluation

Table 12.2 The overlap ratio

	CT	IVT	MIL	TLD	L1T	Ours
pktest02	0.2835	0.7890	0.6842	0.6253	0.7835	0.7890
car4	0.2281	0.8819	0.2421	0.2152	0.3498	0.8121
coupon book	0.6254	0.6068	0.6104	0.3562	0.5022	0.8596
faceocc	0.6419	0.8750	0.6602	0.5514	0.8934	0.8716
race1	0.4708	0.6626	0.4954	0.6480	0.5789	0.7094
cube	0.2951	0.0695	0.2836	0.5862	0.2821	0.9095
egtest01	0.3974	0.5304	0.4391	—	0.3616	0.5797
davidin300	0.4228	0.6251	0.2457	0.5073	0.5320	0.6436
Ave.	0.4206	0.6163	0.4576	0.3240	0.4985	**0.7718**

[10] criterion, $score = \frac{area(R_T \cap R_G)}{area(R_T \cup R_G)}$: an object is regarded as being successfully tracked when the score is above 0:5. Figure 12.10 shows the overlap rates of each tracking algorithm for all the sequences, and Table 12.2 presents the average overlap rates. As indicated by the bold number, our tracker performs favorably against the other algorithms.

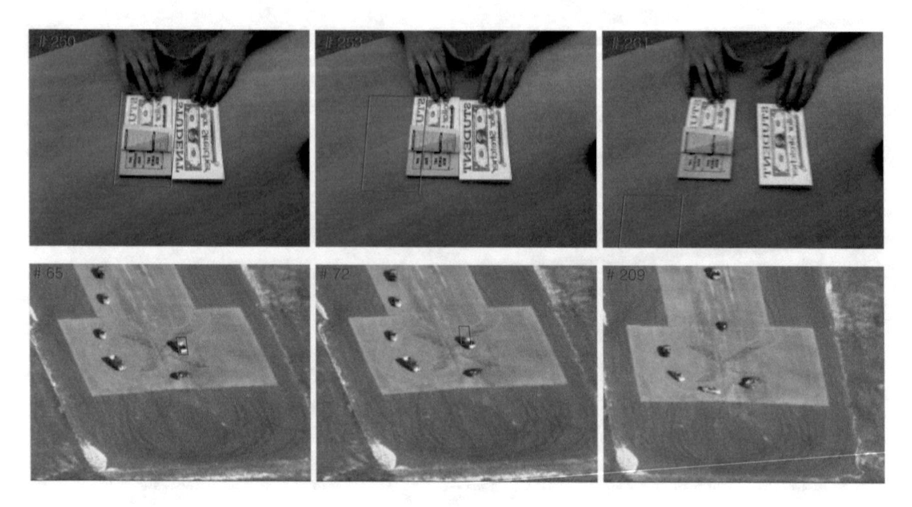

Fig. 12.11 Tracking failures without subspace update

12.5.3 Discussion

Experiments have illustrated that the proposed algorithm works well under complex environments. Particle filter on affine group provides effective search mechanism for visual object tracking, it can effectively handle the appearance caused by pure two-dimensional affine transformation such as sequence *cube*. In addition for geometrical variation, there are many factors which may lead to appearance transformation; template update is particularly important since it is helpful to adapt to target appearance changes. Figure 12.11 gives two failure examples without template update.

For template updating, adaptive step adopted in this algorithm can vary with appearance changes; the strategy makes the subspace well adapted for the appearance variation. Moreover, at each subspace updating, the gradient of the augmented Lagrange function utilizes dual variable to leverage outlier. Model update in IVT does not work well for large appearance update, in that subspaces under different views are not in the same vector space. Template update in L1T is simple, and single false sample in template library may lead to total tracking failure. MIT and CT are based on online classifiers; their tracking windows are fixed and are not directly generalized through affine motion because of their used extended Haar-like features. TLD can run tracker and detector individually and exchange their information by improved online learning mechanism. It merely estimates the target's scale but does not have the ability to accommodate rotation in plane, which does not bring out higher tracking accuracy. TLD method may reinitialize when tracked object disappears and reappears, and it also may be affected by background clutter.

12.6 Conclusion

This chapter has proposed a new incremental tracking algorithm based on ℓ_1 norm approximation and Grassmann subspace update. This model can effectively overcome background clutter or outliers. The tracking problem is regarded as the geometrical particle filter on affine group $Aff(2)$ which provides very effective search framework and in which the state is an affine transformation matrix. In order to tackle occlusion issue, the dual vector is introduced into the likelihood. The subspace update can be considered as an optimization problem on Grassmann manifold. The step size along the geodesic is important; an adaptive step-size strategy is given. The experimental results demonstrate that our tracking performance is superior to the other state-of-art trackers under many challenging tracking situations.

References

1. Absil PA, Mahony R, Sepulchre R (2009) Optimization algorithms on matrix manifold. Princeton University Press, Princeton
2. Absil PA, Mahony R, Sepulchre R (2010) Optimization on manifolds: methods and applications. In: Recent advances in optimization and its applications in engineering. Springer, Berlin, pp 125–144
3. Avidan S (2007) Ensemble tracking. IEEE Trans Pattern Anal Mach Intell 29(2):261–271
4. Babenko B, Yang MH, Belongie S (2011) Robust object tracking with online multiple instance learning. IEEE Trans Pattern Anal Mach Intell 33(8):1619–1632
5. Balzano L, Nowak R, Recht B (2010) Online identification and tracking of subspaces from highly incomplete information. In: 48th Annual Allerton conference on communication, control, and computing. IEEE, New York, pp 704–711
6. Boyd S, Vandenberghe L (2004) Convex optimization. Cambridge University Press, Cambridge
7. Boyd S, Parikh N, Chu E, Peleato B, Eckstein J (2011) Distributed optimization and statistical learning via the alternating direction method of multipliers. Found Trends Mach Learn 3(1): 1–122
8. Edelman A, Arias TA, Smith ST (1998) The geometry of algorithms with orthogonality constraints. SIAM J Matrix Anal Appl 20(2):303–353
9. Elgammal A, Lee CS (2009) Tracking people on a torus. IEEE Trans Pattern Anal Mach Intell 31(3):520–538
10. Everingham M, Van Gool L, Williams CK, Winn J, Zisserman A (2010) The Pascal visual object classes (voc) challenge. Int J Comput Vis 88(2):303–338
11. Grabner H, Grabner M, Bischof H (2006) Real-time tracking via online boosting. In: British Machine Vision Conference (BMVC), vol 1, pp 47–56
12. He J, Balzano L, Lui J (2011) Online robust subspace tracking from partial information. Preprint. arXiv:1109.3827
13. He J, Balzano L, Szlam A (2012) incremental gradient on the Grassmannian for online foreground and background separation in subsampled video. In: IEEE Conference on Computer Vision and Pattern Recognition (CVPR). IEEE, New York, pp 1568–1575
14. Khan ZH, Gu IYH (2013) Nonlinear dynamic model for visual object tracking on Grassmann manifolds with partial occlusion handling. IEEE Trans Cybern 43(6):2005–2019

15. Klein S, Pluim JP, Staring M, Viergever MA (2009) Adaptive stochastic gradient descent optimization for image registration. Int J Comput Vis 81(3):227–239
16. Kwon J, Park FC (2010) Visual tracking via particle filtering on the affine group. Int J Robot Res 29:198–217
17. Liao X, Carin HL, Carin L, Liu Q, Stack JR (2009) Semisupervised multitask learning. IEEE Trans Pattern Anal Mach Intell 31(2):1074–1086
18. Liu R (2014) Research on tracking algorithm for visual target based on statistical learning. Ph.D. thesis, Shanghai Jiao Tong University
19. Mei X, Ling H (2009) Robust visual tracking using $l1$ minimization. In: 12th IEEE international conference on computer vision (ICCV). IEEE, New York, pp 1436–1443
20. Porikli F, Tuzel O, Meer P (2006) Covariance tracking using model update based on lie algebra. In: IEEE Computer Society conference on computer vision and pattern recognition (CVPR), pp 728–735
21. Quattoni A, Carreras X, Collins M, Darrell T (2009) An efficient projection for l1, infinity regularization. In: International conference on machine learning, pp 857–864
22. Ross DA, Lim J, Lin RS, Yang MH (2008) Incremental learning for robust visual tracking. Int J Comput Vis 77(1):125–141
23. Srivastava A, Klassen E (2004) Bayesian and geometric subspace tracking. Adv Appl Probab 36(1):43–56
24. Wang D, Lu H, Yang MH (2012) Online object tracking with sparse prototypes. IEEE Trans Image Process 22(1):314–325
25. Zhang K, Zhang L, Yang MH (2012) Real-time compressive tracking. In: European conference on computer vision (ECCV). Springer, Berlin, pp 864–877.
26. Zhang T, Ghanem B, Liu S, Ahuja N (2012) Robust visual tracking via multi-task sparse learning. In: IEEE conference on computer vision and pattern recognition (CVPR). IEEE, New York, pp 2042–2049
27. Zhou SK, Chellappa R, Moghaddam B (2004) Visual tracking and recognition using appearance-adaptive models in particle filters. IEEE Trans Image Process 13(11):1491–1506

Chapter 13
Image Fusion Using the Quincunx-Sampled Discrete Wavelet Frame

13.1 Introduction

Image fusion combines information in multi-source images to obtain a more accurate description about the same scene. These source images may be captured from different image sensors or the same sensor in different modes. Image fusion technique was firstly applied in military area. Recently, with the development of high-speed processors and imaging sensors, image fusion techniques gradually used in civilian fields, such as digital camera [39], remote sensing [12, 20], medical imaging [6, 8, 37], security and surveillance [17], and intelligent robot [2].

According to Abidi and Gonzalez's fusion architecture [1], the fusion process can be conducted on different levels: signal, pixel, feature, and symbolic levels. This study focuses on the pixel-level fusion, which has great potential in various applications. At this level, the fused result is a single gray or color image, which is generated by fusing individual pixels from various sources images and is more useful for human visual or machine perception than any single source image.

During the last 20 years, various pixel-level image fusion techniques have been proposed, from the simplest weighted averaging [37] to the pyramid algorithms [3, 4, 26, 27] and more advanced wavelets methods [18, 22, 30]. A general image fusion scheme based on multiscale decomposition has been developed. As shown in Fig. 13.1, the scheme can be divided into three steps: (1) Source images are decomposed separately with a multiscale decomposition method; (2) The decomposed coefficients of source images are combined by using a certain rule, the so-called fusion rule; (3) The fused image is obtained from the combined coefficients with the multiscale reconstruction which is the inverse transformation of the multiscale decomposition in the first step.

Recently, the discrete wavelet transform (DWT) has been used in the multiscale scheme to replace pyramid decomposition. It has shown that the DWT-based fusion algorithms generally can achieve better results than the pyramid-based methods

Z. Jing et al., *Non-Cooperative Target Tracking, Fusion and Control*,
Information Fusion and Data Science, https://doi.org/10.1007/978-3-319-90716-1_13

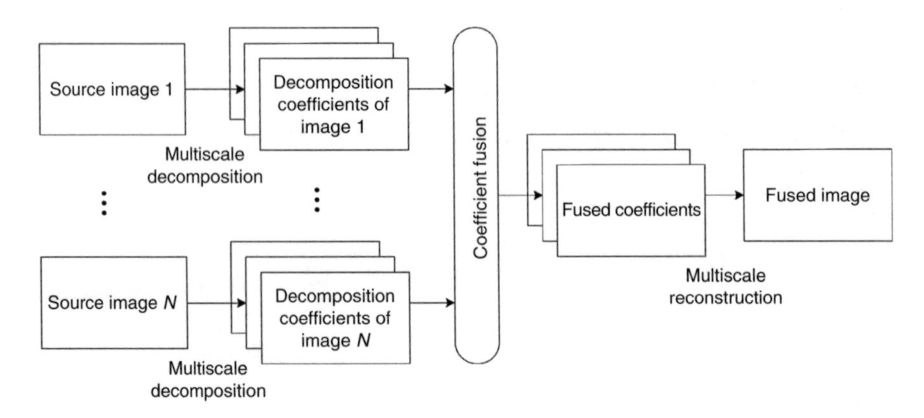

Fig. 13.1 Multiscale-decomposition-based image fusion scheme

[7, 11, 16, 19, 39]. However, the DWT lacks shift invariance [13, 19, 25, 36] due to the down-sampling steps in the decomposition, which may introduce false information in the fused images. To overcome this, Rockinger [23] proposed a fusion algorithm based on an Undecimated Discrete Wavelet Transformation (UDWT) [10], which removes the down-sampling steps in the standard wavelet decomposition. The UDWT solves the problem of shift invariance at the cost of introducing redundancy. Given an image decomposed by an I-level UDWT, the amount of decomposition coefficients is $3I + 1$ times of the amount of the original pixels in the image. These highly redundant coefficients will directly increase the computational complexity and storage burden of fusion algorithm. Therefore, the UDWT fusion is not suitable for dealing with the large amounts of data. As most image fusion tasks require real-time or quasi-real-time processing, the efficiency of fusion algorithms is very important for practical applications. Many excellent algorithms, including the UDWT, are difficult to practice. Efficiency improvement has become an important aspect for improving image fusion algorithms.

In order to improve efficiency, Hill et al. [10] introduce the Double-Tree Complex Wavelet Transform (DTCWT) [13] into the multiscale fusion scheme. The DTCWT is a low-redundancy and near-shift-invariance wavelet transform. Compared with the UDWT, the DTCWT introduces modest redundancy by increasing an additional filtering tree, which generates only four times redundancy for a 2D transformation. However, the filter banks used by the DTCWT are designed much more strictly, such that most of the frequency aliases from different trees can be cancelled each other in every branch. The short FIR filters are difficult to satisfy these extra constraints besides such usual constraints as orthogonality, linear phase (symmetry), and regularity. In addition, the coefficients from the two trees, serving as the real part and imaginary part respectively, must be combined together to form a shift-invariant magnitude representation, which increases the burden of subsequent fusion processing, and the situation will become rather complex in multidimensional cases.

To deal with the challenges of shift invariance and redundancy, we propose a fusion algorithm based on the Quincunx-Sampling Discrete Wavelet Frame (QSDWT) [36], which is built by using the replaceability of the multidimensional Perfect Reconstruction Filter Bank (PRFB) [35]. The QSDWF, similar to the DTCWT, is a low-redundancy and near-shift-invariance wavelet transform. However, the QSDWF employs common filter banks, which is easier to implement than the DTCWT. Besides, the QSDWF has intermediate scales and yields real decomposition coefficients with a higher frequency resolution, which will benefit the subsequent fusion processing. The presentation of this chapter is based on the work [34].

The rest of the chapter is organized as follows. Section 13.2 shows the replaceability of sampling matrix in the multidimensional PRFB and gives the replacement condition. It also shows that tight frames can be built by the replacement of sample matrix. In Sect. 13.3, we replace the standard rectangular sampling matrix of the DWT with a quincunx sampling matrix to build the QSDWF. In Sect. 13.4, the QSDWF is incorporated into the multiscale fusion scheme. Section 13.5 gives the experimental results of the QSDWF-based fusion algorithm on various datasets. Section 13.6 is the conclusion.

13.2 Replaceability of Sampling Matrix in Multidimensional PRFB

13.2.1 Multidimensional PRFB

A separable mode is generally adopted by the wavelets or filter banks when they deal with multidimensional data, i.e., the filtering and resampling operations are conducted in each dimension separately. This mode is simple and easy to implement, however, it is essentially one-dimensional processing. Recently, multi-dimensional non-separable systems have been developed and it is shown that real multi-dimensional processing is more reasonable for the multidimensional data [14, 28, 29, 32]. Perfect reconstruction conditions for a general (separable or non-separable) multidimensional filter bank with an arbitrary sampling matrix are studied in [32]. In [14], the multidimensional filter bank theory is extended to the wavelet domain and non-separable multidimensional wavelet bases are built.

Figure 13.2 shows a general multidimensional filter bank system with an arbitrary sampling matrix. The system is critically sampled, i.e., $D = |\det(D)|$. We define the vectors of analysis and synthesis filter banks as

$$h(\omega) = [H_0(\omega), H_1(\omega), \ldots, H_{D-1}(\omega)]^T \tag{13.1}$$

$$g(\omega) = [G_0(\omega), G_1(\omega), \ldots, G_{D-1}(\omega)]^T \tag{13.2}$$

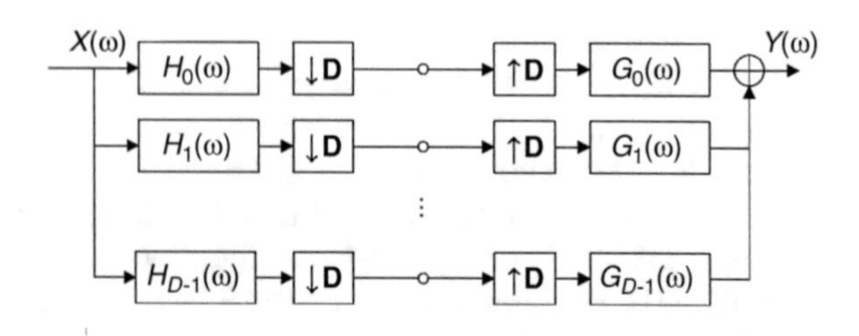

Fig. 13.2 A general multidimensional d-channel filter bank

According to the basic multi-rate formula [32], the output of the system is written as

$$Y(\omega) = \frac{1}{D} g(\omega)^T H(\omega) x(\omega), \tag{13.3}$$

where x is the Aliasing-Component (AC) vector of the input signal and H is the AC matrix of the analysis filter bank

$$x(\omega) = [X(\omega - 2\pi D^{-T} k_0), X(\omega - 2\pi D^{-T} k_1), \ldots, X(\omega - 2\pi D^{-T} k_{D-1})]^T, \tag{13.4}$$

$$H(\omega) = [h(\omega - 2\pi D^{-T} k_0), h(\omega - 2\pi D^{-T} k_1), \ldots, h(\omega - 2\pi D^{-T} k_{D-1})]^T, \tag{13.5}$$

where k_d $(d = 0, 1, \ldots, D - 1)$ are the D sampling points contained by the fundamental parallelepiped [5] of D^T

$$k_d \in \Phi_D = \{k \in \mathbb{Z}^n | D^{-T} k \in [0, 1)^n\} \tag{13.6}$$

and $k_0 = 0$.

From (13.3), the perfect reconstruction condition [32] is obtained. Perfect reconstruction is achieved iff

$$g(\omega)^T H(\omega) = p \exp(j\omega^T q)\underbrace{[1, 0, \ldots, 0]}_{D}, \tag{13.7}$$

where p is a power factor and q is an integral shift vector (without loss of generality, we assume $p = D$ and $q = 0$).

13.2.2 Replacement Condition

In [35], we show the replaceability of sampling matrix in a multidimensional PRFB. Without loss of perfect reconstruction, redundancy can be introduced into the filter bank by replacing the sampling matrix.

We now use $D^{'}$ to replace D, the output of the system is written as

$$Y(\omega) = \frac{1}{D^{'}} g(\omega)^T H^{'}(\omega) x^{'}(\omega) \tag{13.8}$$

where $D^{'} = |\det(D^{'})|$ and

$$x^{'}(\omega) = [X(\omega - 2\pi D^{'-T} k_0^{'}), X(\omega, -2\pi D^{'-T} k_1^{'}), \ldots, X(\omega - 2\pi D^{'-T} k_{D-1}^{'})]^T \tag{13.9}$$

$$H^{'}(\omega) = [h(\omega - 2\pi D^{'-T} k_0^{'}), h(\omega, -2\pi D^{'-T} k_1^{'}), \ldots, h(\omega - 2\pi D^{'-T} k_{D-1}^{'})]^T \tag{13.10}$$

with $k_d^{'} \in \Phi_{D'}$ and $k_0^{'} = 0$.

The two AC matrices, $H^{'}$ and H, have the same first-column, which is the alias-free version of the analysis bank because $k_0 \equiv k_0^{'} \equiv 0$. If all aliased versions of the analysis bank contained by $\mathbf{H}^{'}$ (from the second column to the last column) are exactly contained by H, from (13.7) we can derive

$$g(\omega)^T \mathbf{H}^{'}(\omega) = D\underbrace{[1, 0, \ldots, 0]} \tag{13.11}$$

which means that the replaced system is still perfectly reconstructed.

For convenience, we define the set

$$\Lambda_D = \{D^{-T} k | k \in \Phi_D\} \tag{13.12}$$

The elements of the set correspond to the columns of the AC matrix. Next we have to find when $\Lambda_{D'}$ is a subset of Λ_D. It is proved in [35] that given two nonsingular integer matrices D and $D^{'}$, there is $\Lambda_{D'} \subseteq \Lambda_D$ iff $D^{'-1} D$ is an integer matrix. Therefore, we obtain the following result:

Replacement Condition: The sampling matrix D of a PRFB can be replaced by $D^{'}$ without loss of perfect reconstruction if D is left divisible by $D^{'}$.

13.2.3 Wavelet Framework

It is well known that an n-dimensional orthogonal PRFB constructs an orthogonal basis of $\ell_2(\mathbb{Z}^n)$ [14]. We show that the new, replaced PRFB constructs a tight frame of $\ell_2(\mathbb{Z}^n)$ in [36].

The equivalent synthesis filters for the replaced system after i iterations ($i = 1, 2, \ldots$) can be written as

$$G_d^{'(i)}(\omega) = G_d[(D^{'T})^{i-1}\omega] G_0^{'(i-1)}(\omega) \quad \text{with} \quad G_0^{'(0)}(\omega) = 1 \tag{13.13}$$

The discrete frame functions, similar to the wavelet basis functions, are represented as

$$\varphi'_{i,d,s}(n) = \alpha g_d^{'(i)}(n - D'^i s), \quad n \in \mathbb{Z}^n \tag{13.14}$$

where $g_d^{'(i)}$ are the impulse responses of $G_d^{'(i)}$, and $\alpha = (D'/D)^{\frac{i}{2}}$ is chosen to offset the power increase resulting from the change of sampling.

An overcomplete set is then obtained as

$$\Gamma = \{\varphi'_{i,d,s}(n), \varphi'_{I,0,s}(n) | i = 1, 2, \ldots, I; d = 1, 2, \ldots, D-1; s \in Z^n\} \tag{13.15}$$

We show that if the original PRFB is orthogonal and D' satisfies the replacement condition, the set Γ constitutes a tight frame in $\ell_2(\mathbb{Z}^n)$.

A tight frame in a Hilbert space will lead to a representation, analogous to that associated with an orthogonal basis. Specifically for Γ, there is $\forall x(n) \in \ell_2(\mathbb{Z}^n)$,

$$\begin{aligned} x(n) = &\sum_{i=1}^{I} \sum_{d=1}^{D-1} \sum_{s \in \mathbb{Z}^2} < \varphi'_{i,d,s}(n), x(n) > \varphi'_{i,d,s}(n) \\ &+ \sum_{s \in \mathbb{Z}^2} < \varphi'_{I,0,s}(n), x(n) > \varphi'_{I,0,s}(n) \end{aligned} \tag{13.16}$$

The above analysis can be extended to a biorthogonal PRFB, where a dual frame [9] is constructed by the replacement of sampling matrix, and a more general representation can be obtained:

$$\begin{aligned} x(n) = &\sum_{i=1}^{I} \sum_{d=1}^{D-1} \sum_{s \in \mathbb{Z}^2} < \tilde{\varphi}'_{i,d,s}(n), x(n) > \varphi'_{i,d,s}(n) \\ &+ \sum_{s \in \mathbb{Z}^2} < \tilde{\varphi}'_{I,0,s}(n), x(n) > \varphi'_{I,0,s}(n) \end{aligned} \tag{13.17}$$

where $\tilde{\varphi}'_{i,d,s}$ is the analysis wavelet basis formed by iterating the analysis filters.

The replaced tight (or dual) frame is redundant, and the redundancy ratio is bounded for most cases ($1 < D' < D$)

$$r_f = (D-1) \sum_{i=1}^{I} D'^{-i} + D'^{-I} \leq \frac{D-1}{D'-1} \tag{13.18}$$

13.3 Quincunx-Sampling Discrete Wavelet Frame

The replaceability of sampling matrix provides a simple approach to build redundant PRFB and wavelet frames. In this section, the rectangular sampling matrix D_R of the standard two-dimensional DWT is replaced with the quincunx-sampling matrix D_Q, where

$$D_Q = \begin{pmatrix} 1 & 1 \\ 1 & -1 \end{pmatrix}, \quad D_R = \begin{pmatrix} 2 & 0 \\ 0 & 2 \end{pmatrix} \tag{13.19}$$

Since D_R is left divisible by D_Q, the replaced system is redundant and perfectly reconstructed, which constructs a wavelet frame, i.e., QSDWF. Figure 13.3 illustrates a two-level QSDWF decomposition to the Lena image.

13.3.1 Replacement Matrix Selection

The undecimated version is the only choice in one dimension where the original sampling matrix is a scalar $D = 2$. However, in two dimensions the diagonal matrix D_R is divisible by any integer matrix with determinant ± 2. Here D_Q is selected for the following reasons:

1. All eigenvalues of the matrix have magnitude greater than one, assuring that the corresponding wavelet frame is dilated in all dimensions.
2. The second power of the matrix is a diagonal matrix ($D_Q^2 = D_R$), implying that a rectangular sampling is achieved after every other iteration step.

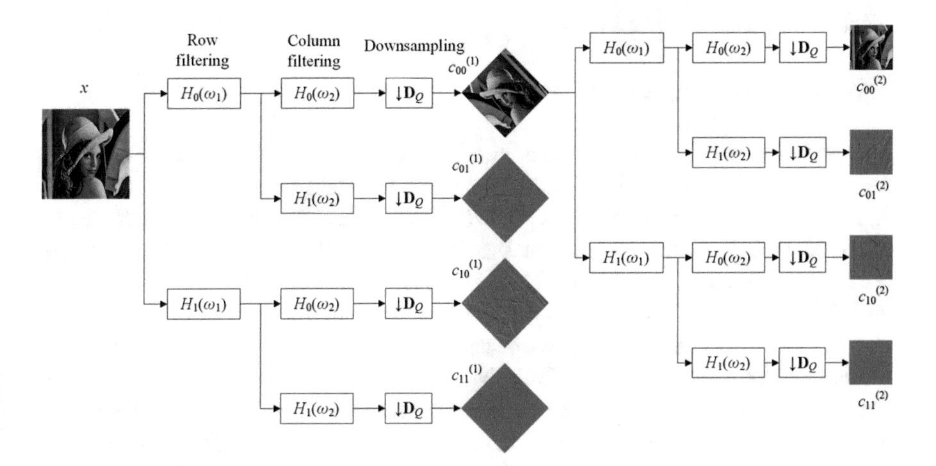

Fig. 13.3 Two-level QSDWF decomposition

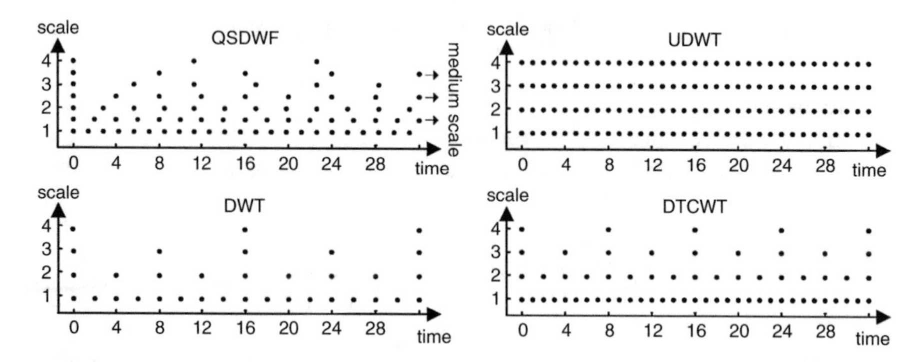

Fig. 13.4 Ideal time-frequency localization diagrams for various wavelet transforms

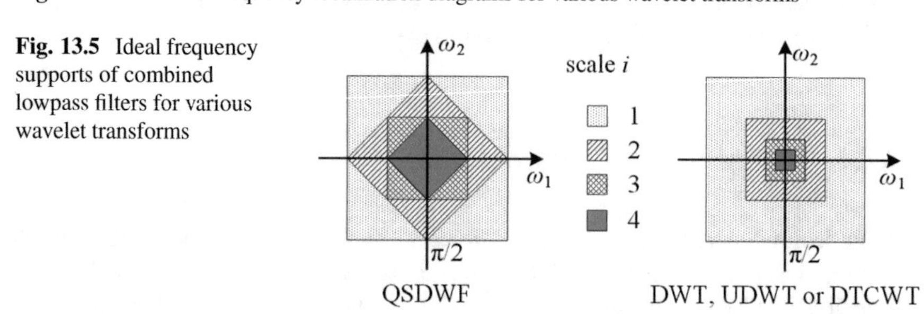

Fig. 13.5 Ideal frequency supports of combined lowpass filters for various wavelet transforms

3. D_Q can be extended to three or more dimensions [15]. The extended matrix has the same determinant as the two-dimension case and still has the above two advantages.

13.3.2 Features of QSDWF

13.3.2.1 Time-Frequency Resolution

The replacement of the sampling matrix causes interesting changes in time-frequency characteristic. Figure 13.4 compares the ideal time-frequency sampling distributions [24] formed by the QSDWF, DWT, UDWT, and DTCWT. It can be observed that the sampling rate of the QSDWF on the time axis (time-domain resolution) decreases with the order of $\sqrt{2}$. The UDWT remains unchanged. Both the DWT and DTCWT decrease by 2, however the sampling rate of the DTCWT is always twice that of the DWT at each scale.

On the other hand, the QSDWF has intermediate scales, whose frequency-domain resolution (sampling rate on the scale axis) is higher than the other three wavelets. This can be understood by examining the frequency supports of equivalent lowpass filters. As shown in Fig. 13.5, the basic spectrum for the QSDWF rotates by 45° and shrinks by a factor $\sqrt{2}$ after every iteration step, while the other three shrink directly by a factor 2. The frequency shrink of the QSDWF is more gradual, leading a higher frequency resolution.

13.3.2.2 Shift Invariance

The QSDWF, benefiting from the higher time-domain resolution, has better shift invariance than the DWT. The aliasing energy ratio (AER) [13] is used to measure the shift invariance. We extend the AER index to the multidimensional case. For a multidimensional DWT, the AER in the subband $\{i, j\}$ is

$$R_d^{(i)} = \frac{\sum_k E\left\{G_d^{(i)}(\omega)H_d^{(i)}\left(\omega - \frac{2k\pi}{2^i}\right)\right\}}{E\{G_d^{(i)}(\omega)H_j^{(i)}(\omega)\}} \tag{13.20}$$

where $k \in \Phi_{D^i}$ and $k \neq 0$, $E\{X\}$ represents the energy of the transfer function X. For the QSDWF, there is $k \in \Phi_{D_Q^i}$, and the corresponding filters are replaced by $G_d^{'(i)}$ and $H_d^{'(i)}$ in (13.20). The AER represents the ratio of the total energies of the frequency-aliasing transfer function and the non-aliasing transfer function. A smaller AER means a better shift invariance. If the AER is equal to zero, the subband is absolute shift invariance.

Figure 13.6 shows the AER of the QSDWF and the DWT at various scales based on the Haar, DB2, and DB4 filter banks, where DBn denotes n-order Daubechies orthogonal wavelet. For comparison, the results of the DTCWT with the 6-tap Q-shift filters [13] (whose AER at the first level is zero, equivalent to the UDWT) are also given in the figure. The QSDWF and the DTCWT generate comparable results,

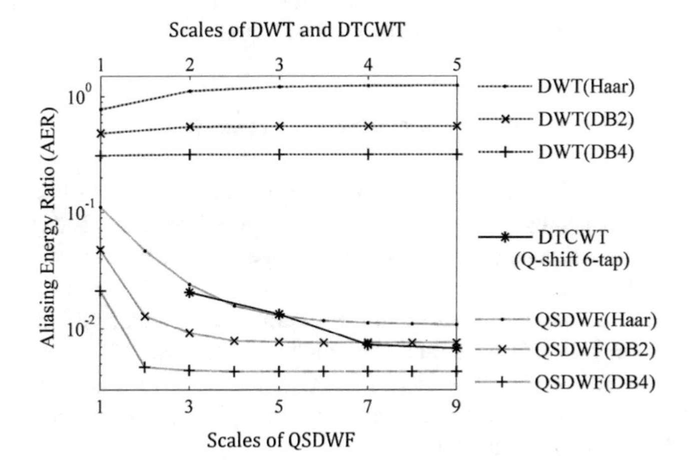

Fig. 13.6 Aliasing energy ratios for various wavelet transforms

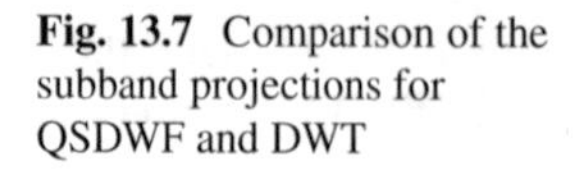

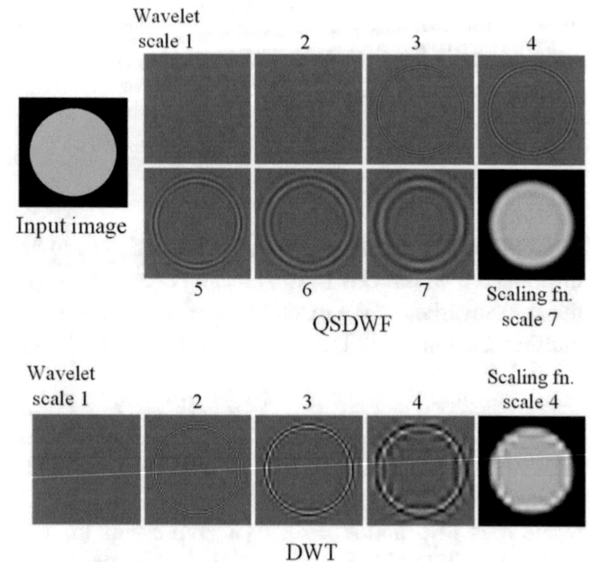

Fig. 13.7 Comparison of the subband projections for QSDWF and DWT

Table 13.1 Redundancy and computational complexities for various wavelet transforms

	DWT	UDWT	DTCWT	QSDWF
Redundancy ratio	1	$3I+1$	4	$3-2^{1-I}$
Complexity	$O(N)$	$O(N\log N)$	$O(N)$	$O(N)$

much better than those generated by the DWT. In addition, the shift invariance can be improved to some extent by using higher order filters. In order to illustrate the improvement of shift invariance, we project a disc image onto the wavelet subbands and the scaling function subband formed by the QSDWF and DWT with the DB2 filter bank. Near-perfect circular arcs are generated by the QSDWF, contrasting with the severely distorted arcs generated by the DWT (Fig. 13.7).

13.3.2.3 Redundancy and Computational Complexity

Let N denote the size of the image and I be the number of the decomposition scale. Table 13.1 shows the redundancy and computational complexity of various transforms in two dimensions. For the same image, the decomposition coefficients of the UDWT are $3I$ times more than those of the DWT, and the redundancy of the UDWT increases infinitely with the number of decomposition scale. The QSDWF and the DTCWT have limited redundancy. The QSDWF generates at most three times decomposition coefficients, less than the DTCWT and far less than the UDWT. In terms of computational complexity, the QSDWF is the same as the DWT and the DTCWT, better than UDWT.

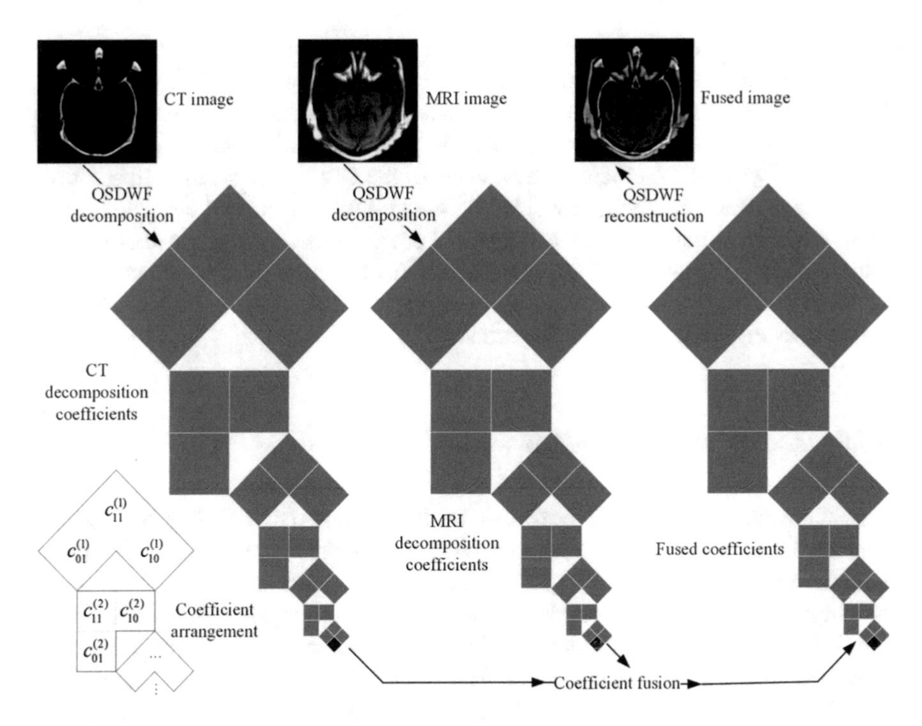

Fig. 13.8 Image fusion scheme based on the QSDWF

13.4 Image Fusion Algorithm

The QSDWF can be easily incorporated into the multiscale fusion scheme (see Fig. 13.1). The QSDWF has near shift invariance, similar as the DTCWT. However, the redundancy of the QSDWF is lower than that of the DTCWT, and the QSDWF is easier to implement than the DTCWT because it employs common filter banks and yields real decomposition coefficients. In addition, the QSDWF has a higher frequency resolution, such that the source images can be fused on finer subbands. Figure 13.8 illustrates the QSDWF fusion algorithm on medical images. The fusion algorithm consists of three steps: decomposition, fusion, and reconstruction, consistent with those in Fig. 13.1. The wavelet coefficients, normally contain edge, texture, and other important information, are fused with a fusion rule. The scale function coefficients at the last scale are usually combined by weighted average. In this study, the simplest Choose-Max (CM) rule [23] is employed for wavelet coefficients. Let $c_{A/B}$ be the wavelet coefficients of the source images A/B, and c_F be the fused coefficient. The CM rule can be written as:

$$c_F(m,n) = \begin{cases} c_A(m,n)\ , & \text{if } |c_A(m,n)| > |c_B(m,n)| \\ c_B(m,n)\ , & \text{else.} \end{cases} \quad (13.21)$$

The wavelet coefficients correspond to the detail information, and their absolute values are the most intuitive measures for the intensity of the detail information. More complex region-based rules [38] such as Windows Based Verification (WBV) rules [16] can also be used, however, the complex rules will increase computational burden and weaken the efficiency advantage of the QSDWF algorithm. In addition, the decomposition coefficients yielded by the QSDWF have good consistency due to good shift invariance, such that the useful coefficients can be selected efficiently with a simple fusion rule. Our experiments (in the next section) show that the performance improvement caused by the complex fusion rule is trivial, which in some cases may degrade fusion performance.

13.5 Experimental Results

The static and dynamic performances of the proposed QSDWF algorithm are evaluated with various evaluation indices. The experimental results are compared with the DWT [16], UDWT [23], and DTCWT [10] algorithms. All source images are 8-bit grayscale and are strictly registered. All fusion algorithms are tested with the CM (default) and WBV rules. The DWT, UDWT, and DTCWT adopt the same 4-level decomposition, and the QSDWF adopts an equivalent 7-level decomposition due to intermediate scales. The DTCWT employs the 14-tap Q-shift filter bank [13], the other wavelets use the DB2 filters.

13.5.1 RMSE Analysis Based on Simulated Images

Since the ideal fused images are unknown, it is difficult to quantitatively evaluate fusion algorithms. The Root Mean Square Error (RMSE) [39] based on the simulated image is commonly used for quantitative evaluation, which includes three steps: (1) An image L is selected as the ideal image; (2) Different regions in the ideal image are blurred with a Gaussian kernel to artificially generate two (or more) source images which is out-of-focus; (3) The source images are fused, and the RMSE between the fused image F and the ideal image L is computed

$$\mathrm{RMSE} = \sqrt{\frac{1}{MN}\sum_{m=1}^{M}\sum_{n=1}^{N}[F(m,n) - L(m,n)]^2} \tag{13.22}$$

where M and N denote the size of the image.

Figure 13.9a shows the ideal image, Fig. 13.9b and c are the out-of-focus images generated by blurring the left and right halves of the ideal image, respectively, and Fig. 13.9d–g shows the difference images between the ideal image and the fused images yielded by various algorithms. It is observed from the difference images that there are visible distortions in the DWT fused image. The DTCWT generates

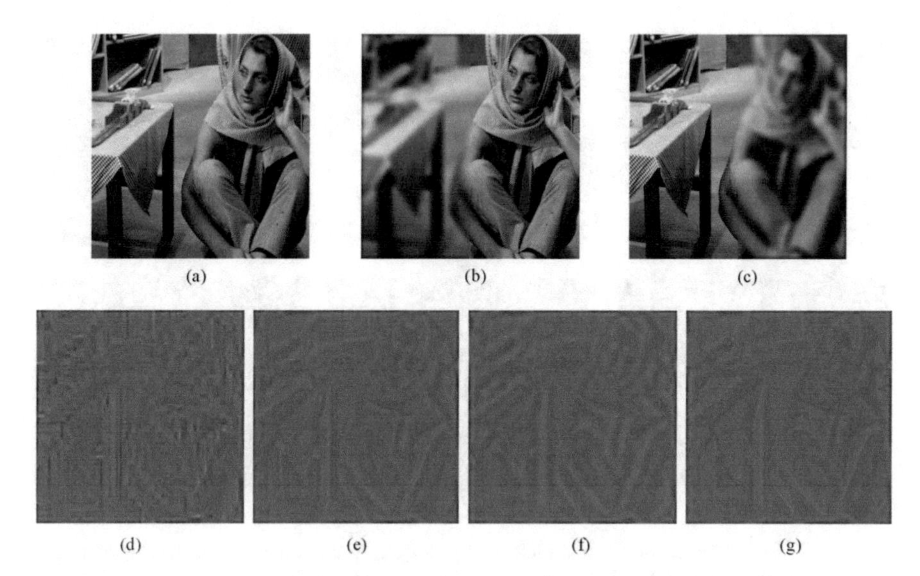

Fig. 13.9 Simulated multi-focus images and difference images for various algorithms. (**a**) Ideal image. (**b**) Left blurred image. (**c**) Right blurred image. (**d**) Diff. image for DWT. (**e**) Diff. image for UDWT. (**f**) Diff. image for DTCWT. (**g**) Diff. image for QSDWT

Table 13.2 RMSE for simulated multi-focus image fusion

Fusion rule	DWT	UDWT	DTCWT	QSDWF
CM	4.297	2.524	2.904	2.407
WBV	3.491	2.623	3.165	2.467

distortions only near the boundary (between focus and out-of-focus regions). The QSDWF yields the best result with very few distortions and the image differences are even.

Table 13.2 compares the RMSE of the four algorithms. The QSDWF with both fusion rules achieves the best results. Although the QSDWF is not as good as the UDWT in terms of shift invariance, the QSDWF achieves better results than the UDWT due to the higher frequency-domain resolution of the QSDWF. In addition, the WBV rule can significantly improve the fusion performance of the DWT, however, it is invalid to the three (nearly) shift invariant algorithms.

13.5.2 Real Image Qualitative and Quantitative Evaluation

Three real image datasets are used for qualitative and quantitative evaluation. Figure 13.10 shows the fused results of the real multi-focus image dataset, where the source images A and B focus on the car and book, respectively. Figure 13.11 shows the fused results of the medical image dataset. Figure 13.12 shows the fusion results of the remote-sensing image set, where the source images A and B are the infrared and low-light images, respectively.

Fig. 13.10 Fusion results on the real multi-focus images. (**a**) Source image A. (**b**) Source image B. (**c**) DWT fused. (**d**) UDWT fused. (**e**) DTCWT fused. (**f**) QSDWF fused

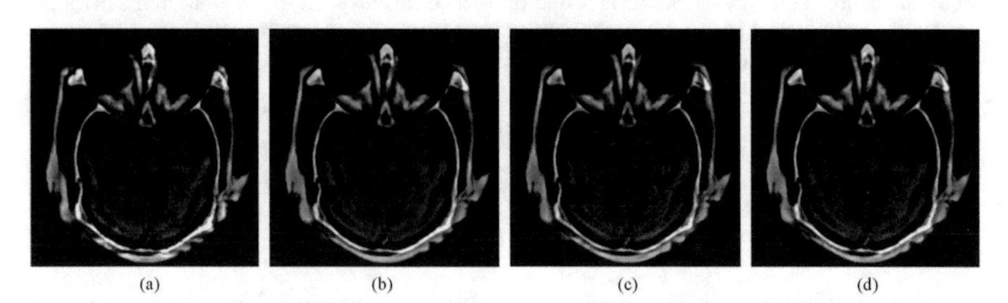

Fig. 13.11 Fused images of the medical CT and MRI images. (**a**) DWT fused. (**b**) UDWT fused. (**c**) DTCWT fused. (**d**) QSDWT fused

13.5.2.1 Subjective Qualitative Evaluation

The three (nearly) shift-invariant fusion algorithms perform better than the DWT algorithm on all three datasets. For the multi-focus images, the DTCWT generates slight distortions at the boundary of focus and out-of-focus, while the QSDWF and the UDWT perform well in the whole image. For the medical images, the DWT yields obvious ringing distortions near the curved edges, while the three (nearly) shift-invariant algorithms effectively eliminate these distortions. The QSDWF and the UDWT perform slightly better than the DTCWT. In fact, it is difficult to distinguish the differences between the QSDWF and UDWT fused images just through visual inspection. The results on the remote sensing dataset also confirm the above analysis.

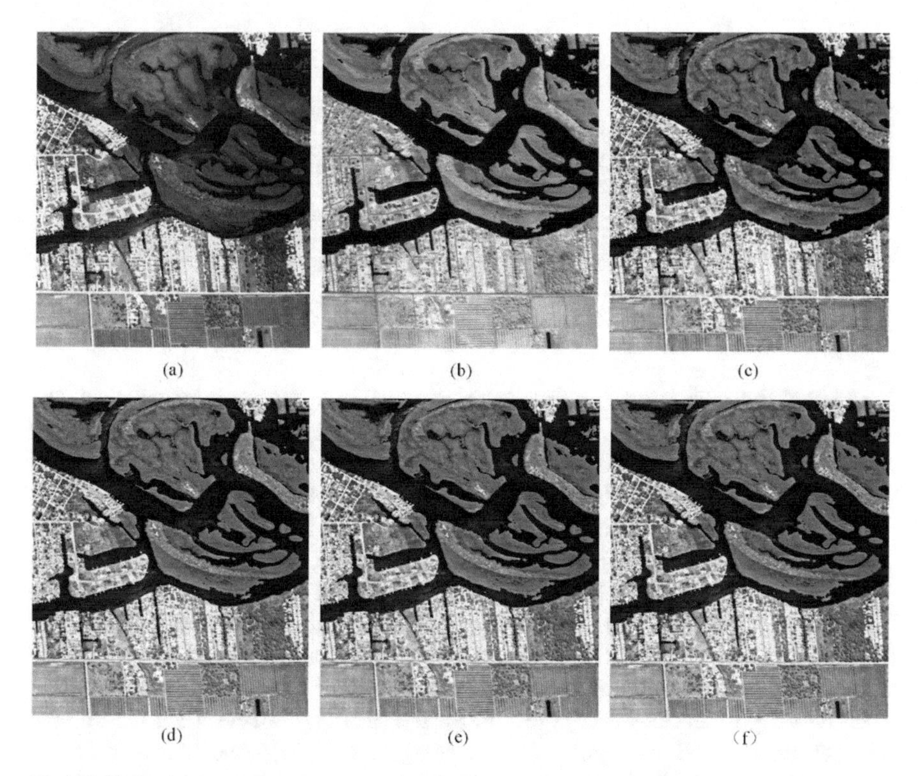

Fig. 13.12 Fusion results on the remote sensing images. (**a**) Source image A. (**b**) Source image B. (**c**) DWT fused. (**d**) UDWT fused. (**e**) DTCWT fused. (**f**) QSDWF fused

13.5.2.2 Objective Quantitative Evaluation

Quantitative evaluation on real image sets is difficult due to unknown ideal fused images. There is currently no widely accepted evaluation method. In this study, two complementary indices: Mutual Information (MI) [21] and Edge Preservation (EP) [33] are used for quantitative evaluation.

The MI is a concept in information theory, which is used to measure the statistical correlation between two random variables [31]. The MI index in the image fusion field can be used to measure the similarity between the joint random variable of the source images (A, B) and the random variable of the fused image F:

$$\mathrm{MI}_{AB/F} = P(A, B) + P(F) - P(A, B, F) \tag{13.23}$$

where $P(*)$ denotes the (joint) information entropy and can be calculated from the normalized (joint) histogram. A larger MI means higher pixel similarity between the source images and the fused image and thus better fusion performance.

Table 13.3 MI and EP evaluations on real image datasets

	Fusion	MI				EP			
Datasets	Rule	DWT	UDWT	DTCWT	QSDWF	DWT	UDWT	DTCWT	QSDWF
Multi-focus	CM	4.562	5.220	5.046	5.186	0.636	0.683	0.682	0.683
	WBV	4.831	5.278	5.030	5.251	0.666	0.689	0.682	0.692
Medical	CM	1.858	2.179	1.923	2.128	0.600	0.711	0.666	0.700
	WBV	1.701	2.161	1.755	2.058	0.456	0.710	0.533	0.640
Remote sensing	CM	2.880	3.152	3.101	3.111	0.537	0.600	0.584	0.591
	WBV	2.840	3.148	3.098	3.097	0.520	0.590	0.575	0.582

The MI indicates the ability of fusion algorithm to transfer pixel information. The EP index, which evaluates the ability of transferring edge information, is also used in our study. The EP method extracts edges from the source images and the fused image, and then calculates the edge preservation. The weighted edge preservation is used for quantitative evaluation (more details are given in [33]).

Table 13.3 gives the MI and EP results on the three real image datasets. The results are consistent with those obtained by the RMSE analysis and the qualitative analysis: (1) The three (nearly) shift-invariant algorithms are obviously better than the DWT algorithm. The QSDWF and UDWT are very close, slightly better than the DTCWT; (2) For the multi-focus images, the WBV rule is efficient for promoting the DWT algorithm, however is invalid for the (nearly) shift-invariant algorithms; (3) For the medical and remote-sensing images, the WBV rule is invalid for all four algorithms. The locations of important information contained by medical and remote sensing images are usually intertwined between source images, and there are no obvious dividing lines, like in the multi-focus images. Consequently, the adjacent pixels used by the WBV may be unreliable for fusion.

13.5.3 Dynamic Performance Analysis

Fusion systems usually need to process successive image sequences. Calculating the above performance indices frame-by-frame is essentially an intra-frame (static) evaluation, which is not enough for evaluating image sequence fusion. The fusion algorithm without shift invariance may introduce inconsistent information in inter-frame, resulting in flicker and jitter in the fused image sequences. The human visual system is particularly sensitive to these dynamic disturbances when the fused sequences are continually rendered. Therefore, inter-frame (dynamic) performance is very important for a practical fusion algorithm.

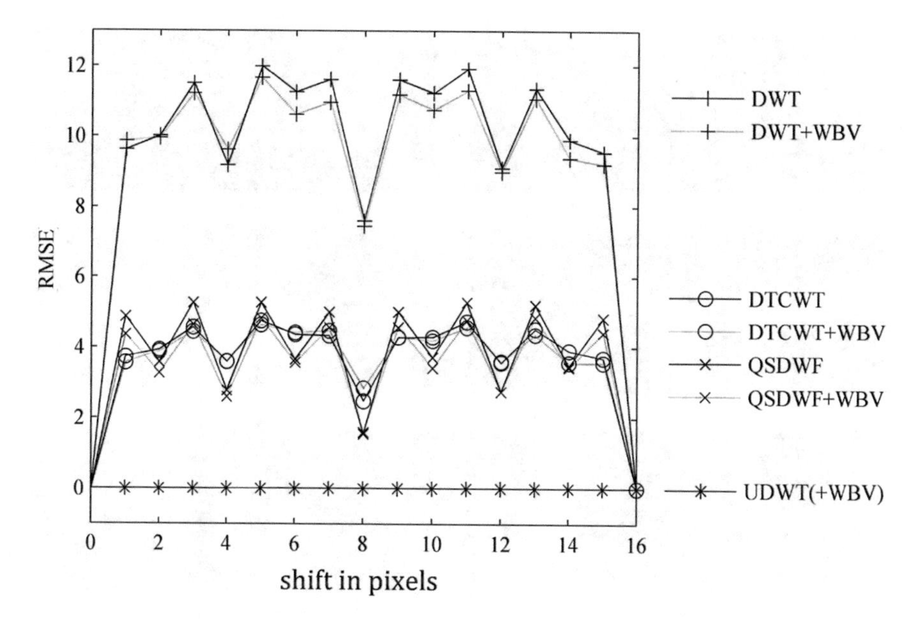

Fig. 13.13 Shift errors (RMSE) for various algorithms

13.5.3.1 Shift Error Analysis on Simulated Image Sequences

In [23], artificial-generated image sequences are used to evaluate the dynamic performance. First, two static source images are fused to obtain a reference image. Second, the source images are synchronously shifted and fused again. The fused image is finally shifted back to the original position and compared with the reference image by calculating the RMSE of the overlapped region.

The remote sensing images shown in Fig. 13.12 are used for the shifting experiment. Figure 13.13 shows the RMSE curves for the four algorithms with the CM (default) and WBV rules. The UDWT has absolute shift invariance with the shift error of zero. The shift errors of the QSDWF and the DTCWT are small and clearly better than those of the DWT. In addition, the WBV rule slightly improves the shift invariance of the DWT, however, has little effect on the DTCWT and the QSDWF.

13.5.3.2 True Sequence Qualitative and Quantitative Evaluation

Real visible and infrared image sequences are also used for dynamic performance evaluation, as shown in Fig. 13.14. Each sequence includes 200 frames. Three (nearly) shift-invariant algorithms perform better than the DWT algorithm in terms

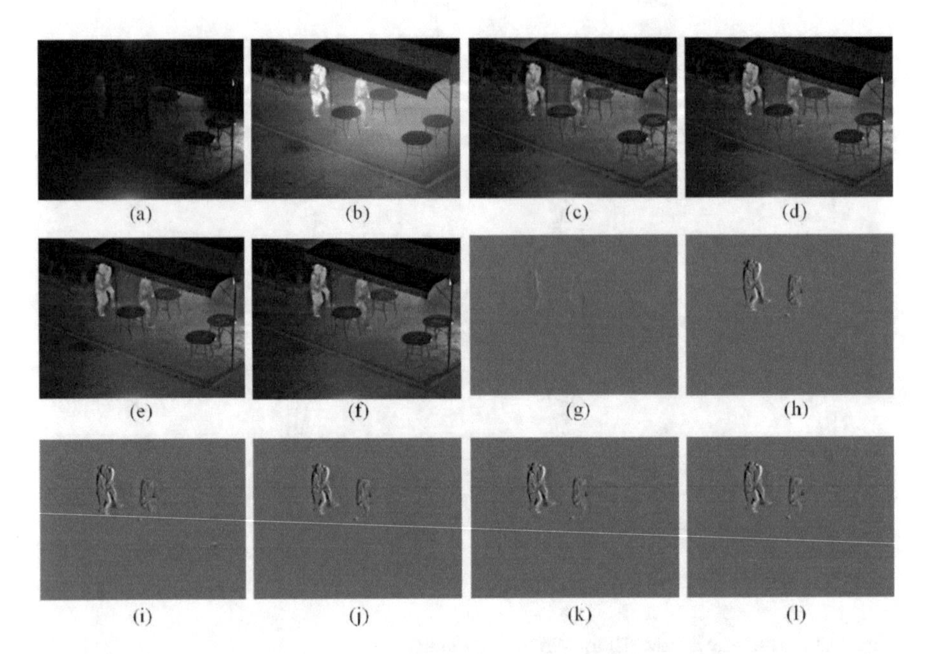

Fig. 13.14 Fusion results of the visible light (VL) and infrared (IR) image sequences. (**a**) VL frame. (**b**) IR frame. (**c**) DWT fused. (**d**) UDWT fused. (**e**) DTCWT fused. (**f**) QSDWF fused. (**g**) IFD for VL. (**h**) IFD for IR. (**i**) IFD for DWT. (**j**) IFD for UDWT. (**k**) IFD for DTWCT. (**l**) IFD for QSDWF

Table 13.4 Average IFD-MI and speed for the visible light and infrared image sequence fusion

	MI				EP			
Fusion rule	DWT	UDWT	DTCWT	QSDWF	DWT	UDWT	DTCWT	QSDWF
CM	1.358	1.839	1.624	1.723	11	0.8	3.0	5.8
WBV	1.333	1.808	1.504	1.692	4.9	0.5	2.4	3.0

of stability and consistency. To illustrate this, the Inter-Frame Difference (IFD) images of each fused sequence are calculated, as shown in Fig. 13.14g–l. There are visible distortions in the IFD of the DWT fused sequence, which cannot be explained by the corresponding IFD of the source sequences. The three (nearly) shift-invariant algorithms effectively eliminate these inter-frame distortions.

In [23], the MI of IFD (IFD-MI) is used to evaluate dynamic performance of real image sequences, which can be calculated by substituting the corresponding inter-frame difference images into (13.23). The left-half of Table 13.4 gives the average IFD-MI over 199-frame visible and infrared difference image sequences. The QSDWF algorithm performs slightly better than the DTCWT, slightly poorer than the UDWT. The three (nearly) shift invariant algorithms are obviously better than the DWT. The WBV rule is useless to improve the inter-frame performance.

13.6 Conclusion

A new image fusion algorithm based on the quincunx-sampled discrete wavelet frame is proposed. The replaceability of sampling matrix in multidimensional PRFB is demonstrated. The replaced filter bank keeps perfect reconstruction if the new sampling matrix exactly left divides the original matrix, and the iterated filters form a tight (or dual) frame in the $\ell_2(\mathbb{Z}^n)$ space. The rectangular sampling matrix of the standard two-dimensional discrete wavelet transform is replaced with the quincunx sampling matrix, which builds a discrete wavelet frame (QSDWF) of higher frequency-domain resolution, lower redundancy, and near shift invariance. The new built wavelet frame is incorporated into the multiscale image fusion scheme. The static and dynamic performances of the new fusion algorithm is evaluated on simulated and real image datasets. The experimental results show that the proposed fusion algorithm performs better than the standard wavelet algorithm. It yields high-quality fused images, close to that yielded by the best undecimated discrete wavelet algorithm. However, the proposed fusion algorithm is much more efficient than the undecimated algorithm, it is also faster than another well-established nearly shift-invariant algorithm based on the double tree complex wavelet transform.

References

1. Abidi MA, Gonzalez RC (1992) Data fusion in robotics and machine intelligence. Academic, Cambridge
2. Broussard RP, Rogers SK, Oxley ME, Tarr GL (1999) Physiologically motivated image fusion for object detection using a pulse coupled neural network. IEEE Trans Neural Netw 10(3): 554–563
3. Burt PJ, Adelson EH (1985) Merging images through pattern decomposition. Appl Digit Image Process VIII 575:173–181
4. Burt PJ, Kolczynski RJ (1993) Enhanced image capture through fusion. In: Fourth international conference on computer vision. IEEE, Piscataway, pp 173–182
5. Cassels JWS (2012) An introduction to the geometry of numbers. Springer, Berlin
6. Chavez P, Sides SC, Anderson JA, et al (1991) Comparison of three different methods to merge multiresolution and multispectral data-landsat TM and SPOT panchromatic. Photogramm Eng Remote Sens 57(3):295–303
7. Chipman LJ, Orr TM, Graham LN (1995) Wavelets and image fusion. In: International conference on image processing, vol 3. IEEE, Piscataway, pp 248–251
8. Choi M, Kim RY, Nam MR, Kim HO (2005) Fusion of multispectral and panchromatic satellite images using the curvelet transform. IEEE Geosci Remote Sens Lett 2(2):136–140
9. Daubechies I (1992) Ten lectures on wavelets. SIAM, Philadelphia
10. Hill PR, Canagarajah CN, Bull DR (2002) Image fusion using complex wavelets. In: British machine vision conference, pp 1–10
11. Jiang X, Zhou L, Gao Z (1996) Multispectral image fusion using wavelet transform. In: Photonics China'96. International society for optics and photonics, pp 35–42
12. Joseph V. Hajnal DLH (2001) Medical image registration, vol 392. CRC Press/Taylor & Francis Group, Boca Raton/Oxford

13. Kingsbury N (2001) Complex wavelets for shift invariant analysis and filtering of signals. Appl Comput Harmon Anal 10(3):234–253
14. Kovacevic J, Vetterli M (1992) Nonseparable multidimensional perfect reconstruction filter banks and wavelet bases for r/sup n. IEEE Trans Inf Theory 38(2):533–555
15. Kovacevic J, Vetterli M (1993) FCO sampling of digital video using perfect reconstruction filter banks. IEEE Trans Image Process 2(1):118–122
16. Li H, Manjunath B, Mitra SK (1994) Multi-sensor image fusion using the wavelet transform. In: IEEE international conference image processing, vol 1. IEEE, Piscataway, pp 51–55
17. Li J, Pan Q, Yang T, Mei Cheng Y (2004) Color based grayscale-fused image enhancement algorithm for video surveillance. In: Third international conference on image and graphics (ICIG). IEEE, Piscataway, pp 47–50
18. Mallat SG (1989) A theory for multiresolution signal decomposition: the wavelet representation. IEEE Trans Pattern Anal Machine Intel 11(7):674–693
19. Pajares G, De La Cruz JM (2004) A wavelet-based image fusion tutorial. Pattern Recogn Lett 37(9):1855–1872
20. Qu G, Zhang D, Yan P (2001) Medical image fusion by wavelet transform modulus maxima. Opt Express 9(4):184–190
21. Qu G, Zhang D, Yan P (2002) Information measure for performance of image fusion. Electron Lett 38(7):313–315
22. Rioul O (1993) A discrete-time multiresolution theory. IEEE Trans Signal Process 41(8): 2591–2606
23. Rockinger O (1997) Image sequence fusion using a shift-invariant wavelet transform. In: International conference on image processing, vol 3. IEEE, Piscataway, pp 288–291
24. Selesnick IW (2006) A higher density discrete wavelet transform. IEEE Trans Signal Process 54(8):3039–3048
25. Simoncelli EP, Freeman WT, Adelson EH, Heeger DJ (1992) Shiftable multiscale transforms. IEEE Trans Inf Theory 38(2):587–607
26. Toet A (1989) Image fusion by a ratio of low-pass pyramid. Pattern Recogn Lett 9(4):245–253
27. Toet A (1989) A morphological pyramidal image decomposition. Pattern Recogn Lett 9(4):255–261
28. Vaidyanathan P (1987) Perfect reconstruction QMF banks for two-dimensional applications. IEEE Trans Circuits Syst 34(8):976–978
29. Vetterli M (1984) Multi-dimensional sub-band coding: some theory and algorithms. Signal Process 6(2):97–112
30. Vetterli M, Herley C (1992) Wavelets and filter banks: theory and design. IEEE Trans Signal Process 40(9):2207–2232
31. Viola P, Wells WM (1997) Alignment by maximization of mutual information. Int J Comput Vis 24(2):137–154
32. Viscito E, Allebach JP (1991) The analysis and design of multidimensional FIR perfect reconstruction filter banks for arbitrary sampling lattices. IEEE Trans Circuits Syst 38(1): 29–41
33. Xydeas C, Petrovic V (2000) Objective image fusion performance measure. Electron Lett 36(4):308–309
34. Yang B (2008) Research on wavelet-based pixel-level image fusion algorithms. Ph.D. thesis, Shanghai Jiao Tong University
35. Yang B, Jing Z (2007) Replacing resamplers in filter banks without losing perfect reconstruction. Electron Lett 43(16):864–865
36. Yang B, Jing Z (2007) A simple method to build oversampled filter banks and tight frames. IEEE Trans Image Process 16(11):2682–2687

37. Yonghong J (1998) Fusion of landsat TM and SAR image based on principal component analysis. Remote Sens Technol Appl 13(1):46–49
38. Zhang Z, Blum RS (1997) Region-based image fusion scheme for concealed weapon detection. In: Proceedings of the 31st annual conference on information sciences and systems, pp 168–173
39. Zhang Z, Blum RS (1999) A categorization of multiscale-decomposition-based image fusion schemes with a performance study for a digital camera application. Proc IEEE 87(8): 1315–1326

Chapter 14
Multi-Focus Image Fusion Using Pulse Coupled Neural Network

14.1 Introduction

It's nearly impossible for charge-coupled devices (CCD), such as camera and light optical microscope, to take an image with all objects in focus when there are multiple objects in the situation. This is because the depth-of-focus of optical lenses is limited. Generally, most objects in one image are blurred except for those that are focused. One method to get an image in which all objects are clear is the multi-focus image fusion technology. Several images are taken in different focal setup from the same viewpoint and then fused into one desired image by the fusion algorithm. Multi-focus image fusion is an important pre-processing tool. The resulting clear image can contribute humans or machines to better perception and further process.

Currently, there are two representative types of algorithms in multi-focus image fusion. They are the methods based on multiscale decomposition (MSD) and the ones based on image blocks. MSD methods have been successfully applied in image fusion field [1–3], while they have their limitations. For example, DWT algorithm is not shift-invariant due to the down-sampling procedure. If the source images are not registered ideally, the algorithm would lose its efficacy. Discrete wavelet frame transform (DWFT) was proposed to alleviate this problem [4]. However, DWFT is very complicated so its computation efficiency may be low. Besides, most MSD methods fuse the images in the transformed domain. As a result, frequently, the pixel information of the source images cannot be preserved very well.

Another kind of multi-focus image fusion method is based on image blocks. Li et al. [5] used this concept to design an effective algorithm. The idea is to construct the resulting image by jointing the clear image blocks from source images. Each source image is divided into several image blocks in the same way and the clarity of each block is measured by a certain evaluating method. Using clearer blocks can compose a fused image with better characteristic of clarity. A large number of

Z. Jing et al., *Non-Cooperative Target Tracking, Fusion and Control*,
Information Fusion and Data Science, https://doi.org/10.1007/978-3-319-90716-1_14

evaluating methods are proposed to measure the clarity of blocks. Li et al. [5] used spatial frequency (SF) as the measure to choose the blocks. The algorithm can solve the shift-variant problem, which may be difficult for the MSD methods.

Recently, pulse coupled neural network (PCNN) model was used to deal with image fusion problem [6, 8–10]. Miao and Wang [10] improved PCNN and applied it in multi-focus image fusion. The source images are used as the external stimulus of PCNN. The parameter, linking strength β, is calculated by the energy of image gradient. Based on the image blocks concept, the algorithm constructs the fused image by comparing the PCNN outputs.

In this chapter, a multi-focus image fusion algorithm is proposed. The energy of image Laplacian and PCNN is used to measure the clarity of image blocks. By selecting the clearer image blocks, the method can generate a multi-focus image. The presentation of this chapter is based on the work in [7, 20].

The rest of this chapter is organized as follows. The multi-focus image fusion method based on image blocks is briefly introduced in Sect. 14.2. In Sect. 14.3, PCNN model used in our algorithm is described. The scheme of our proposed algorithm is given in Sect. 14.4. The determinations of the parameters are analyzed in Sect. 14.5. Comprehensive experiments are showed in Sect. 14.6. In Sect. 14.7, the chapter is concluded.

14.2 Multi-Focus Image Fusion Method Based on Image Blocks

With respect to the algorithms based on image blocks, clarity evaluation methods are significant and may affect the results greatly. Spatial frequency (SF) is used in the work of Li et al. [5] to measure the clarity of image blocks. Energy of image gradients is another frequently used method [11].

Eskicioglu and Fisher [12] first introduced SF. For an image $I_{K\times L} = f(x, y)$, SF is defined as:

$$\mathrm{SF} = \sqrt{(\mathrm{RF})^2 + (\mathrm{CF})^2}, \tag{14.1}$$

where RF and CF are the row frequency and column frequency, respectively,

$$\mathrm{RF} = \sqrt{\frac{1}{K \times L}\sum_{x=1}^{K}\sum_{y=2}^{L}[f(x, y) - f(x, y-1)]^2}, \tag{14.2}$$

$$\mathrm{CF} = \sqrt{\frac{1}{K \times L}\sum_{y=1}^{L}\sum_{x=2}^{K}[f(x, y) - f(x-1, y)]^2}. \tag{14.3}$$

The definition of energy of image gradient (EOG) is:

$$\mathrm{EOG} = \sum_x \sum_y (f_x^2 + f_y^2), \tag{14.4}$$

where

$$f_x = f(x+1, y) - f(x, y), \tag{14.5}$$

$$f_y = f(x, y+1) - f(x, y). \tag{14.6}$$

Comparing the definition of SF and EOG from Eqs. (14.1)–(14.6), we can find the SF is a modified version of EOG.

The clarity evaluation methods, can also be called focus measures, are studied extensively in the autofocusing field. In the literature [13, 14], the focus measures are proposed. The measures have a maximum value for the best focused image and they will decrease when the extent of defocus increases. In this chapter, the energy of image Laplacian (EOL) is used to evaluate image clarity. As shown in the previous literature [15, 16], EOL has a better performance than SF and EOG in the general case.

EOL is defined by:

$$EOL = \sum_x \sum_y (f_{xx} + f_{yy})^2 \tag{14.7}$$

where

$$\begin{aligned} f_{xx} + f_{yy} = &-f(x-1, y-1) - 4f(x-1, y) - f(x-1, y+1) \\ &-4f(x, y-1) + 20f(x, y) \\ &-4f(x, y+1) - f(x+1, y-1) \\ &-4f(x+1, y) - f(x+1, y+1) \end{aligned} \quad . \tag{14.8}$$

The schematic diagram of the methods based on image blocks is displayed in Fig. 14.1. The clarity evaluation including SF, EOG, and EOL is calculated by the pixel values of the source images. Procedures of the algorithm consist of the following steps:

1. Source images are divided into blocks. To simplify the problem, the number of source images is set as two. Denote the image by A and B. Accordingly, the image blocks are denoted by A_i and B_i.
2. For each block, its clarity is computed by the clarity evaluation. The results of A_i and B_i are denoted by M_i^A and M_i^B, respectively.
3. Compare the clarity of two corresponding blocks A_i and B_i. The fused image C can be constructed by the blocks with larger value of clarity, i.e.,

$$C_i = \begin{cases} A_i, & M_i^A > M_i^B \\ B_i, & \text{otherwise} \end{cases} \tag{14.9}$$

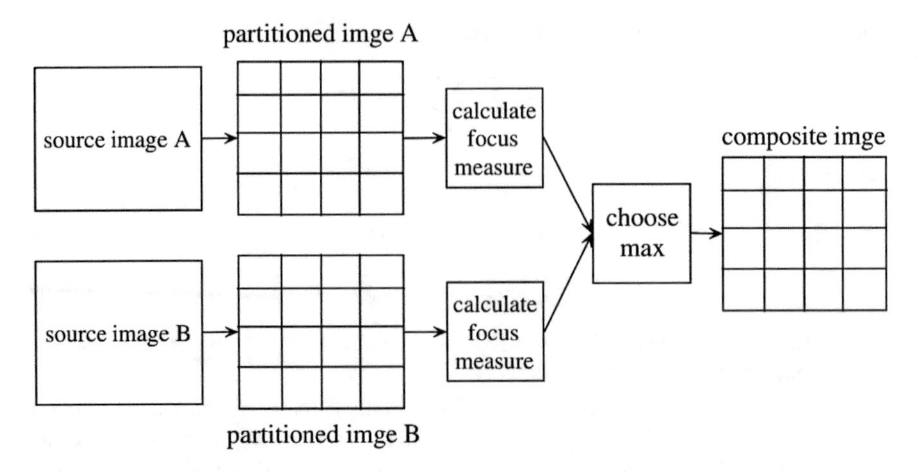

Fig. 14.1 Schematic diagram of the methods based on image blocks

Table 14.1 Evaluation of different focus measures with different block sizes by RMSE ("clock")

	Block size						
Focus measures	4 × 4	4 × 8	8 × 8	8 × 16	16 × 16	16 × 32	32 × 32
SF	2.4506	1.7742	1.5248	1.4121	0.8330	0.7278	0.6506
EOG	2.4505	1.7742	1.5248	1.4121	0.8330	0.7278	0.6506
EOL	1.6318	1.3035	1.0983	0.9208	0.7758	0.7913	0.7084

4. Compute root mean square error (RMSE) for the fused image with the ground truth image to evaluate the resulting image.

RMSE is frequently used to evaluate the quality of the processed image. It is defined as:

$$\text{RMSE} = \sqrt{\frac{\sum_x \sum_y [R(x, y) - F(x, y)]^2}{M \times N}} \tag{14.10}$$

where R and F represent the reference image (ground truth) and fused image respectively.

Table 14.1 shows some experiment results of the multi-focus fusion methods based on image blocks. The experiments are conducted by processing images in Fig. 14.4b and c. Different block sizes and different clarity evaluation methods are set in the experiments. The effect of this type of methods can be briefly summarized in Table 14.1. It is shown that EOL can provide better performance than SF and EOG.

14.3 PCNN Model

PCNN is proposed by the observation of cat's primary visual cortex [17]. A neuron of the network consists of three parts: the receptive fields, the modulation product, and the pulse generator. Figure 14.2 shows the structure of PCNN neuron model. There are two kinds of channel in the receptive field. One is F channel, which receives the feeding input F_{ij}. The other is L channel, which accepts the linking input L_{ij}. The external stimulus signals are input into neuron N_{ij} by F channel and the outputs from adjacent neurons are input by L channel. A PCNN neuron can be described by the following equations:

$$F_{ij}(n) = \exp(-1/\alpha_F)F_{ij}(n-1) + S_{ij} + V_F \sum_{kl} M_{ijkl}Y_{kl}(n-1) \tag{14.11}$$

$$L_{ij}(n) = \exp(-1/\alpha_L)L_{ij}(n-1) + V_L \sum_{kl} W_{ijkl}Y_{kl}(n-1) \tag{14.12}$$

$$U_{ij}(n) = F_{ij}(n)[1 + \beta L_{ij}(n)] \tag{14.13}$$

$$\theta_{ij} = \begin{cases} \exp(-1/\alpha_T)\theta_{ij}(n-1), & \text{if } Y_{ij}(n-1) = 0 \\ V_T, & \text{otherwise} \end{cases} \tag{14.14}$$

$$Y_{ij}(n) = \begin{cases} 1, & \text{if } U_{ij}(n) > \theta_{ij}(n) \\ 0, & \text{otherwise} \end{cases} \tag{14.15}$$

U_{ij}, called internal activity, is generated by F_{ij} and L_{ij} together. Compare U_{ij} with dynamic neuronmime threshold θ_{ij} and then output Y_{ij} could be obtained. If U_{ij} is big enough, the neuron will produce a pulse signal.

α_F, α_L, and α_T are the decay constants of the PCNN neuron. V_F, V_L, and V_T are the magnitude scaling terms. M and W are the synaptic weight in feeding field and linking field, respectively. Constant β represents the linking strength between neurons.

PCNN can be used in image processing. The structure of PCNN is single layer. It is similar with a 2D matrix. A neuron is connected with its surrounding neurons. The structures and functions of neurons in the network are identical. The number of neurons is the same as the number of pixels in one source image. There is a one-to-one correspondence between the neurons and pixels.

For a neuron N_{ij}, the pixel value S_{ij} is chosen as the input of F channel, i.e.,

$$F_{ij}(n) = S_{ij} \tag{14.16}$$

In the common case, the output of a PCNN neuron can be computed by Eq. (14.15). But we rectify the equation as Eq. (14.17),

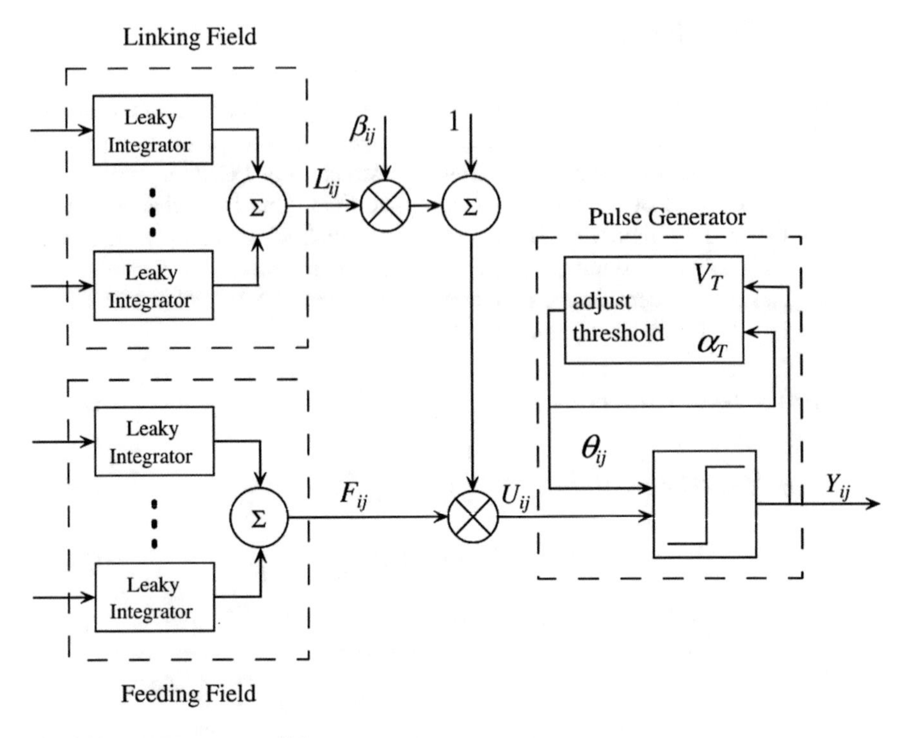

Fig. 14.2 PCNN neuron model

$$Y_{ij}(n) = \begin{cases} U_{ij}(n), & \text{if } U_{ij}(n) > \theta_{ij}(n) \\ 0, & \text{otherwise} \end{cases} \tag{14.17}$$

We denote the whole output of the PCNN as O. It is a 2D matrix and also can be seen as an image $[O_{ij}]_{P \times Q}$. The element O_{ij} equals the sum of the outputs of neurons N_{ij}, i.e.,

$$O_{ij} = \sum_n Y_{ij}(n) \tag{14.18}$$

The PCNN used in our algorithm can be described as follows. Some annotations are listed first.

The external stimulus (feeding input) F is a matrix with dimensions $P \times Q$, which can be obtained by Eq. (14.16). The elements of F equal the corresponding values of the input image. The linking input L is also a matrix with dimensions $P \times Q$. It can be computed by:

$$L = \exp(-1/\alpha_L) * L + V_L * work, \tag{14.19}$$

where $work = (Y\text{conv2}K)$. Y is an output matrix, which records all the outputs of neurons. K is a kernel matrix, whose size is $(2r+1) \times (2r+1)$. Θ is a threshold matrix. U is a matrix which records every internal activity. O is the output matrix of PCNN. $F, L, work, \Theta, U$ and O are matrices with the same size. ".*" indicates array multiplication and "conv2" represents 2D convolution operation.

Then the pseudo code is shown as follows:

1. $F = S, L = U = Y = 0, \Theta = 1$.
   K : The center element of K is one. The other elements are $1/d$, where d is the distance from the center $d \leq r$.
2. Select the number of iterations np. Initial n, $n = 1$.
3. $work = (Y\text{conv2}K)$;
   $L = \exp(-1/\alpha_L) * L + V_L * work$;
   If $Y_{ij} = 0$, then $\Theta_{ij} = \exp(-1/\alpha_T) * \Theta_{ij}$, else $\Theta_{ij} = V_T$;
   $U = F. * (1 + \beta * L)$;
   If $U_{ij} > \Theta_{ij}$, then $Y_{ij} = U_{ij}$, else $Y_{ij} = 0$, $(1 \leq i \leq P, 1 \leq j \leq Q)$;
   $O = O + Y$
   $n = n + 1$
4. If $n > np$, go to step 5, else go back to step 3.
5. output O as the result value of PCNN.

Detailed implementation of PCNN can refer to the literature [18].

14.4 Proposed Multi-Focus Image Fusion Method

Based on the aforementioned PCNN model, a method for multi-focus image fusion is proposed. It has the following steps:

1. Divide the source images A and B into $P \times Q$ image blocks, respectively. Denote the image blocks in A as A_{ij}, and the image blocks in B as B_{ij}. The index i, j represents the position in the image. The size of each image block is chosen as 8×8.
2. Calculate the clarity of each image block A_{ij} in A by EOL. Then generate the feature map S^A. The size of S^A is $P \times Q$. The element S^A_{ij} equals the clarity of corresponding image block A_{ij}. By the same way, the feature map S^B of source image B can also be generated.
3. Normalize S^A and S^B into region [0, 1]. Input the normalized feature maps into PCNN model as the external stimulus. S^A and S^B are operated independently. The output from S^A is denoted as O^A and the output from S^B is denoted as O^B.
4. Compare the output of two PCNN O^A and O^B. The results will generate the image blocks C_{ij} in fused image. C_{ij} can be calculated by:

$$C_{ij} = \begin{cases} A_{ij}, & \text{if } O^A_{ij} > O^B_{ij} \\ B_{ij}, & \text{otherwise} \end{cases} \tag{14.20}$$

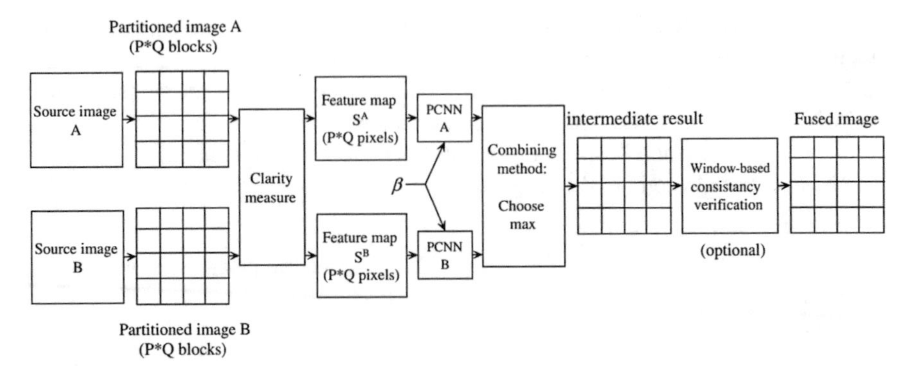

Fig. 14.3 The schematic diagram of our proposed algorithm

5. Use consistency verification (3×3 window) to fine-tune the result. For example, in the fused image, if a block comes from A while most blocks (>50%) surrounding it come from B, then the block should be reset by choosing the one from B.

Figure 14.3 shows the schematic diagram of our proposed algorithm. It can be noted that the PCNNs don't need training. It is assumed that the source images should be registered well.

14.5 Parameters for PCNN

Several parameters in PCNN need to be determined in a certain application. In the problem of multi-focus image fusion, the parameters are:

1. PCNN iteration times: $np = 300$;
2. Decay term for linking: $\alpha_L = 1.0$;
3. Decay term for threshold: $\alpha_T = 5.0$;
4. Magnitude scaling term for linking: $V_L = 0.2$;
5. Magnitude scaling term for threshold: $V_T = 20.0$
6. Radius of linking field: $r = 13.0$.

The linking strength β is a significant parameter in PCNN model. βs in two networks $PCNN^A$ and $PCNN^B$ are set as the same value. In the general case, β is chosen by 0.2. But in our method, we propose an algorithm to determine the value of β.

Consider an independent PCNN neuron N_{ij}, which is not connected by other neurons. N_{ij} has a stable external stimulus input S_{ij}. According to the concept of PCNN, when the N_{ij} has no linking input (i.e., $\beta = 0$), the neuron will generate the pulse periodically. The pulsing period $T(S_{ij})$, which is also called natural period of neuron, can be obtained by [19]:

$$T(S_{ij}) = \alpha_T \ln\left(\frac{V_T}{S_{ij}}\right) \tag{14.21}$$

With regard to the proposed algorithm, we assume $\beta = 0$. The clarity of image block (EOL) is used to be the external stimulus, Thus we can get the natural period of N_{ij}^A: $T(S_{ij}^A) = \alpha_T \ln(\frac{V_T}{S_{ij}^A})$. Similarly, $T(S_{ij}^B)$, the natural period of N_{ij}^B, equals $\alpha_T \ln(\frac{V_T}{S_{ij}^B})$. Under the setting of iteration times np, the number of pulses from N_{ij}^A is $np/T(S_{ij}^A) = np/[\alpha_T \ln(\frac{V_T}{S_{ij}^A})]$, where $[\cdot]$ is the ceiling operator. Because $\beta = 0$, $U_{ij} = S_{ij}$ and then:

$$Y_{ij}(n) = \begin{cases} S_{ij}, & \text{if } S_{ij} > \theta_{ij}(n) \\ 0, & \text{otherwise} \end{cases} \tag{14.22}$$

According to (14.18), the output O_{ij} can be obtained by:

$$O_{ij}^A = S_{ij}^A \times np/T(S_{ij}^A) = S_{ij}^A \times np/\left[\alpha_T \ln\left(\frac{V_T}{S_{ij}^A}\right)\right] \tag{14.23}$$

$$O_{ij}^B = S_{ij}^B \times np/T(S_{ij}^B) = S_{ij}^B \times np/\left[\alpha_T \ln\left(\frac{V_T}{S_{ij}^B}\right)\right] \tag{14.24}$$

Generally, assume $S_{ij}^A > S_{ij}^B$, then we have:

$$\alpha_T \ln\left(\frac{V_T}{S_{ij}^A}\right) < \alpha_T \ln\left(\frac{V_T}{S_{ij}^A}\right); \; O_{ij}^A < O_{ij}^B.$$

Therefore, when $\beta = 0$, Eq. (14.20) becomes:

$$C_{ij} = \begin{cases} A_{ij}, & \text{if } S_{ij}^A > S_{ij}^B \\ B_{ij}, & \text{otherwise} \end{cases} \tag{14.25}$$

It can be obviously concluded that when $\beta = 0$, the proposed method is identical to Li's method [5].

We conduct a group of experiments to find the influence of β. Figure 14.4 shows the experimental images. Figure 14.4a is the ground truth image. Figure 14.4b and c are the source images. The goal of multi-focus image fusion is to construct a clear image by using the source images. Choosing different value of β, we can get

Fig. 14.4 The experimental images "clock." (**a**) Reference image (ground truth). (**b**) Source images A. (**c**) Source images B

different resulting image. Figure 14.5 shows the decision mask of this example when $\beta = 0$. Figure 14.5a is the ideal decision mask and Fig. 14.5b shows the decision mask obtained by the algorithm. Each block shown in Fig. 14.5 represents an 8×8 image block coming from the source images. The white block represents the block comes from Fig. 14.4b and the black block is from Fig. 14.4c. Comparing Fig. 14.4a and 14.5b, we can find there also exist some mistakes.

Suppose D_{ij} is a white block that is a wrong decision in Fig. 14.5b and $O_{ij}^A > O_{ij}^B$. And assume a neighbor of N_{ij}, which is denoted as N_{pq}, is decided correctly, i.e., D_{pq} is a correct decision. If we can find a suitable β which makes N_{ij}^B be captured by N_{pq}^B, the output O_{ij}^B will increase and then the inequation $O_{ij}^A < O_{ij}^B$ will hold. Thus D_{ij} will become the right decision.

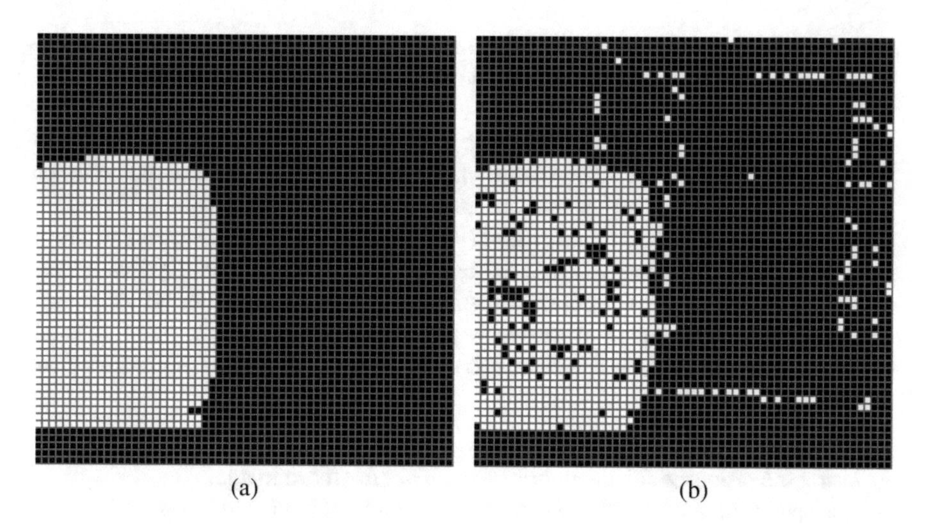

Fig. 14.5 Decision mask of the example "clock." (**a**) Ideal decision mask. (**b**) Proposed method result ($\beta = 0$)

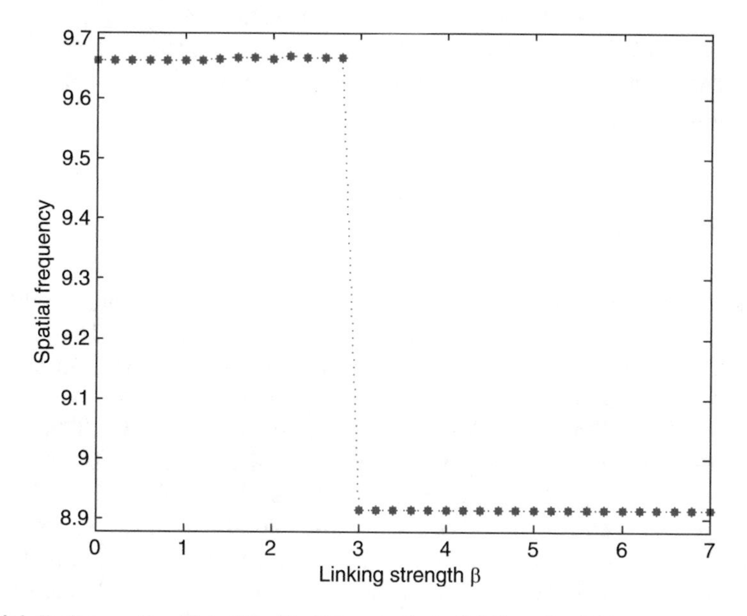

Fig. 14.6 Fusion results of "clock" with different values of β ($0 \geq \beta \geq 7.0$)

Figure 14.6 shows the results evaluation (SF value) of the proposed algorithm with different values of β. It is shown that the result will become worse significantly when β is larger than a certain value. In this case, when $\beta \geq 3.0$, the SF value would decrease to 8.9145. Note that the SFs of two source images are 5.5912 and 8.9145, respectively. 8.9145 is just the larger SF value. As a result, it can be concluded that when β is greater than a specified threshold, the fusion result will become worse and the SF value of the resulting image will equal to the maximum value of the SFs of source images.

According to the discussion above, we propose an iterative method to determine the suitable value of β. Zheng et al. [21] introduced an iterative method for image fusion. The iterative method can determine the optimized value of a tunable parameter. In detail, image quality index (IQI) is used to evaluate the quality of fused image and then the value of IQI is fed back to the fusion method. By this tuning method, the parameters are adjusted to make algorithm have better performance. In this chapter, an iterative algorithm is proposed to find a proper β. SF is used to measure the quality of fused image. The iterative algorithm is described as follows:

1. Initialize β: $\beta = 0$;
2. Using the proposed fusion algorithm to obtain the fused image;
3. Use spatial frequency (SF) to measure the quality of the fused image;
4. If $SF > SF_{max}$, then $SF_{max} = SF$, $\beta_{max} = \beta$, where SF_{max} records the maximum value of SF in the iteration process. If the SF value equals the maximum value of the ones of source images, it indicates that the value of β has exceeded the threshold. The iteration procedure should be completed, so go to step 5. Otherwise, increase β, such as $\beta = \beta + 0.2$, and then go back to step 2;
5. The fine-tuned β is obtained (β_{max}).

14.6 Experimental Evaluations

In this section, experiments are conducted to evaluate our proposed multi-focus image fusion algorithm. Three sets of 256-level images are used. They are clock (size 480 × 480), disk (size 480 × 640), and lab (size 480 × 640). In clock and disk sets, reference images and source images are given, while "lab" set just gives its source images. The spatial frequency of these images is listed in Table 14.2. In the experiments, our method is compared with DWT-based method, Li's method [5] and Miao's method [10]. The fusion results are listed in Table 14.3 (evaluated by RMSE). Miao's method also uses PCNN to fuse multi-focus images. It uses clarity values of pixels to determine the value of β. Li's method is a classic multi-focus image fusion method based on image blocks. For Li's method, SF and EOL are used to measure the clarity of image blocks, respectively. With respect to the size of image blocks, our proposed method and Li's method use blocks with 8 × 8. The threshold in our method is set to 0. In DWT-based method, "DB4" wavelet is used with decomposition level of 5.

Image set of "clock" is shown in Fig. 14.4. The fusion results of DWT-based method, Miao's method and proposed method are shown in Fig. 14.7. Intuitively, the

Table 14.2 Spatial frequency of images in "clock," "disk," "lab"

	Reference image	Source image	
		Image *A*	Image *B*
"clock"	9.6800	5.5912	8.9145
"disk"	15.4671	14.4524	6.9317
"lab"	–	8.6104	10.8338

Table 14.3 Evaluation of fusion results (RMSE)

	"clock"	"disk"
DWT-based method	0.7559	1.5710
Li's method (SF)	0.9130	2.2515
Li's method (EOL)	0.7198	1.7066
Miao's method	0.7024	3.3428
Our proposed method	0.5365 ($\beta = 2.2$)	1.4621 ($\beta = 4.6$)

(a) (b) (c)

Fig. 14.7 Fusion results of "clock." (**a**) DWT-based method. (**b**) Miao's method. (**c**) Proposed method ($\beta = 2.2$)

fused image that is obtained by DWT-based method has low overall image quality. The blurry region of the source images would be clearer in the fused image while the clear region of the source images will be blurred. With regard to Miao's method and our proposed method, the results of them are better than the result of DWT-based method in the "clock" example. The fused image of our proposed method can better retain the original information of the source images and the block effect in it is not obvious. As mentioned in Sect. 14.5, the SF results of fused images by choosing different β are shown in Fig. 14.6. When $\beta \geq 3.0$, the SF decreases to 8.9145. SF value reaches its maximum 9.6694 when $\beta = 2.2$. Therefore, in "clock" image set, we choose $\beta = 2.2$ and get the fused image (Fig. 14.7c). Its RMSE is 0.5365.

Figure 14.8 describes the experiment results of "disk" image set. Figure 14.8d to f are the fused image obtained by DWT-based method, Miao's method and proposed method. In Fig. 14.8g, we can find when $\beta \geq 4.8$, the SF of fused image will decrease to 14.4524. Figure 14.8h is a part of Fig. 14.8g with $0 \geq \beta \geq 4.6$. When $\beta = 4.6$, the result reaches its maximum value 15.4004. Figure 14.9 shows the results of "lab" set. In this example, we choose for our method.

Figure 14.10 shows the differences between fused image and source image. From these difference images, we can conclude that our proposed method is the best in these fusion methods. The important information is retained well and the boundary is trimmed. The fusing result of Miao's method is not ideal. The information of "3" in the clock is missed. With respect to DWT-based method, the head region of the resulting image is bad because position of the head changes in the two source images. DWT is not shift-invariant so the method can't handle the situation of moving object. From the above experiments, it is shown that our proposed method is better. DWT-based method may blur the clear region in the source images and it is not shift-invariant. Miao's method has good performance in "clock" but it's worse in "disk" and "lab."

14.7 Conclusion

In this chapter, we propose a multi-focus image fusion method based on image blocks and PCNN model. EOL is used to measure the clarity of image blocks. The method is inspired by the neuro network of biologic vision. With regard to the problem of determining the value of β in PCNN, an iteration method is proposed to deal with this problem. By analyzing the characteristics of PCNN, we deduce that our proposed method is identical with Li's method when $\beta = 0$. According to the evaluation results of experiments, our proposed method shows better performance than traditional methods.

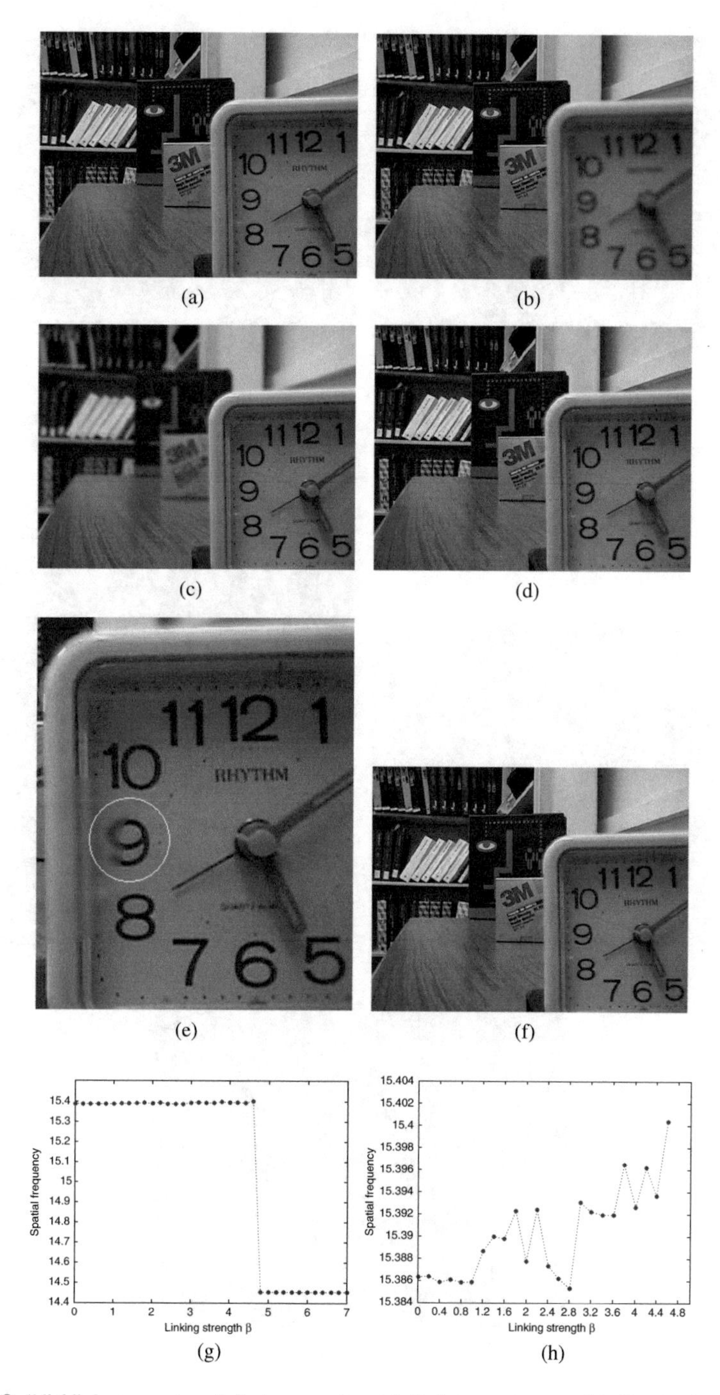

Fig. 14.8 "disk" image set and fusion results. (**a**) Reference image. (**b**) Source images A. (**c**) Source images B. (**d**) DWT-based method. (**e**) Miao's method (a detailed part of fused image). (**f**) Proposed method ($\beta = 4.6$). (**g**) SF of fused image with different values of $\beta (0 \geq \beta \geq 7.0)$. (**h**) Part of g, SF of fused image $(0 \geq \beta \geq 4.6)$

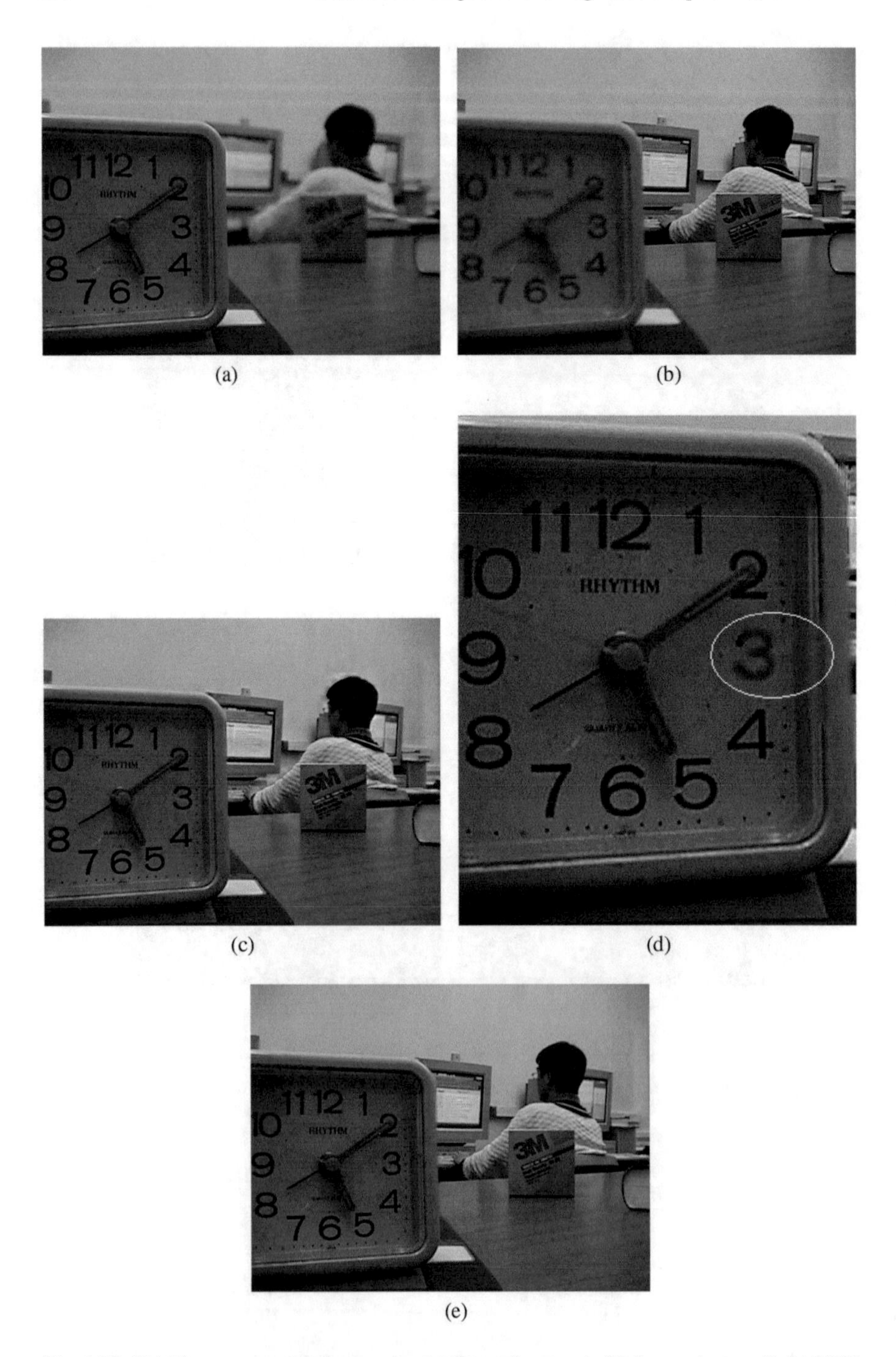

Fig. 14.9 "lab" image set and fusion results. (**a**) Source images A. (**b**) Source images B. (**c**) DWT-based method. (**d**) Miao's method (a detail part of fused image). (**e**) Proposed method ($\beta = 2.2$)

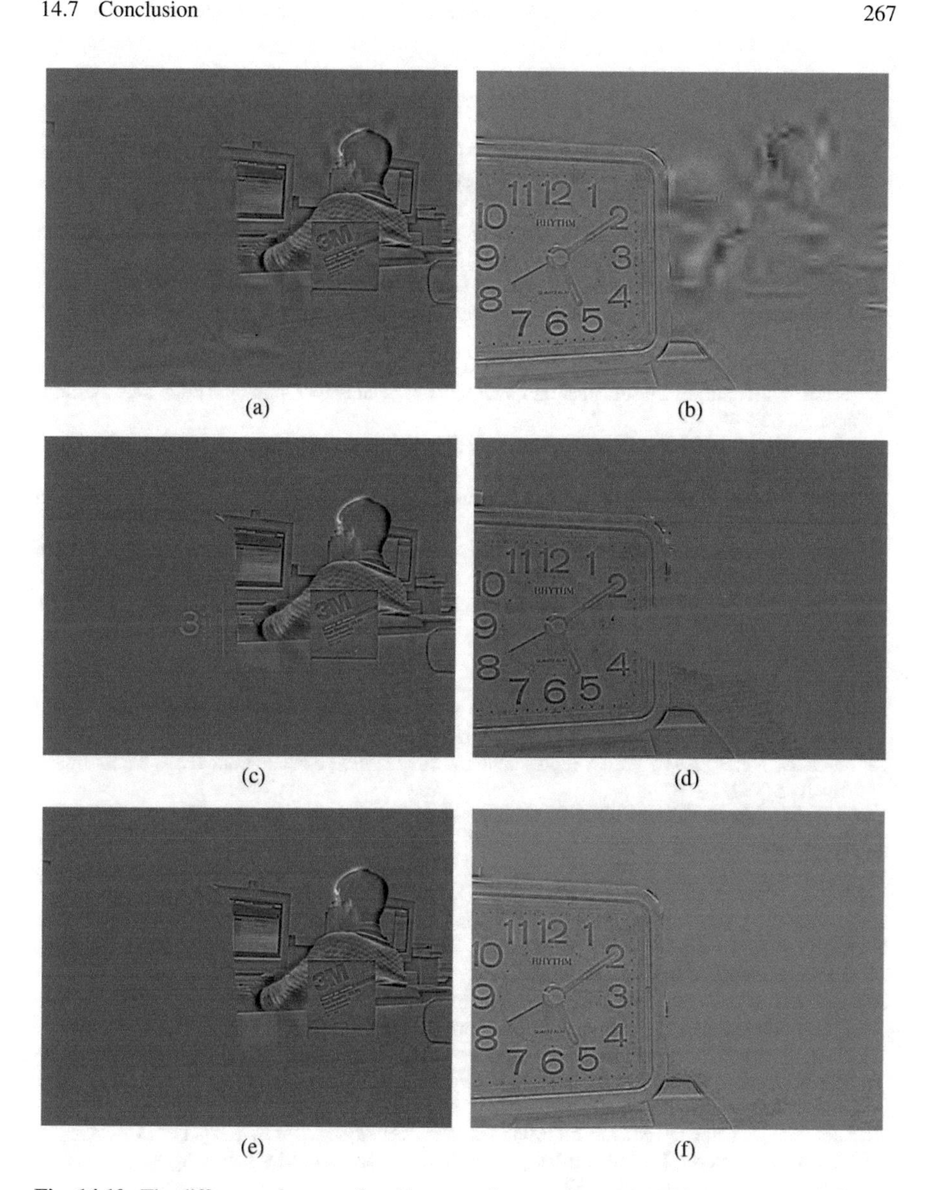

Fig. 14.10 The differences between fused image and source image. (**a**) Differences between DWT and Fig. 14.9a. (**b**) Differences between DWT and Fig. 14.9b. (**c**) Differences between Miao's method and Fig. 14.9a. (**d**) Differences between Miao's method and Fig. 14.9b. (**e**) Differences between proposed method and Fig. 14.9a. (**f**) Differences between proposed method and Fig. 14.9b

References

1. Broussard RP, Rogers SK, Oxley ME, Tarr GL (1999) Physiologically motivated image fusion for object detection using a pulse coupled neural network. IEEE Trans Neural Netw 10(3): 554–563
2. Burt PJ, Kolczynski RJ (1993) Enhanced image capture through fusion. In: Fourth international conference on computer vision. IEEE, Piscataway, pp 173–182
3. Eckhorn R, Reitboeck HJ, Arndt M, Dicke P (1990) Feature linking via synchronization among distributed assemblies: simulations of results from cat visual cortex. Neural Comput 2(3): 293–307
4. Eltoukhy HA, Kavusi S (2003) Computationally efficient algorithm for multifocus image reconstruction. In: Electronic imaging. International Society for Optics and Photonics, Bellingham, pp 332–341
5. Eskicioglu AM, Fisher PS (1995) Image quality measures and their performance. IEEE Trans Commun 43(12):2959–2965
6. Huang W, Jing Z (2007) Evaluation of focus measures in multi-focus image fusion. Pattern Recogn Lett 28(4):493–500
7. Huang W, Jing Z (2007) Multi-focus image fusion using pulse coupled neural network. Pattern Recogn Lett 28(9):1123–1132
8. Johnson JL, Padgett ML (1999) PCNN models and applications. IEEE Trans Neural Netw 10(3):480–498
9. Kinser JM (1997) Pulse-coupled image fusion. Opt Eng 36(3):737–742
10. Krotkov E (1987) Focusing. Int J Comput Vis 1:223–237
11. Kuntimad G, Ranganath HS (1999) Perfect image segmentation using pulse coupled neural networks. IEEE Trans Neural Netw 10(3):591–598
12. Laine A, Fan J (1996) Frame representations for texture segmentation. IEEE Trans Image Process 5(5):771–780
13. Li H, Manjunath B, Mitra SK (1995) Multisensor image fusion using the wavelet transform. Graph Models Image Process 57(3):235–245
14. Li S, Kwok JT, Wang Y (2001) Combination of images with diverse focuses using the spatial frequency. Inf Fusion 2(3):169–176
15. Li M, Cai W, Tan Z (2005) Pulse coupled neural network based image fusion. In: Wang J, Liao XF, Yi Z (eds) Advances in neural networks – ISNN 2005. Lecture notes in computer science, vol 3497. Springer, Berlin
16. Ligthart G, Groen FC (1982) A comparison of different autofocus algorithms. In: Proceedings of the sixth international conference on pattern recognition, pp 597–600
17. Miao Q, Wang B (2005) A novel adaptive multi-focus image fusion algorithm based on PCNN and sharpness. In: Defense and security. International Society for Optics and Photonics, Bellingham, pp 704–712
18. Subbarao M, Choi TS, Nikzad A (1993) Focusing techniques. Opt Eng 32(11):2824–2836
19. Toet A, Van Ruyven LJ, Valeton JM (1989) Merging thermal and visual images by a contrast pyramid. Opt Eng 28(7):287–789
20. Wang W (2008) Research on pixel-level image fusion. Ph.D. thesis, Shanghai Jiao Tong University
21. Zheng Y, Essock EA, Hansen BC (2005) Advanced discrete wavelet transform fusion algorithm and its optimization by using the metric of image quality index. Opt Eng 44(3):037003

Chapter 15
Evaluation of Focus Measures in Multi-Focus Image Fusion

15.1 Introduction

The majority of imaging systems of practical interest has a finite depth of field (DOF), i.e., confined focal length. Generally, the objects with the DOF appear sharp in the captured photograph while other objects tend to be blurred. A possible way to overcome this problem is to combine the images taken from the same viewpoint under different focal settings. Most of the established techniques consider this problem in practical situations often to critically rely on focus measures. One major concern is to find a metric able to deliver reliable focus area in a reasonable way.

Multi-focus image fusion aims to obtain an all-in-focus image for preserving repetitive textures and details of the observed scene. Multi-focus image fusion has emerged as a key research field of image fusion. It has been applied to many different fields such as remote sensing, biomedical imaging, and computer vision. With recent technological advances, the availability of several images corresponding to different focal lengths provides new degrees of freedom, leading to increased interest in exploiting them efficiently. Thus, leveraging the natural diversity of multi-focus images has led to many algorithmic advances. A key problem of multi-focus image fusion is to identify the focused areas.

In the past two decades, many fusion techniques have been developed, e.g., multi-resolution image fusion [1, 2, 7, 11]. However, many of the multi-resolution approaches, such as discrete wavelet transform (DWT) algorithm, suffer from shift-variant. This shortcoming may lead to the variations in the distribution of energy between DWT coefficients at different scales. For example, if the camera is moving or the source images are not registered, the fusion performance of those algorithms will deteriorate. An alternative way to handle this problem is taking advantage of the discrete wavelet frame transform [12] (DWFT) or dual tree complex wavelet transform [4] (DT-CWT). DWFT and DT-CWT are shift-invariant. However, it

Z. Jing et al., *Non-Cooperative Target Tracking, Fusion and Control*,
Information Fusion and Data Science, https://doi.org/10.1007/978-3-319-90716-1_15

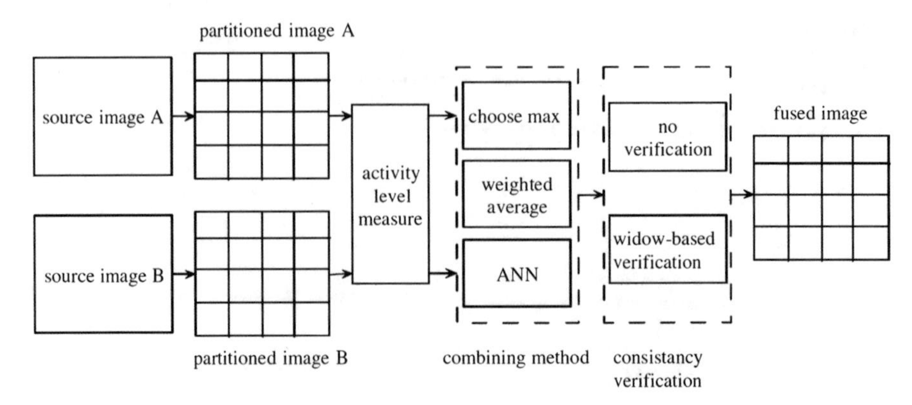

Fig. 15.1 A generic scheme for fusing multi-focus images based on the selection of image blocks from source images

should be noted that DWFT and DT-CWT are complicated and time consuming. Beyond this, these transformation methods exhibit high redundancy in the output information.

In literatures [6, 7], Li et al. introduced a new fusion method based on the selection of image blocks from source images. The basic idea of this method is to choose the clearer image blocks from source images to construct the fused image. A feature selection method by consistency verification was applied. Especially, the fusion coefficient according to the neighboring coefficients was determined by using a majority filter. Figure 15.1 illustrates a generic schematic diagram for this approach.

It is apparent that this method can reduce the artifacts introduced by the shift variant property of DWT. The image blocks selection method demonstrated impressive capabilities in image fusion. And its implementation is computationally simple and can be extended to a real-time image fusion system.

Activity-level measurement has been proposed to determine the quality of a given part of the input images. It consists of many fusion rules, such as choose max, weighted average, artificial neural networks (ANN) as well. Among these schemes, spatial frequency is one of the successful criteria to measure the clarity of image blocks that have been proposed in the literature [6, 7]. Energy of image gradients was also utilized to measure the activity. As shown in Fig. 15.1, it can be easily seen that the measure of clarity plays an important role in this kind of fusion method. A better measure results in a superior fusion performance. Thus, it is important to explore the capability of the image clarity measures in the field of multi-focus image fusion. The image clarity measures, namely, focus measures, are deeply studied in the field of autofocusing. A detailed discussion of this topic can be found in several papers [9, 10]. Focus measures are usually divided into spatial domain focus measure and frequency domain focus measure. But following literature [14], frequency domain focus measures will not be used in a real-time system as their complexity makes it difficult to produce fast algorithms [6, 14].

Spatial domain focus measures have a simple physical interpretation which makes them quite popular. Thus, we focused our investigations on several spatial domain focus measures, such as variance, energy of image gradient as well.

We give a brief description of focus measures in Sect. 15.2. Then, evaluations of several focus measures will be presented in Sects. 15.3, and 15.4 will give some concluding remarks. The presentation of this chapter is based on the work [5, 13].

15.2 Focus Measures

Focus measure is defined as a quantity to evaluate the sharpness of a pixel locally. Some general requirements for a focus measure are provided in literatures [6, 14]. Specially, the maximum of a focus measure implies that the best focused image is generated. And it generally decreases as the defocus increases. Therefore, in view of the subject above, the focused image areas of the source images must produce maximum focus measures, the defocused areas must produce minimum focus measures in contrast. In this context, there is a strong need for examining the invariance and persistence of the focus measure.

A typical focus measure should satisfy these requirements:

1. Independent of image content.
2. Monotonic with respect to blur.
3. The focus measure must be unimodal, that is, it must have one and only one maximum value.
4. Large variation in value with respect to the degree of blurring.
5. Minimal computation complexity.
6. Robust to noise.

These requirements are also useful to multi-focus image fusion.

A number of different focus functions are assessed in this section. Let $f(x, y)$ be the gray-level intensity of pixel (x, y).

1. **Variance:** The variance of all the pixels is a measure of how the pixel values are dispersed around its mean μ. Higher variance indicates that the test image has a better contrast. And, the smaller variance implies the gray-level (intensity) values of all the pixels are close to the mean. The expression for the image $f(x, y)$ is given by

$$\text{Variance} = \frac{1}{M \times N} \sum_{x} \sum_{y} (f(x, y) - \mu)^2 \tag{15.1}$$

where $\mu = \frac{1}{M \times N} \sum_x \sum_y f(x, y)$ denotes the mean of all pixels. And, $M \times N$ stands for the size of image.

2. **Energy of image gradient (EOG):** This focus measure relies on the computation of image gradient, which is given as follows:

$$\mathrm{EOG} = \sum_x \sum_y (f_x^2 + f_y^2), \tag{15.2}$$

where $f_x = f(x+1, y) - f(x, y)$ and $f_y = f(x, y+1) - f(x, y)$ denote the partial derivatives of $f(x, y)$ along the horizontal and vertical directions respectively.

3. **Tenenbaum's algorithm (Tenengrad):** This focus measure is obtained by summing the squared responses of the vertical and horizontal components with a Sobel convolution mask. It is defined as follows:

$$\mathrm{Tenengrad} = \sum_{x=2}^{M-1} \sum_{y=2}^{N-1} [\nabla S(x, y)]^2 \quad \text{for} \quad \nabla S(x, y) > T, \tag{15.3}$$

where T is a discrimination threshold value, and $\nabla S(x, y)$ is the Sobel gradient magnitude value by

$$\nabla S(x, y) = [\nabla S_x(x, y)^2 + \nabla S_y(x, y)^2]^{1/2}, \tag{15.4}$$

where $\nabla S_x(x, y)^2$, $\nabla S_y(x, y)^2$ are the outcomes of the convolution of the image with the following convolution mask:

$$\nabla S_x(x, y) = \begin{pmatrix} -1 & 0 & +1 \\ -2 & 0 & +2 \\ -1 & 0 & +1 \end{pmatrix}, \quad \nabla S_y(x, y) = \begin{pmatrix} -1 & -2 & -1 \\ 0 & 0 & 0 \\ +1 & +2 & +1 \end{pmatrix}. \tag{15.5}$$

Equivalently, the gradient magnitude value can be represented as follows:

$$\nabla S_x(x, y) = \{-[f(x-1, y-1)] + 2f(x-1, y) + f(x-1, y+1)] \\ +[f(x+1, y-1) + 2f(x+1, y) + f(x+1, y+1)]\}, \tag{15.6}$$

$$\nabla S_y(x, y) = \{+[f(x-1, y-1)] + 2f(x, y-1) + f(x+1, y-1) \\ +[f(x-1, y+1) + 2f(x, y+1) + f(x+1, y+1)]\}. \tag{15.7}$$

4. **Energy of Laplacian of the image (EOL):** This focus measure employs Laplacian operator to analyze high spatial frequencies associated with image border sharpness. It can be computed by

$$\mathrm{EOL} = \sum_x \sum_y (f_{xx} + f_{yy})^2 \tag{15.8}$$

where $f_{xx} + f_{yy}$ is defined as follows:

$$\begin{aligned} f_{xx} + f_{yy} = & -f(x-1, y-1) - 4f(x-1, y) - f(x-1, y+1) \\ & -4f(x, y-1) + 20f(x, y) \\ & -4f(x, y+1) - f(x+1, y-1) \\ & -4f(x+1, y) - f(x+1, y+1) \end{aligned} \quad . \tag{15.9}$$

5. **Sum-modified-Laplacian (SML):** This focus measure [9] is proposed by Nayar. It relies on the introduction of the modified Laplacian (ML), which avoids the cancellation of second derivatives in the horizontal and vertical directions that have opposite signs. The approximation of ML is computed as follows:

$$\begin{aligned} \nabla^2_{\mathrm{ML}} f(x, y) = & |2f(x, y) - f(x - \mathrm{step}, y) - f(x + \mathrm{step}, y)| \\ & + |2f(x, y) - f(x, y - \mathrm{step}) - f(x, y + \mathrm{step})|, \end{aligned} \tag{15.10}$$

where step is a variable spacing between the pixels, which utilized to accommodate for possible variations in the size of texture elements. In our experiments, 'step' is always set as 1.

At last, SML at a point (i, j) is obtained by

$$\mathrm{SML} = \sum_{i=x-N}^{i=x+N} \sum_{j=y-N}^{j=y+N} \nabla^2_{\mathrm{ML}} f(i, j) \quad \text{for } \nabla^2_{\mathrm{ML}} f(i, j) \geq T. \tag{15.11}$$

where T is a discrimination threshold value. The parameter N determines the window size used to compute the focus measure. In most cases, N is set to be $3\times$ or 5×5, i.e., $N = 1$ or $N = 2$.

6. **Spatial Frequency (SF) [3]:** It should be noted that this measure is based on the energy of image gradient (EOG). A higher SF value implies that the images are richer in structural information, such as textures, edge. SF presented here is used to compare with other focus measures in the following experiments.

Spatial frequency is defined by

$$\mathrm{SF} = \sqrt{(\mathrm{RF})^2 + (\mathrm{CF})^2}, \tag{15.12}$$

where RF and CF are the row frequency

$$\mathrm{RF} = \sqrt{\frac{1}{M \times N} \sum_{x=1}^{M} \sum_{y=2}^{N} [f(x, y) - f(x, y-1)]^2} \tag{15.13}$$

and column frequency, respectively.

$$\mathrm{CF} = \sqrt{\frac{1}{M \times N} \sum_{x=2}^{M} \sum_{y=1}^{N} [f(x, y) - f(x-1, y)]^2} \tag{15.14}$$

15.3 Evaluation of Focus Measures

It is important to analyze the behavior and performance of existing focus measures. In the field of autofocusing, a typical evaluation experiment can be found in [10]. The process for evaluating focus measures can be divided into three basic steps: (1) place an object in front of a camera; (2) take a picture of the object at each lens position; (3) compute focus measures for these images. Among these criteria, the monotonicity, magnitude of slope, and smoothness are used to evaluate the focus measures. However, there are some differences in the experiments of multi-focus image fusion. First, there are always two or three source images. Second, in one source image, only those objects within the depth of field of the camera are focused, while other objects are blurred. The basic idea underlying our experiments is to evaluate the focus measures' capability of distinguishing focused image blocks from defocused image blocks. In other words, we aim at calculating the values of the focus measures in the context of image blocks.

A detailed description of the evaluation experiment will be given in Sect. 15.3.1.

15.3.1 Experimental Setup

In order to simplify the evaluation method, we only consider two source images. The process of evaluating focus measures consists of the following steps:

1. Decompose the two source images into blocks. Denote the i-th image block pair by A_i and B_i, respectively.
2. Compute the focus measure or SF of each block, and denote the results of A_i and B_i by M_i^A and M_i^B, respectively.
3. Compare the focus measure or SF of two corresponding blocks A_i and B_i, and construct the i-th block C_i of the composite image as

$$C_i = \begin{cases} A_i, & M_i^A > M_i^B \\ B_i, & \text{otherwise.} \end{cases} \tag{15.15}$$

4. Compute root mean square error (RMSE) for the composite image with a reference image.

Figure 15.2 shows a schematic diagram for evaluating the focus measures and spatial frequency by using images with 256 gray levels.

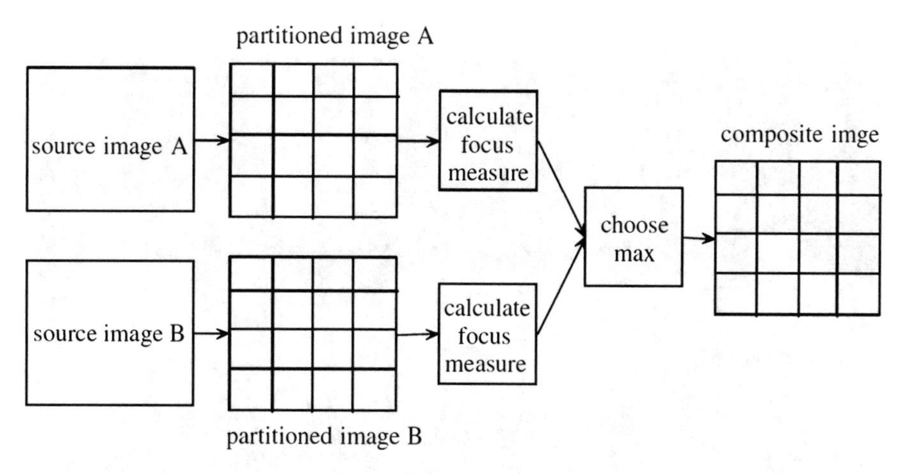

Fig. 15.2 Schematic diagram for evaluating the focus measures in multifocus image fusion

Here, RMSE is utilized to evaluate the performance of the focus measures. Though, mutual information (MI) and the percentage of correct decisions (PCD) have also been used to evaluate the performance of the focus measures, the results are similar to RMSE. Therefore, the outcomes of MI and PCD are not presented here.

RMSE is defined as

$$\text{RMSE} = \sqrt{\frac{\sum_x \sum_y [R(x, y) - F(x, y)]^2}{M \times N}}, \tag{15.16}$$

where R and F denote reference image and composite image, respectively, with size $M \times N$ pixels.

No consistency verification was used in our experiments. It should be noted that consistency verification will disturb the results obtained by focus measures.

15.3.2 Experimental Results

In this section, some experiments were conducted on several sets of images. Only the results on three sets of images will be presented here (two sets of 256-level gray images, namely 'clock' and 'disk', and a set of color images). Their sizes are 480×480, 640×480, and 640×480, respectively. Experiments were implemented on a Celeron M 1.40 GHz computer with 512 MB RAM. The simulation software is Matlab 7.0.

Figure 15.3a–c show the reference image (focused everywhere) and the source images of 'clock'. The fusion result based on SML is displayed in Fig. 15.3d.

Fig. 15.3 Example of composite image, namely fused image, by using SML (block size 32 × 32). (a) Reference image (focused everywhere); (b) source image (focus on the right); (c) source image (focus on the left); (d) fused image using SML (threshold $T = 0$)

Figure 15.4a–d show the images for *disk*. In this chapter, we just foucs on the fusion results using SML.

It should be noted that the method proposed in Sect. 15.3.1 for evaluating the focus measures can be easily extended to handle color images. Consider the color source images presented in Fig. 15.5b and c. Denote Fig. 15.5b and c by A and B, respectively. We use Eq. (15.17) to obtain the intensity components of Fig. 15.5b and c, namely A^I and B^I.

$$I = (R + G + B)/3 \tag{15.17}$$

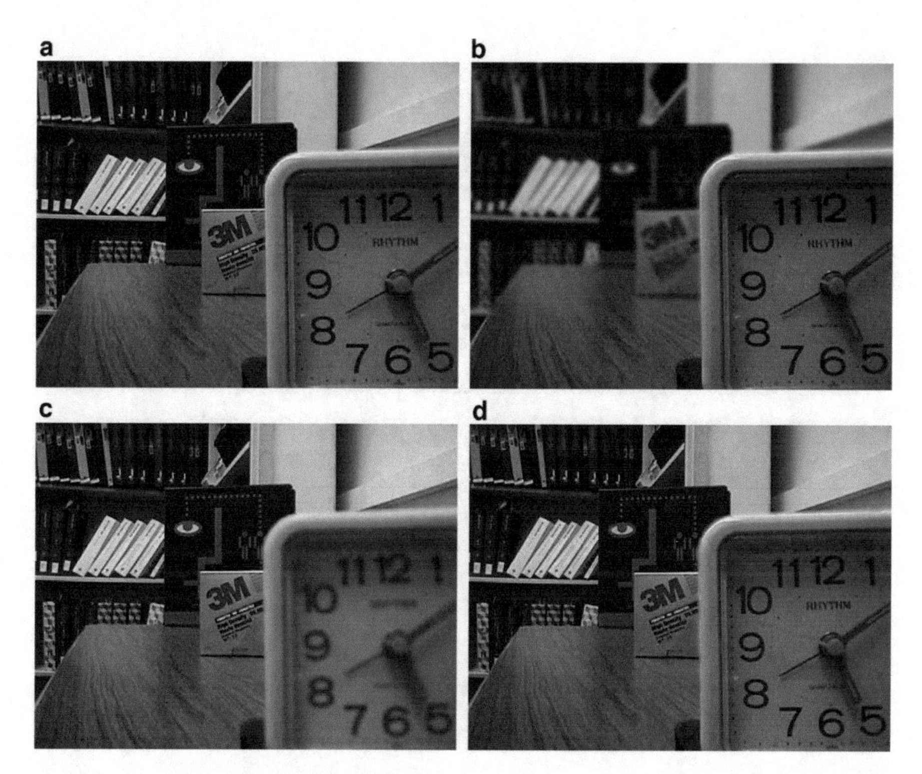

Fig. 15.4 Example of composite image by using SML (block size 8 × 16). (**a**) Reference image (focused everywhere); (**b**) source image (focus on the right); (**c**) source image (focus on the left); (**d**) fused image using SML (threshold $T = 5$)

where I is the intensity component of the color image. The color image has three channels R, G, and B. Decompose A, B, A^I, and B^I into blocks. Compute the focus measure for each block of A^I and B^I, and denote the results of A_i^I and B_i^I by $M_i^{A_I}$ and $M_i^{B_I}$, respectively. We use the following expression to obtain the composite color image,

$$\begin{cases} C_i^R = A_i^R, C_i^G = A_i^G, C_i^B = A_i^B & M_i^{A_I} > M_i^{B_I}, \\ C_i^R = B_i^R, C_i^G = B_i^G, C_i^B = B_i^B & \text{otherwise.} \end{cases} \tag{15.18}$$

where X_i^R, X_i^G, and X_i^B ($X = C$, A or B) are the i-th image blocks of the R, G, and B channels, respectively, of the color image X.

The composite color image of Fig. 15.5b and c by using SML is presented in Fig. 15.5d.

Table 15.1 shows the quantitative comparison of the focus measures by RMSE obtained on processing images in Fig. 15.3b and c. Table 15.2 shows the comparison obtained on processing images in Fig. 15.4b and c. In Table 15.3, RMSE are

Fig. 15.5 Example of composite image by using SML (block size 16 × 16). (**a**) Reference image (focused everywhere); (**b**) source image (focus on the right); (**c**) source image (focus on the left); (**d**) fused image using SML (threshold $T = 2$)

Table 15.1 Evaluation of different focus measures with different block sizes by RMSE (Fig. 15.3b and c)

	Focus measures					
Block size	Variance	EOG	Tenengrad	EOL	SML	SF
4 × 4	3.01	2.45	1.88, $T = 87$	1.63	1.13, $T = 6$	2.45
4 × 8	2.34	1.77	1.51, $T = 87$	1.30	1.01, $T = 7$	1.77
8 × 8	2.05	1.52	1.24, $T = 86$	1.10	0.85, $T = 6$	1.52
8 × 16	1.67	1.41	1.04, $T = 88$	0.92	0.78, $T = 6$	1.41
16 × 16	1.06	0.83	0.77, $T = 77$	0.78	0.77, $T = 9$	0.83
16 × 32	0.73	0.73	0.74, $T = 41$	0.79	0.76, $T = 4$	0.73
32 × 32	0.65	0.65	0.65, $T = 0$	0.71	0.65, $T = 0$	0.65

calculated between the intensity component of the reference image (Fig. 15.5a) and the intensity component of the composite images obtained by using different focus measures. Although the analysis of Tenengrad and SML included the thresholds from 0 to 200, Tables 15.1, 15.2, and 15.3 show only the best results.

Table 15.2 Evaluation of different focus measures with different block sizes by RMSE (Fig. 15.4b and c)

	Focus measures					
Block size	Variance	EOG	Tenengrad	EOL	SML	SF
4×4	3.87	2.84	4.24, $T = 5$	1.86	1.90, $T = 0$	2.84
4×8	3.61	2.44	3.06, $T = 60$	1.82	1.77, $T = 4$	2.44
8×8	3.37	2.25	2.66, $T = 48$	1.71	1.73, $T = 4$	2.25
8×16	2.68	2.03	2.20, $T = 80$	1.54	1.48, $T = 5$	2.03
16×16	2.04	1.67	1.68, $T = 5$	1.52	1.56, $T = 5$	1.67
16×32	1.92	1.57	1.58, $T = 5$	1.58	1.54, $T = 5$	1.57
32×32	1.85	1.52	1.52, $T = 6$	1.52	1.52, $T = 32$	1.52

Table 15.3 Evaluation of different focus measures with different block sizes by RMSE (Fig. 15.5b and c)

	Focus measures					
Block size	Variance	EOG	Tenengrad	EOL	SML	SF
4×4	2.15	1.64	2.47, $T = 0$	1.11	1.09, $T = 0$	1.64
4×8	1.92	1.41	1.99, $T = 0$	0.95	0.97, $T = 1$	1.41
8×8	1.82	1.31	1.69, $T = 7$	0.93	0.90, $T = 1$	1.31
8×16	1.71	1.22	1.52, $T = 4$	0.90	0.97, $T = 2$	1.22
16×16	1.28	0.85	1.03, $T = 4$	0.73	0.71, $T = 2$	0.85
16×32	1.23	0.88	0.98, $T = 4$	0.76	0.76, $T = 2$	0.88
32×32	1.07	0.76	0.80, $T = 0$	0.81	0.81, $T = 2$	0.76

Table 15.4 Average implementation time(s) of different focus measures with different block sizes (Fig. 15.3b and c)

	Focus measures					
Block size	Variance	EOG	Tenengrad	EOL	SML	SF
4×4	1.3280	0.9060	0.9220, $T = 87$	0.9060	0.9530, $T = 6$	0.9530
4×8	0.6870	0.4840	0.5150, $T = 87$	0.5000	0.5310, $T = 7$	0.5000
8×8	0.3750	0.2660	0.3120, $T = 86$	0.2960	0.3280, $T = 6$	0.2810
8×16	0.2180	0.1560	0.2180, $T = 88$	0.1720	0.2180, $T = 6$	0.1710
16×16	0.1250	0.0930	0.1720, $T = 77$	0.1250	0.1710, $T = 9$	0.1090
16×32	0.0780	0.0620	0.1400, $T = 41$	0.0930	0.1400, $T = 4$	0.0620
32×32	0.0620	0.0460	0.1250, $T = 0$	0.0780	0.1400, $T = 0$	0.0470

The average implementation time of Table 15.1 is shown in Table 15.4. Though the average implementation times of Tables 15.2 and 15.3 have also been considered in our experiments, the results are similar to that of Table 15.4 and will not be presented here.

As can be seen from Tables 15.1, 15.2, and 15.3, the major thresholds of the best results of SML are located on interval [0,10]. On the contrary, the best thresholds of Tenengrad are decentralized.

After analyzing Tables 15.1, 15.2, and 15.3, we can find that the larger the block size, the smaller the RMSE. We believe that a larger image block gives more information for measuring the block's focus or clarity. However, using a block size too large is undesirable [8].

It can be seen that the focus measures are classified in the following groups of decreasing performance, ordered by the values of RMSE: (1) SML; (2) EOL; (3) EOG; (4) SF; (5) Tenengrad; (6) Variance.

Besides, the performance of SF is similar to the performance of EOG in our experiments. It should be noted that Tenengrad is superior to EOG and SF in Table 15.1. However, in Tables 15.2 and 15.3, Tenengrad is inferior to EOG and SF. Therefore, Tenengrad, EOG, and SF are arranged in the same group. The focus measures can also be listed as follows, ordered as implementation time increase: (1) EOF; (2) SF; (3) EOL; (4) Tenengrad; (5) SML; (6) Variance.

Tenengrad had been found to be the best focus measure, until SML was introduced. In our experiments Tenengrad is inferior to EOL and SML, when the execution time is not included in the evaluation.

Although SML has not been proved to be theoretically sound, it performs well in our experiments. SML's capability of distinguishing focused image blocks from defocused image blocks ranks first in our experiments. It should also be noted that SML achieves good results even with small image blocks, such as 4×8 or 8×8 image blocks. However, SML needs more implementation time than many other focus measures.

15.4 Conclusion

In this chapter, several focus measures were compared according to focus measures' capability of distinguishing clear image blocks from blurred image blocks. A general way to evaluate the performance of the focus measures is presented. Experiment results indicate that SML can provide better performance than other focus measures. This measure involves an approximation of second-derivative information in the underlying natural scene. These experiments suggest that using SML or EOL replaces SF as the measure of images clarity. Meanwhile, it can be seen that SML offers a nice way to incorporate the significant correlations in neighboring pixels.

Acknowledgements The authors would like to thank the anonymous reviewers for their valuable suggestions. This work is jointly supported by National Natural Science Foundation of China (60375008), EXPO Technologies Special Project of National Key Technologies R&D Program (2004BA908B07), Shanghai World EXPO Technologies Special Project (04DZ05807), China Ph.D. Discipline Special Foundation (20020248029), China Aviation Science Foundation (02D57003), Aerospace Supporting Technology Foundation (2003-1.3 0 2).

References

1. Burt P, Adelson E (1983) The Laplacian pyramid as a compact image code. IEEE Trans Commun 31(4):532–540
2. Burt PJ, Kolczynski RJ (1993) Enhanced image capture through fusion. In: Proceedings of the fourth international conference on computer vision. IEEE, Piscataway, pp 173–182
3. Eskicioglu AM, Fisher PS (1995) Image quality measures and their performance. IEEE Trans Commun 43(12):2959–2965
4. Hill, PR, Canagarajah CN, Bull DR (2002) Image fusion using complex wavelets. In: 13th British machine vision conference, pp 1–10. Citeseer
5. Huang W, Jing Z (2007) Evaluation of focus measures in multi-focus image fusion. Pattern Recogn Lett 28(4):493–500
6. Krotkov E (1987) Focusing. Int J Comput Vis 1:223–237
7. Li H, Manjunath B, Mitra SK (1994) Multi-sensor image fusion using the wavelet transform. In: IEEE international conference on image processing (ICIP), vol 1. IEEE, Piscataway, pp 51–55
8. Li S, Kwok JT, Wang Y (2001) Combination of images with diverse focuses using the spatial frequency. Inf Fusion 2(3):169–176
9. Nayar SK, Nakagawa Y (1994) Shape from focus. IEEE Trans Pattern Anal Mach Intell 16(8):824–831
10. Subbarao M, Choi TS, Nikzad A (1993) Focusing techniques. Opt Eng 32(11):2824–2836
11. Toet A, Van Ruyven LJ, Valeton JM (1989) Merging thermal and visual images by a contrast pyramid. Opt Eng 28(7):789–792
12. Unser M (1995) Texture classification and segmentation using wavelet frames. IEEE Trans Image Process 4(11):1549–1560
13. Wang W (2008) Research on pixel-level image fusion. Ph.D. thesis, Shanghai Jiao Tong University
14. Yeo T, Ong S, Sinniah R et al (1993) Autofocusing for tissue microscopy. Image Vis Comput 11(10):629–639

Part IV
Spacecraft Control for Tracking

Chapter 16
Dynamic Optimal Sliding-Mode Control for Active Satellite Tracking

16.1 Introduction

To guarantee the continuity of inter-satellite tracking, for example, for tracking satellite with noncooperative maneuver, spacecraft control is sometimes necessary for the chaser satellite. In this chapter, we consider a control-based follow-up target tracking scheme for active satellite tracking. The scheme is composed of a robust tracking algorithm and a follow-up control law.

At first, a relative motion model is described by using osculating reference orbit (ORO) and applied with the RAREKF discussed in Chap. 8 in order to yield an ORO-based tracking algorithm. After that, a dynamic optimal sliding-mode control (DOSMC) method developed on dynamic optimal sliding surface (DOSS) is given as the follow-up control law for both orbit and attitude control of the chaser.

We use three numerical examples to illustrate the advantage of the ORO-based tracking and control method and to verify the effectiveness of the presented control-based follow-up tracking scheme.

The chapter is organized as follows. Firstly, we derive the ORO relative motion model and attitude dynamic model. Then, we present the ORO-based target tracking algorithm, followed by the six-DOF DOSMC law. Simulations are demonstrated finally. The presentation of this chapter is primarily based on the work [5, 6, 16].

16.2 Control-Based Satellite Tracking

In traditional problems of inter-satellite tracking, spacecraft is usually passive. Recently, maneuver capability has become one of the critical spacecraft properties. Active target with potential maneuvers should be paid more attention.

Z. Jing et al., *Non-Cooperative Target Tracking, Fusion and Control*,
Information Fusion and Data Science, https://doi.org/10.1007/978-3-319-90716-1_16

As a developed type of inter-satellite tracking, active satellite tracking focuses on unknown target maneuvers, model uncertainties, and external disturbances. It is more suitable for using adaptive and robust tracking approaches.

For disturbed circumstances, the optimal gain of the traditional Kalman filtering algorithm may lead to divergence due to the lack of filtering robustness. The adaptive and robust algorithms become necessary to acquire better tracking performance, by online identifying noise covariance [15] or self-adjusting filter gains [10].

Xiong [12] provided an adaptive robust extended Kalman filter (AREKF) by introducing a switched attenuation factor to tune gains adaptively. However, due to the existence of nonlinear modeling errors, the switching structure between optimal and robust filtering modes works improperly, staying on robust status with excessive cost of optimality. It weakens the meaning of introducing the switching mechanism and increases the tracking errors.

In order to let the switching structure of the AREKF work properly, a redundant AREKF [6], RAREKF, shown in Chap. 8 is developed. The RAREKF introduces a redundant factor to obtain redundancy to modeling errors and disturbances, achieving more precise tracking results for less cost of filtering optimality.

16.2.1 Six-DOF Follow-up Tracking

Adaptive and robust tracking method is a feasible way for active satellite tracking. However, it is unsuitable for long-term tracking. The tracking process becomes questionable with time lasting. This is because when the target satellite maneuvers, relative range between chaser and target varies correspondingly, so nonlinear error of relative motion model is changed. Consequently, tracking results can hardly be kept at a stable level and probably diverge if the relative distance and modeling error are large enough.

Therefore, for long-term active satellite tracking, it is necessary to consider the chaser satellite control during tracking in order to stabilize the relative motion and avoid tracking divergence. In addition, chaser attitude control is also needed to keep the target in sight. This scheme is named six-DOF follow-up tracking.

16.2.2 Problems of the Tracking Scheme

For implementing the scheme, two points should be considered.

First, tracking and control are related closely. Tracking results give information for control decision, whereas control law provides a feasible platform for long-term tracking. Furthermore, the coupling will be strengthened by the potential maneuvers of active target. This is because when the target maneuvers, the chaser satellite may change trajectory according to tracking results, leading to uncertain relative motion. In the meantime, reference coordination in relative motion model is usually built by using a static reference orbit (SRO). As a result, if relative motion moves away from SRO, modeling error must increase, reducing the tracking precision. That will weaken the quality of chaser control and probably fail long-term tracking. Therefore, the coupling between tracking and control facilitates the transition of modeling error through SRO, so it should be considered in establishment of the follow-up tracking scheme. Unfortunately, tracking and control are rarely studied together in previous works.

Second is the requirement of designing a follow-up control law for the chaser. For space operations with disturbances, robustness and fuel cost are more important than others for chaser control. Sometimes they are conflicting. Sliding-mode control is a well-known method with robustness, having been applied into station-keeping [13] and formation reconfiguration [7]. Efforts have also been made to optimize the sliding surface [8, 9, 11, 13] and the control gain [14]. However, because the cost functions in those works only include inner system states, but without real control vector, they are not the solution with optimal fuel, even if they are called the optimal sliding-mode control (OSMC) [8, 9, 11].

16.2.3 Control Method

In this chapter, we solve the problem above from model and control.

First is the relative motion model. An osculating orbit of the chaser is adopted as the reference to generate a reference frame. Although the chaser needs maneuver, the model built on osculating reference orbit (ORO) can follow up the movement and describe the relative motion more accurately than the SRO model. Although there is the coupling of tracking and control, negative effect of nonlinear error transition can be gradually weakened because the modeling error of the ORO model is restrained at a low level. In this chapter, we will present the ORO-based state and measurement models. Also, we applied the models to RAREKF and yielded an ORO-based robust tracking algorithm. The algorithm is suitable for long-term tracking even when both target and chaser are maneuvering.

Second is the chaser control law. Considering disturbances and fuel optimality, we developed a dynamic optimal sliding-mode control (DOSMC) law established by introducing dynamic optimum into sliding surface. Notice that the sliding surface of the typical OSMC is a hyper-curve in reduced dimensional space when the weight matrix of the optimum index is time-invariant. Such surfaces are hard to induce the system states to follow the dynamic fuel-optimal solution in full dimensional space. Hence, a dynamic optimal sliding surface (DOSS) is built by using a desired state trajectory caused by a virtual optimal LQR and forms the DOSMC. The DOSMC is robust globally and can be regarded as an extended version of LQR. We applied the DOSMC into both orbit and attitude control of the chaser and considered gravity gradient torque [13] into the attitude dynamics.

16.3 Modeling of Relative Motion

Assume that two satellites (the chaser and the target) move on proximity orbits. The target has potential maneuver unknown to the chaser. To keep the relative range and sight angles for onboard sensors to guarantee the tracking continuity, the chaser has maneuver capability to control orbit and attitude. We use laser and microwave radar as the onboard measurement device, and regard each satellite as a point mass. The relative motion model and the chaser attitude model are presented below.

16.3.1 The ORO Model

The osculating chaser orbit, representing instantaneous orbital motion of the chaser, is chosen as reference orbit. The position vectors are described in the local vertical and local horizontal (LVLH) coordinate system of the reference orbit. The relative range is denoted as $\rho = [\rho_x, \rho_y, \rho_z]^T$ along radial, in-track, and cross-track. Subscript C or T means chaser or target. The dynamic model of relative motion, defined in the LVLH frame of the ORO, is written as [13]

$$\rho = -2\omega_c \times \dot{\rho} - \dot{\omega}_C \times \rho - \omega_T \times (\omega_T \times \rho) + (\mu/r_C^3)[r_C - (r_C/r_T)^3 r_T] - U_C + U_D \tag{16.1}$$

where ω, r, and U represent orbital angular rate, position, and acceleration in the Earth centered frame (ECF). μ is the geocentric gravitational constant. The disturbances, including natural perturbations and other bounded uncertainties such as U_T, the target maneuver, are represented by U_D.

Notice the small relative range between the satellites compared with geocentric distance of the chaser, that is, $r \ll r_T$, the higher-order terms of Taylor series expansion of $r_T = r + r_C$ are negligible. Equation (16.1) can be rewritten as follows [2, 14]

$$\dot{X} = AX - BU_C + BU_D \tag{16.2}$$

$$X = \begin{bmatrix} \rho \\ \dot{\rho} \end{bmatrix}; \quad A = \begin{bmatrix} A_{11} & A_{12} \\ A_{21} & A_{22} \end{bmatrix}; \quad X = \begin{bmatrix} B_1 \\ B_2 \end{bmatrix}$$

where $U = [u_x, u_y, u_z]^T$, $A_{11} = B_1 = O_{3\times3}$, $A_{12} = B_2 = I_{3\times3}$. Matrices A_{21} and A_{22} are time-variant and expressed as

$$A_{21} = \begin{bmatrix} \omega_C^2 + 2\mu/r_C^3 & \dot{\omega}_C & 0 \\ -\dot{\omega}_C & \omega_C^2 - \mu/r_C^3 & 0 \\ 0 & 0 & -\mu/r_C^3 \end{bmatrix}; \quad A_{22} = \begin{bmatrix} 0 & 2\omega_C & 0 \\ -2\omega_C & 0 & 0 \\ 0 & 0 & 0 \end{bmatrix}$$

In Eq. (16.2), the state transfer matrix can be derived with different types of coordinates. We use DOE-based transfer function because nonlinear modeling error has the weakest propagation in the state space of orbital elements [1]. Define the orbital element set as $\sigma = [a, \theta, i, u_1, u_2, \Omega]^T$. The argument of latitude is $\theta = u + f$. $u_1 = e\cos u$ and $u_2 = e\sin u$. The difference set of the satellites is $\delta\sigma = \sigma_C - \sigma_T$. If $U_D = 0$, the transition function of $\delta\sigma$ and the geometric transformation function between X and $\delta\sigma$ are

$$\delta\sigma(t) = \varphi(t, t-\delta)\delta\sigma(t-\delta) \tag{16.3}$$

$$X(t) = \phi(t)\delta\sigma(t) \tag{16.4}$$

in which $\delta > 0$ is small enough. φ and ϕ are specified in [3]. Then, for arbitrary t_1 and $t_2 = t_1 + \delta$, state transfer function satisfies

$$X(t_2) = \Phi(t_2, t_1)X(t_1) \tag{16.5}$$

where the transfer matrix $\Phi(t_2, t_1) = \phi(t_2)\varphi(t_2, t_1)\phi^{-1}(t_1)$.

ORO-based dynamic model of relative motion is expressed by Eq. (16.2) or (16.5). Compared with the traditional SRO model, the orbital parameters and states in the ORO model vary with both the orbit position and the osculating orbit of the chaser. That is to say, A and Φ are derived a cluster of time-variant Keperian orbits, or a cluster of non-Keplerian orbits due to the potential chaser control. As a result, the body-fixed reference frame can help update orbital parameters in real time by using navigation system so that the nonlinear propagation of the modeling error can be restrained. Therefore, the ORO model has much higher precision than the SRO model.

To implement a tracking algorithm, it is necessary to derive an iterative form of Eq. (16.5). Note Φ_k as the transfer matrix in the k-th sampling time span, $X_k(k)$ and $X_k(k+1)$ as the states at the start and the end of the time period. Considering the sampling frequency is high, we may have

$$X_k(k+1) = \Phi_k(k+1, k)X_k(k) \tag{16.6}$$

where $\Phi_k(k+1, k) = \phi_k(k+1)\varphi_k(k+1, k)\phi_k^{-1}(k)$. Additionally, the reference orbit is updated at each connection time point, as shown in Fig. 16.1. Rotation of coordinates is required to combine the transfer functions together. The rotation matrix from LVLH to ECF can be written as

$$I = \begin{bmatrix} \cos\Omega_C & -\sin\Omega_C & 0 \\ \sin\Omega_C & \cos\Omega_C & 0 \\ 0 & 0 & 1 \end{bmatrix} \begin{bmatrix} 1 & 0 & 0 \\ 0 & \cos i_C & -\sin i_C \\ 0 & \sin i_C & \cos i_C \end{bmatrix} \begin{bmatrix} \cos\theta_C & -\sin\theta_C & 0 \\ \sin\theta_C & \cos\theta_C & 0 \\ 0 & 0 & 1 \end{bmatrix} \tag{16.7}$$

Then, $\rho_k(k+1)$, the end state of the k-th period, can be matched with $\rho_{k+1}(k+1)$, the initial state of the next period by using

$$\rho_{k+1}(k+1) = I^T(k+1)I(k)\rho_k(k+1) \tag{16.8}$$

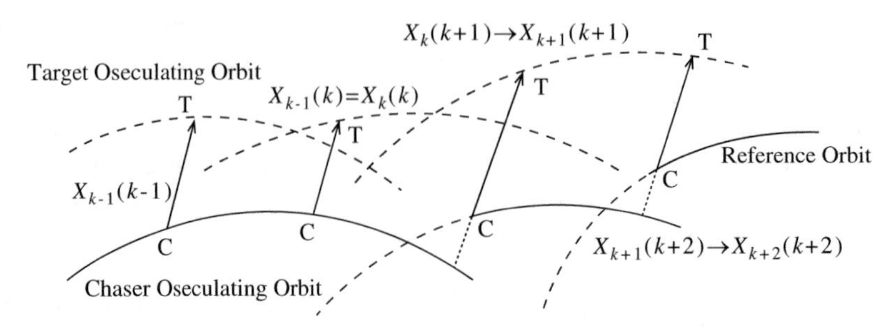

Fig. 16.1 Osculating reference orbit

Thus, we may get the state connection function as

$$X_{k+1}(k+1) = I_R(k+1,k)X_k(k+1) \tag{16.9}$$

where

$$I_R(k+1,k) = \begin{bmatrix} I^T(k+1)I(k) & 0 \\ \dot{I}^T(k+1)I(k) + I^T(k+1)\dot{I}(k) & I^T(k+1)I(k) \end{bmatrix}$$

Then, an iterative form of the ORO relative motion model can be described by Eqs. (16.6) and (16.9). If no chaser maneuver appears, it will degenerate to the SRO model.

16.3.2 Coupled Attitude Dynamics

The attitude dynamics of the chaser with thrust control is governed by [13]

$$\dot{\omega}_A = -J^{-1}\tilde{\omega}_A J\omega_A + J^{-1}N_{GGT} + J^{-1}M \tag{16.10}$$

$$N_{GGT} = 3\mu(\tilde{r}_C^e J r_C^e)/r_C^3 \tag{16.11}$$

where $J \in \mathbb{R}^{3\times 3}$, $\omega_A = [\omega_{Ax}, \omega_{Ay}, \omega_{Az}]^T$, N_{GGT}, and $M = [M_x, M_y, M_z]^T$ denote mass-dependent inertia, attitude angular rate, gravity gradient torque, and external torques, respectively. r_C^e indicates the unit position vector. In this chapter, vectors with an over tilde represent the skew symmetric matrices.

The attitude kinematic model based on quaternion can be expressed as

$$\dot{q}_A = \frac{1}{2}\begin{bmatrix} -\tilde{\omega}_A & \omega_A \\ -\omega_A^T & 0 \end{bmatrix} q_A \tag{16.12}$$

where $q_A = [q, q_0]^T$ and $q = [q_1, q_2, q_3]^T$. Their derivatives can be transformed into Euler angular rate by

$$\begin{bmatrix} \omega_A \\ 0 \end{bmatrix} = 2 \begin{bmatrix} q_0 I_{3\times3} - \tilde{q} & -q \\ q^T & q_0 \end{bmatrix} \dot{q}_A \tag{16.13}$$

Let $X_A = [q_A, \omega_A]^T$. The attitude dynamics of the chaser can be written as

$$\dot{X}_A = f_A(X_A, N_{GGT}, t) + B_A M + B_A M_D \tag{16.14}$$

where f_A is the dynamics function derived by Eqs. (16.10), (16.12), and $B_A = [O_{4\times3}, J^{-1}]^T$. M_D denotes attitude disturbances. We note $f_A = [f_{A1}, f_{A2}]^T$, where $f_{A1} \in \mathbb{R}^{4\times1}$ and $f_{A2} \in \mathbb{R}^{3\times1}$. As pointed out in [13], the coupling between orbital and attitude motion is exactly due to the existence of f_A.

16.4 ORO-Based Robust Tracking Algorithm

The tracking algorithm discussed in this section uses robust filtering to adapt the potential target maneuver and the ORO model to consider the follow-up control of the chaser. Such an ORO-based robust tracking method is expected to be available for two maneuvering satellites.

Here, we use RAREKF [6] as the robust filtering method to acquire stable and accurate tracking results in the existence of modeling errors and disturbances.

Denote the measurement vector in the reference frame at the $(k+1)$-th period as $Y_{k+1}(k+1)$, the noise as $v_{k+1}(k+1)$. The connection functions of neighboring measurements are

$$Y_k(k+1) = I^T(k) I(k+1) Y_{k+1}(k+1) \tag{16.15}$$

$$v_k(k+1) = I^T(k) I(k+1) v_{k+1}(k+1) \tag{16.16}$$

and the measurement equation at the k-th time span is

$$Y_k(k+1) = H_k(k+1) X_k(k+1) + v_k(k+1) \tag{16.17}$$

where $H \in \mathbb{R}^{3\times6}$. Then, ORO-based measurement model is described by Eqs. (16.15)–(16.17).

For the k-th iterative period, RAREKF has the following standard steps:

1. Prediction. One-step state prediction and covariance are

$$X_k(k+1|k) = \Phi_k(k+1, k) X_k(k|k) \tag{16.18}$$

$$P_k(k+1|k) = \Phi_k(k+1, k) P_k(k|k) \Phi_k^T(k+1, k) \tag{16.19}$$

2. Switching. Prediction covariance matrix is replaced with an adaptive switching function as

$$\Sigma_k(k+1|k) = \begin{cases} (P_k^{-1}(k+1|k) - \gamma^{-2}L_k^T(k+1)L_k(k+1))^{-1}, & \bar{P}_{Yk}(k+1) > \alpha P_{Yk}(k+1) \\ P_k(k+1|k), & \text{otherwise.} \end{cases} \tag{16.20}$$

where γ and α are tunable attenuation factor and redundancy factor. Let

$$L_k(k+1) = \gamma\left(P_k^{-1}(k+1|k) - \left[\left(\operatorname{diag}\frac{\bar{P}_{Yk}(k+1)}{P_{Yk}(k+1)} \otimes I_2\right)\right]^{-1} \alpha P_k^{-1}(k+1|k)\right)^{1/2} \tag{16.21}$$

in which $\otimes$ represents Kronecker product. P_Y and $\bar{P}_Y$ denote estimated and real measurement noise covariance, respectively. Take

$$\bar{P}_{Yk}(k+1) = \begin{cases} \hat{Y}_k(k+1)\hat{Y}_k^T(k+1) & k=0 \\ \frac{\rho P_{Yk}(k) + \hat{Y}_k(k+1)\hat{Y}_k^T(k+1)}{\rho+1} & k>0 \end{cases} \tag{16.22}$$

where ρ is the forgetting factor. $\hat{Y}_k$ is the measurement residue.

3. Innovation. Measurement residue and covariance are

$$\hat{Y}_k(k+1) = Y_k(k+1) - H_k(k+1)\hat{X}_k(k+1|k) \tag{16.23}$$

$$P_{Yk}(k+1) = H_k(k+1)\Sigma_k(k+1|k)H_k^T(k+1) + V_k(k+1) \tag{16.24}$$

4. Update. Filtering gain, state estimates, and state covariance matrix are

$$K_k(k+1) = \Sigma_k(k+1|k)H_k^T(k+1)P_{Yk}^{-1}(k+1) \tag{16.25}$$

$$\hat{X}_k(k+1|k+1) = \hat{X}_k(k+1|k) + K_k(k+1)\hat{Y}_k(k+1) \tag{16.26}$$

$$P_k(k+1|k+1) = (\Sigma_k^{-1}(k+1|k) + H_k^T(k+1)V_k^{-1}(k+1)H_k(k+1))^{-1} \tag{16.27}$$

Let $k = k+1$ and update Φ_k. Equations (16.18)–(16.27) complete an iteration cycle.

Nevertheless, to take RAREKF into the ORO model, an additional step for variable matching is required.

5. Match. Measurement residue can be matched by Eq. (16.15), noise covariance by

$$V_k(k+1) = I^T(k)I(k+1)V_{k+1}(k+1)I^T(k+1)I(k) \tag{16.28}$$

Considering Eqs. (16.6) and (16.9), it is easy to get the matching function of state estimates as

$$\hat{X}_k(k|k) = I_R(k, k-1)\hat{X}_{k-1}(k|k) \tag{16.29}$$

and of estimates covariance as

$$P_k(k|k) = \phi_k(k)\phi_{k-1}^{-1}(k-1)P_{k-1}(k|k)\phi_{k-1}^{-1}(k-1)\phi_k^T(k) \tag{16.30}$$

Then, the ORO-based RAREKF tracking algorithm can be fully expressed by Eqs. (16.18)–(16.27), (16.15), and (16.28)–(16.30). The difference from the standard RAREKF is at the step of variable matching, which is necessary for the time-variant reference frame.

16.5 Six-DOF Follow-up Control

For tracking an active target satellite, the chaser should be follow-up controlled due to the limited capability of onboard sensors. The ORO-based tracking algorithm gives a feasible way to tracking under control but does not consider the way of controlling the chaser. In this section, the whole scheme of follow-up tracking is clarified with the solution of follow-up control.

Figure 16.2 figures out the scheme, in which target is deemed as a black box with only relative position measurable by onboard radar of the chaser. There is no other information transformed between satellites.

In the scheme, follow-up control law has six DOFs composed of relative orbit and chaser attitude. The robust tracking algorithm gives the orbit control feedback of relative states, restraining the relative distance in the measurable region. In other words, keep relative trajectory at the desired formation geometry, that is, $X_A \to \bar{X}_A$.

Additionally, the chaser needs attitude control to maintain the target moving in sight. Suppose the attitude sensor is at the in-track direction of the body-fixed frame. The objective of attitude control is to make the in-track axis follow the target sightline, that is, let X_A, the real attitude approximates to the desired $\bar{X}_A$, determined by X and provided by the transformation block.

In Fig. 16.2, the transform function can be derived as follows. Note the unit vector of the in-track axis as X_{In}^e. It has the Euler axis vector $X_{Eu} = \rho \times X_{In}^e$. Then, the relative position can be expressed with attitude quaternion as

$$\begin{bmatrix} q_0 = \cos(\rho/2) & q_1 = -\rho_z \sin(\sigma/2)/\sqrt{\rho_x^2 + \rho_z^2} \\ q_2 = 0 & q_3 = \rho_x \sin(\rho/2)/\sqrt{\rho_x^2 + \rho_z^2} \end{bmatrix} \tag{16.31}$$

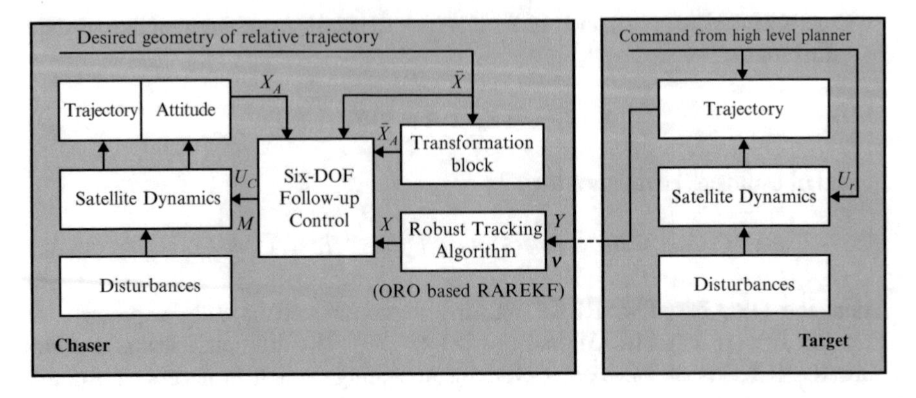

Fig. 16.2 Six-DOF follow-up robust tracking scheme

where $\sigma = \tan^{-1}(\sqrt{\rho_x^2 + \rho_z^2}/\rho_y)$. Denote Eq. (16.31) as $q_A = g_1(\rho, \sigma)$. Taking Eq. (16.13) into consideration, the relation between Euler angular rate and relative state can be written as

$$\omega_A = g_2(q_A)\dot{q}_A \tag{16.32}$$

where

$$g_2 = 2[q_0 I_{3\times3} - \tilde{q} \quad -q]$$

and

$$\begin{bmatrix} \dot{q}_0 = -\dot{\sigma}\cos(\sigma/2)/2 & \dot{q}_1 = \rho_{xz}q_3 - \rho_z\rho_\sigma q_0 \\ \dot{q}_2 = 0 & \dot{q}_3 = -\rho_{xz}q_1 + \rho_x\rho_\sigma q_0 \end{bmatrix}, \rho_{xz} = \frac{\dot{\rho}_x\rho_z - \rho_x\dot{\rho}_z}{\rho_x^2 + \rho_z^2}$$

$$\rho_\sigma = \frac{\dot{\sigma}}{2\sqrt{\rho_x^2 + \rho_z^2}}; \dot{\sigma} = \frac{-\dot{\rho}_y\sqrt{\rho_x^2 + \rho_z^2} - \rho_y(\rho_x\dot{\rho}_x + \rho_z\dot{\rho}_z)(\rho_x^2 + \rho_z^2)^{-3/2}}{\rho_x^2 + \rho_z^2 + \rho_z^2}$$

Equations (16.31) and (16.32) imply that the requirement of chaser attitude is determined by the relative state, explicating the coupling of orbit and attitude in the follow-up control.

16.6 Designing of DOSMC

To design the follow-up control law, we need to consider the following two facets. One is the fuel cost optimization and the other is the potential disturbances. Sliding mode control is a natural way for such a scenario.

In this section, we present a fuel-optimized dynamic optimal SMC and apply it into the six-DOF follow-up control.

16.6.1 Traditional OSMC

Typically, we may define the relative state error $X = X - \bar{X}$, so that the sliding surface can be written as

$$S = \dot{\rho} + \lambda\rho \tag{16.33}$$

where $\lambda \in \mathbb{R}^{3\times3}$ is a diagonal and positive definite weight matrix. Let the reaching law be a switching function with constant gain as

$$\dot{S} = \epsilon \mathrm{sgn}(S) \tag{16.34}$$

Taken into Eq. (16.2), the orbital control law of the chaser can be formulated as

$$U_C = A_{21}\rho + (A_{22} + \lambda)\dot{\rho} + A_{21}\bar{\rho} + A_{22}\dot{\bar{\rho}} + \ddot{\bar{\rho}} + \epsilon \mathrm{sgn}(S) \tag{16.35}$$

where $\epsilon \geq \|U_D\|$ to guarantee the robust stability. The operator $\|\cdot\|$ notes two-norm of real vectors or spectral norm of real matrices.

If λ is given, the sliding surface of SMC will be condensed to a three-dimensional hyper-curve. An optimal sliding surface is developed in [11] and [8] by optimizing λ, where the adopted objective function of such OSMC is

$$J = \int_{t_0}^{t} X^T(\tau)Q(\tau)X(\tau)d\tau \tag{16.36}$$

where Q is positive definite. Applied with the nonsingular transformation below

$$L^T(t)Q(t)L(t) = \begin{bmatrix} Q_{11}(t) & Q_{12}(t) \\ Q_{21}(t) & Q_{22}(t) \end{bmatrix}; Q_{11}, Q_{12}, Q_{21}, Q_{22} \in \mathbb{R}^{3\times3}, L \in \mathbb{R}^{6\times6}$$

into Eq. (16.36), it can be reformed as

$$J = \int_{t_0}^{t} [\rho^T(\tau)Q_{11}(\tau)\rho(\tau) + 2\rho^T(\tau)Q_{12}(\tau)\dot{\rho}(\tau) + \dot{\rho}^T(\tau)Q_{22}(\tau)\dot{\rho}(\tau)]d\tau \tag{16.37}$$

According to the theory of optimal LQR, the optimal λ can be derived by

$$\lambda(t) = Q_{22}^{-1}(t)[A_{12}^T(t)P'(t) + Q_{12}^T(t)] \tag{16.38}$$

in which P' satisfies the following Riccati equation:

$$\dot{P}' + P'A' + A'^T P' - P'B'R'B'^T P' + Q' = 0 \tag{16.39}$$

where $A' = A_{11} - A_{12}Q_{22}^{-1}Q_{12}^T$, $B' = A_{12}$, $Q' = Q_{11} - Q_{12}Q_{22}^{-1}Q_{12}^T$, and $R' = Q_{22}$.

It is clear to see that the sliding surface of the OSMC is optimized by minimizing (16.36). However, the optimized sliding surface does not take control information into account, so it is not fuel optimal. Moreover, the optimal λ equals to a constant matrix when Q is time-invariant. That is to say, the sliding surface will still be a reduced-dimensional curve, but a fuel-optimal state trajectory generally belongs to the full-dimensional state space. Therefore, an SMC law with a full-dimensional optimal sliding surface needs to be designed for fuel optimization.

16.6.2 Dynamic Optimal Sliding Surface

We design the sliding surface by using the state trajectory of a reference model under nominal linear quadratic control. The nominal control is optimal for minimizing its cost function

$$J = \int_{t_0}^{t} [X^T(\tau)Q(\tau)X(\tau) + U_C^T(\tau)R(\tau)U_C(\tau)]d\tau \tag{16.40}$$

where $R \in \mathbb{R}^{3\times3}$ is positive definite. According to Eq. (16.2), the following system

$$\dot{X}(t) = AX(t) - BU_C^*(t) \tag{16.41}$$

is chosen as the referenced model, in which U_C^* is a nominal LQR for tracking control to relative orbit, expressed as [14]

$$U_C^* = K(t)(X(t) - \bar{X}) \tag{16.42}$$

The optimal gain matrix is

$$K \triangleq [K_p \quad K_d] = R^{-1}B^T P \tag{16.43}$$

$$\dot{P} + PA + A^T P - PBRB^T P + Q = 0 \tag{16.44}$$

Actually, Eq. (16.42) is an optimal PD controller with control gain K_p and $K_d \in \mathbb{R}^{3\times3}$.

Then, the state of the controlled system (16.41) is

$$X^*(t, t_0) = \Phi(t, t_0)X^*(t_0) + \int_{t_0}^{t} \Phi(t, \tau)BU_C^*(\tau)d\tau \tag{16.45}$$

where $X^*(t_0)$ is the reference initial state, equaling the real value $X(t_0)$. The expression describes the desired state trajectory with (16.40) minimized. It implies that the optimal relation between position and velocity is time-variant and determined by control, so the nominal trajectory is full-dimensional and dynamic optimal and we take it as the DOSS to optimize the state variable under disturbance.

16.6.3 DOSMC for Orbital Relative Motion

Write the DOSS as a generalized form

$$S = X - X^* \tag{16.46}$$

Note $S = [(S^{\rho})^T, (S^{\dot{\rho}})^T]^T$. Substituting Eq. (16.42) into (16.41), we have

$$\dot{X}^*(t) = AX^*(t) \tag{16.47}$$

$$A = \begin{bmatrix} A_{11} & A_{12} \\ A_{21} & A_{22} \end{bmatrix} \tag{16.48}$$

where $A_{11} = O_{3\times3}$, $A_{12} = I_{3\times3}$, $A_{21} = A_{21} - K_p$, and $A_{22} = A_{22} - K_d$. Design the reaching law as

$$\dot{S} = -\alpha AS - \epsilon K\text{sgn}(S) \tag{16.49}$$

where $\alpha \geq 0$ is tunable to control the reaching speed of states to the sliding surface. The optimal gain matrix is to balance the diversity of orthogonal directions.

Implementing Eqs. (16.46) and (16.49)–(16.2), the dynamic optimal sliding-mode control law of relative orbital motion can be derived with the form

$$U_C(t) = U_C^*(t) + (1 - \alpha)U_C^{'}(t) + \epsilon K_d\text{sgn}(S^{\rho}) \tag{16.50}$$

in which

$$U_C^{'}(t) = A_{21}(\rho - \rho^*) + A_{22}(\dot{\rho} - \dot{\rho}^*) \tag{16.51}$$

Clearly, the equivalent term of Eq. (16.50) is exactly an optimal LQR when $\alpha = 1$. Hence, the expression of DOSMC can be regarded as an extended form of the usual LQR.

Use $V = S^T S/2$ as a Lyaponuv function to obtain

$$\dot{V} = S^T \dot{S} = -|S^T|(\epsilon K + U_d \text{sgn}(S)) - \alpha S^T A \tag{16.52}$$

Matrix $\tilde{A}$ is negative definite for (16.47) is stable, so $\dot{V} \leq 0$ if $\epsilon \geq \|K_d^{-1}\|\|U_D\|$. Because the satellite thrust is finite, the control vector has an upper bound as

$$\|U_C(t)\| \leq U_m \tag{16.53}$$

Then, the switching constant should satisfy

$$\|K_d^{-1}\|\|U_D\| \leq \epsilon \leq \|K_d^{-1}\|[U_m - \|U_C^*\| - (1-\alpha)\|U_C^{'}\|] \tag{16.54}$$

to keep system stable.

For the purpose of acquiring limited reaching time, we take

$$\alpha(S^{\dot{\rho}}, s) = \begin{cases} 0 & S^{\dot{\rho}} \leq s \\ 1 & S^{\dot{\rho}} > s \end{cases} \tag{16.55}$$

where $s \in \mathbb{R}^{3\times 1}$ is a user-defined threshold. If sliding mode is below the threshold, it means that the disturbance is weak, so we need to decrease α to increase reaching speed. Otherwise, α should be increased to approximate to the optimal control for fuel saving.

For further reducing fuel cost, a flying corridor may be designed by loosening the control precision. That is, control is fulfilled only if the state error is out of the acceptable range. Note corridor boundary width as X_{cb}. Equation (16.50) becomes

$$U_C(t) = \begin{cases} U_C^*(t) + (1-\alpha)U_C^{'}(t) + \epsilon K_d \text{sgn}(S^{\dot{\rho}}) & \|\tilde{X}\| > X_{cb} \\ 0 & \|\tilde{X}\| \leq X_{cb} \end{cases} \tag{16.56}$$

Then, the DOSMC for the system (16.2) can be fully described by Eqs. (16.56) and (16.51), satisfying (16.42)–(16.46), (16.48), (16.54), and (16.55).

The DOSMC can achieve two objectives: reduced fuel cost due to fuel-related optimization, and improved robustness caused by DOSS and adaptive reaching law.

16.6.4 DOSS-Based Control for Attitude Cooperation

The sliding surface of the chaser attitude control can be derived from the DOSS of relative orbit by using Eqs. (16.31) and (16.32). Note the attitude error as $X_A = X_A - \bar{X}_A$. The sliding surface of attitude can be written as

$$S_A = X_A - X_A^* \triangleq [(S_A^q)^T, (S_A^{\omega_A})^T]^T \tag{16.57}$$

Based on Eq. (16.14) and the attitude corridor boundary X_{Acb}, we may design attitude cooperative control law as

$$M(t) = \begin{cases} f_A(X_A^*, t) - f_A(X_A, t) - j\epsilon_A \text{sgn}(S_A^{\omega_A}); & \|\tilde{\omega}_A\| > X_{Acb} \\ 0 & \|\tilde{\omega}_A\| \leq X_{Acb} \end{cases} \quad (16.58)$$

where ϵ_A is the constant switching gain.

Consider the attitude thrust has an upper bound denoted as M_m. Based on the Lyaponuv function $V = S_A^T S_A/2$, we may know that the inequity

$$\|M_D\| \leq \epsilon_A \leq \|J^{-1}\|[M_m - \|f_A(X_A^*)\| + \|f_A(X_A)\|] \quad (16.59)$$

should be satisfied to guarantee the system stability. Define maximum sight angular range as σ_m. According to Eq. (16.31), it has the maximal quaternion error as

$$q_{Am} = |g(\rho, \sigma) - g(\rho, \sigma_m)| \quad (16.60)$$

Then, we get two constraint conditions as

$$|\rho| \leq \|\rho\|\sigma_m \quad (16.61)$$

$$|q_A| \leq q_{Am} \quad (16.62)$$

They can guarantee the errors caused by relative orbit and chaser attitude control small enough to keep the target maintaining at the sight range. Usually, Eq. (16.61) can be naturally satisfied due to the relatively large $\|\rho\|$, whereas we need

$$J^{-1}\left[\int_0^t g_2^{-1}(q_A)\tau d\tau\right]^{-1}(q_A - q_A^* - q_{Am})$$

$$\leq \epsilon_A \leq J^{-1}\left[\int_0^t g_2^{-1}(q_A)\tau d\tau\right]^{-1}(q_A - q_A^* + q_{Am}) \quad (16.63)$$

to satisfy Eq. (16.62).

Proof Substitute Eq. (16.58) into Eq. (16.14) to get

$$\dot{\omega}_A(t) = f_{A2}(X_A^*, t) + J\epsilon_A \text{sgn}(S_A^{\omega_A}) \quad (16.64)$$

Considering $\dot{\omega}_A^*(X_A^*, t) = f_{A2}(X_A^*, t)$, it has

$$\omega_A(t) = \omega_A^*(X_A^*, t) + J\epsilon_A \int_0^t \text{sgn}(S_A^{\omega_A}) d\tau \quad (16.65)$$

Clearly, if ϵ_A exists, $q_A \to q_A^*$, then, $g_2(q_A) \to g_2(q_A^*)$. Integrate Eq. (16.65) and apply Eq. (16.32) to obtain

$$q_A(t) = q_A^*(t) + J\epsilon_A \int_0^t g_2^{-1}(q_A^*) \int_0^{\tau_1} \text{sgn}(S_A^{\omega_A}) d\tau d\tau_1 \tag{16.66}$$

Substitute it into (16.62) and notice that $\int_0^t \text{sgn}(S_A^{\omega_A}) d\tau \le t$, then Eq. (16.63) can be drawn.

Thus, the DOSS-based control law for attitude cooperation can be described by Eq. (16.58), with the switching gain satisfying Eqs. (16.59) and (16.63). It is coupled with the DOSMC of relative orbit because the attitude sliding surface is generated by the orbital one and their control objectives are coherent. Then, the six-DOF follow-up control law can be formulated by combining Eqs. (16.56) with (16.58).

Note that for chattering restraint, typically we use a saturation function

$$\text{sat}(S, \eta) = \begin{cases} S/\eta; & S \le \eta \\ \text{sgn}(S); & S > \eta \end{cases} \tag{16.67}$$

to replace the signum function. $\eta \in \mathbb{R}^{3\times 1}$ is the small boundary width.

16.7 Simulations

Three simulation cases are shown in this section. Case A is to compare robust tracking algorithms based on SRO and ORO models for two maneuvering satellites. Case B is to illustrate the advantage of DOSMC over the usual SMC and OSMC methods. In the case, a scenario of hovering control is adopted without considering tracking error and target maneuver. Case C is to verify the effectiveness of the six-DOF follow-up tracking scheme.

The orbital elements of the initial satellite orbits are

$$a_T = a_C = 6678.137\,\text{km}, e_T = e_C = 0.05, i_T = i_C = 28.5^\circ$$

$$\Omega_T = \Omega_C = 28.5^\circ, f_T = f_C = 133.912^\circ, u_T = 0.5^\circ, u_C = 0$$

The two satellites follow in-track formation [4]. The true states of orbital motion are calculated from Keplerian equations with J_2 perturbation. Both satellites have maneuver capability but without information transfer.

Case A Two satellites are supposed to maneuver 200 s from the time 10,300 s along radial and in-track directions with the same magnitude of acceleration of 0.02 m/s^2. The RAREKF is used as the filtering algorithm. Figure 16.3 shows the filtering error based on the SRO and ORO models, respectively. It is clear to see that, after

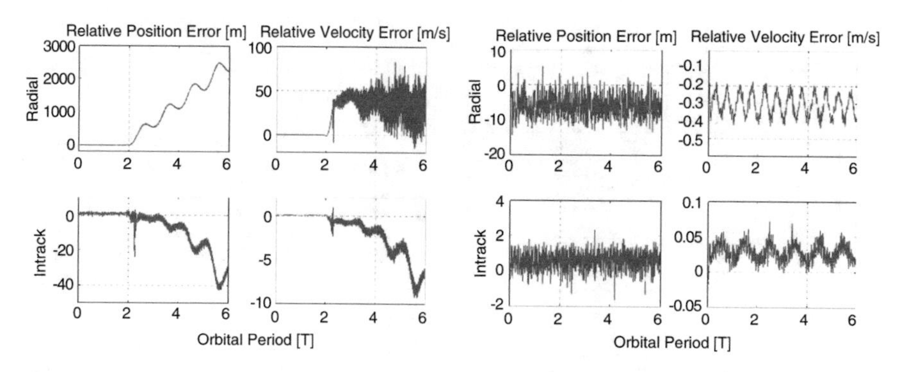

Fig. 16.3 Tracking error of RAREKF based on SRO model (Left) and based on ORO model (Right)

the chaser maneuvers, relative motion moves away from the SRO. The model error increases and worsens the tracking results. The filtering error based on the ORO model gets less effect from the chaser maneuver, showing much higher precision than the SRO model.

Clearly, the RAREKF can reach m-level relative position accuracy under the target maneuver, so the ORO-based RAREKF is much more suitable for tracking between maneuvering satellites.

Case B Consider a scenario of controlling the chaser satellite to hover around a free-flying target. The objective is to keep relative position at the initial state with relative velocity zero. Assume that signal loss by sensor or measurement link fault appears irregular and sometimes unavoidable, starting at the 50 s and lasting 100 s. During the process of disturbance, relative state is stayed at the final feedback value before signal loss.

We take $s = [10^{-7}, 10^{-7}, 10^{-7}]^T$, $\eta = [10^{-3}, 10^{-3}, 10^{-3}]^T$, $Q = I_{6\times6}$, and $R = I_{3\times3}$. λ is chosen arbitrarily in SMC, equaling to $10^{-1}I_{3\times3}$, and taken as Eq. (16.38) in OSMC.

The state error of DOSMC is compared with that of SMC and OSMC methods, showing the results in Fig. 16.4. Clearly, the DOSMC has the smallest deviation from the desired hover position with the best stability.

Figure 16.5 presents the comparison result of fuel cost. For the purpose of simplicity, we normalize the gross velocity increment by taking

$$g(U_C, t) = \frac{G(U_C, t)}{G(U_C^*, t)} \tag{16.68}$$

where $G(U_C, t) = \int_0^t \|U_C(\tau)\|d\tau$ is defined as the total velocity increment. U_C^* is chosen as the norm because of the independence to disturbance.

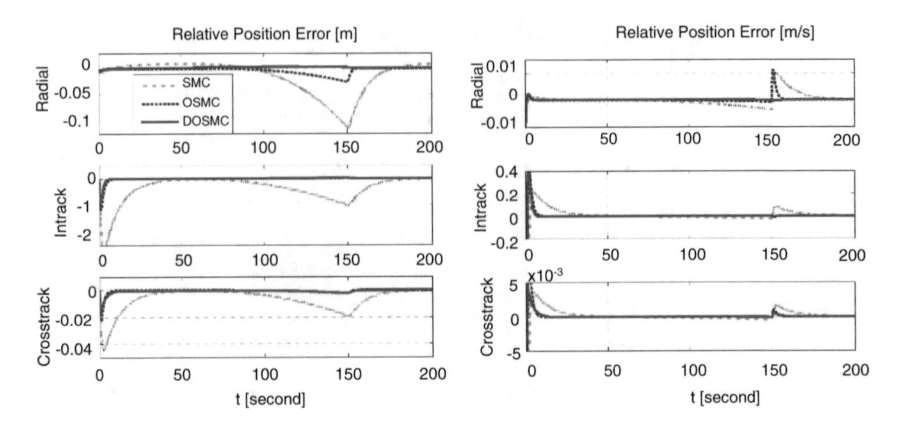

Fig. 16.4 Comparison of the relative state error with different control method

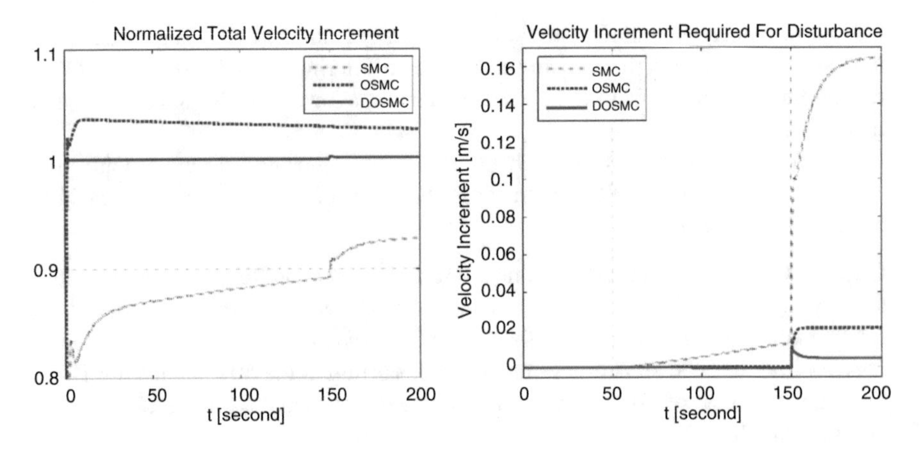

Fig. 16.5 Comparison of the fuel cost with different control method

In the left figure of Fig. 16.5, SMC yields an insufficient thrust making state error augmented greatly. The OSMC achieves better control precision but costs excessive fuel. The DOSMC shows the best approximation to the nominal optimal control, which means the least fuel cost, as illustrated in the right figure.

Case C To verify the six-DOF follow-up tracking scheme, assume that the two satellites work normally. The target is not cooperative with the chaser, maneuvering along radial and in-track with $0.001\ \text{m/s}^2$. Keep the maneuver time the same as Case A. The ORO-based RAREKF is used as tracking algorithm, Eqs. (16.56) and (16.58) based on DOSMC are as six-DOF follow-up control law. The objective is to realize long-term tracking of the target, keeping the desired initial formation.

Figure 16.6 shows the changing of relative orbit. Clearly, if the chaser is flying freely, even a tiny target maneuver may cause large deviation from the initial formation and make itself outside the tracking range. The follow-up control of the chaser can avoid the situation and maintain the relative motion between two satellites.

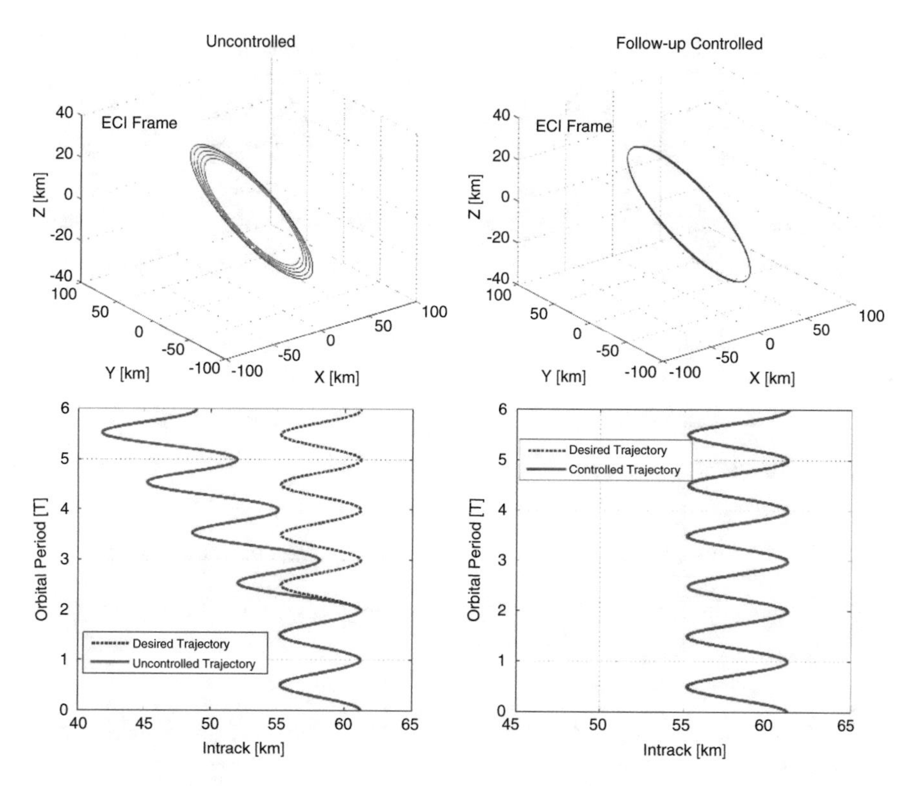

Fig. 16.6 Comparison of the relative orbit change

The six-DOF control law is shown in Fig. 16.7. It is implemented with

$$U_m = 0.001\,\text{m/s}^2,\ M_m = 0.002\,\text{kg}\,\text{m}^2/\text{s}^2,\ J = \text{diag}[60, 70, 40]\,\text{kg}\,\text{m}^2$$

$$X_{cb} = [50\,\text{m}, 60\,\text{m}, 10\,\text{m}, 0.5\,\text{m/s}, 0.2\,\text{m/s}, 0.1\,\text{m/s}]^T,$$
$$X_{Acb} = [0.05, 0.05, 0.05]^T\ \text{mrad/s}$$

Other parameters are the same as in Case B. After six orbital periods, total velocity increment for orbital control is 22.303 m/s and for attitude control is $10.312\,\text{kg}\,\text{m}^2/\text{s}^2$.

The errors of orbital and attitude control are shown in Figs. 16.8 and 16.9. Due to the existence of nonlinear modeling error, relative orbit hardly remains the same after the target maneuvers. That is why we design a flying corridor. The figures indicate that the control accuracy is loosened to the corridor boundary, leading to fuel saved. In Fig. 16.8, real lines present the control result with tracking error, and dashed lines are without considering the tracking error. The small difference implies that if the follow-up control guarantees the stability of the overall tracking-control-integrated system, the tracking and control errors can be deemed as uncoupled.

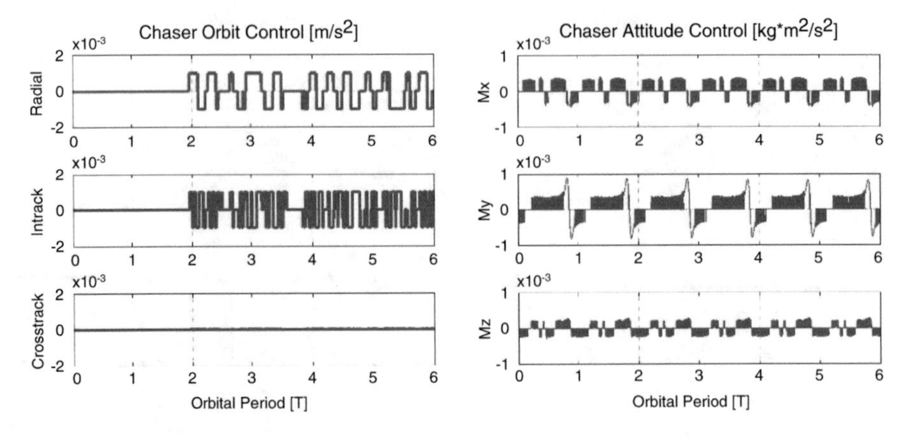

Fig. 16.7 Six-DOF follow-up control

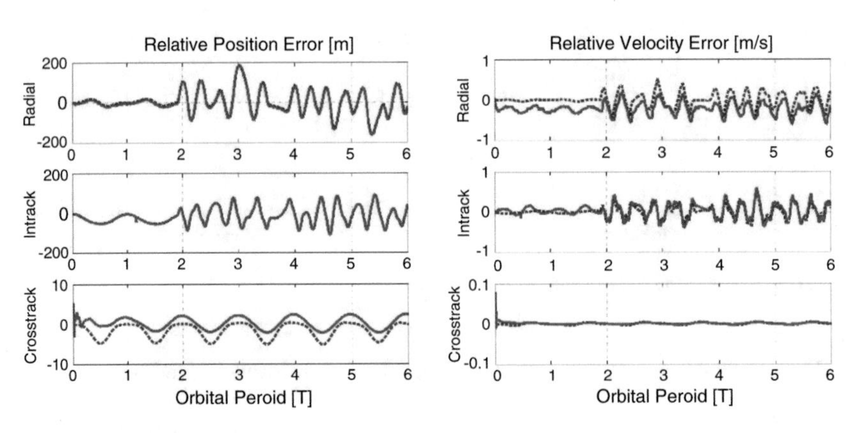

Fig. 16.8 State error of the relative orbital control

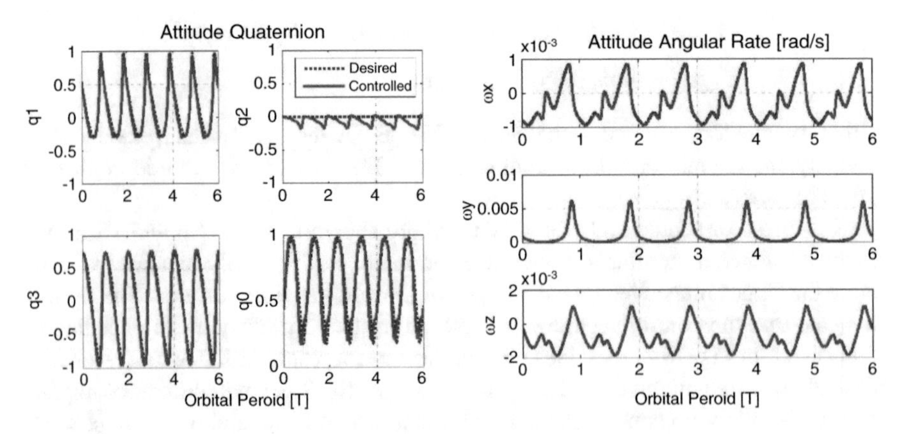

Fig. 16.9 State error of the chaser attitude control

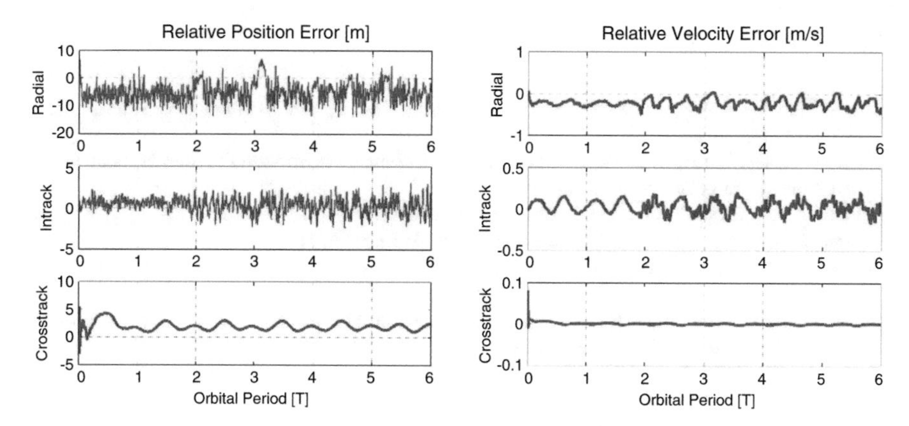

Fig. 16.10 Follow-up tracking error

The control-based tracking error is given in Fig. 16.10. During the entire tracking process, relative position error is at m-level and velocity at dm-level. Thus, we may conclude that the six-DOF follow-up tracking scheme is effective to keep tracking continuity for maneuvering satellites.

16.8 Conclusions

The discussed six-DOF follow-up tracking scheme is developed for long-term tracking of active satellite. It integrates tracking and control into a closed loop and guarantees the tracking continuity when any satellite appears maneuvering. The scheme is mainly composed of a robust tracking algorithm and a six-DOF follow-up control law. The presented ORO-based RAREKF is more suitable for tracking between maneuver satellites than traditional SRO-based methods. Also, the follow-up control law based on DOSMC can help orbital and attitude control obtain better robustness and less fuel cost than usual SMC and OSMC methods. In the scheme, tracking and control support with each other, so the robust tracking algorithm and follow-up control law should be implemented in an integrated framework.

Acknowledgements This work was supported in part by China Aviation Science Foundation (2009ZC57003) and in part by China Natural Science Foundation (No. 60775022 and No. 60674107).

References

1. Alfriend K, Yan H (2005) Evaluation and comparison of relative motion theories. J Guid Control Dyn 28(2):254–261
2. Broucke RA (2003) Solution of the elliptic rendezvous problem with the time as independent variable. J Guid Control Dyn 26(4):615–621
3. Gim DW, Alfriend KT (2003) State transition matrix of relative motion for the perturbed noncircular reference orbit. J Guid Control Dyn 26(6):956–971
4. Lane CM, Axelrad P (2006) Formation design in eccentric orbits using linearized equations of relative motion. J Guid Control Dyn 29(1):146–160
5. Li Y (2010) Autonomous follow-up tracking and control of noncooperative space target. Ph.D. dissertation, Shanghai Jiao Tong University
6. Li Y, Jing Z, Hu S (2010) Redundant adaptive robust tracking of active satellite and error evaluation. IET Control Theory Appl 4(11):2539–2553
7. Liu H, Li J, Hexi B (2006) Sliding mode control for low-thrust earth-orbiting spacecraft formation maneuvering. Aerosp Sci Technol 10(7):636–643
8. Salamci MU, Özgören MK, Banks SP (2000) Sliding mode control with optimal sliding surfaces for missile autopilot design. J Guid Control Dyn 23(4):719–727
9. Schumacher C, Cottrill G, Yeh HH (1999) Optimal sliding mode flight control. In: Guidance, navigation, and control conference and exhibit, p 4002
10. Seo J, Yu MJ, Park CG, Lee JG (2006) An extended robust h_∞ filter for nonlinear constrained uncertain systems. IEEE Trans Signal Process 54(11):4471–4475
11. Sinha A, Miller DW (1995) Optimal sliding-mode control of a flexible spacecraft under stochastic disturbances. J Guid Control Dyn 18(3):486–492
12. Xiong K, Zhang H, Liu L (2008) Adaptive robust extended Kalman filter for nonlinear stochastic systems. IET Control Theory Appl 2(3):239–250
13. Xu Y (2005) Sliding mode control and optimization for six DOF satellite formation flying considering saturation. J Astronaut Sci 53(4):433–443
14. Yoon H, Agrawal BN (2009) Novel expressions of equations of relative motion and control in Keplerian orbits. J Guid Control Dyn 32(2):664–669
15. Yu KK, Watson N, Arrillaga J (2005) An adaptive Kalman filter for dynamic harmonic state estimation and harmonic injection tracking. IEEE Trans Power Delivery 20(2):1577–1584
16. Yuankai L, Zhongliang J, Shiqiang H (2011) Dynamic optimal sliding-mode control for six-DOF follow-up robust tracking of active satellite. Acta Astronaut 69(7):559–570

Chapter 17
Optimal Dynamic Inversion Control for Spacecraft Maneuver-Aided Tracking

17.1 Introduction

In this chapter, we address a generalized problem of inter-satellite tracking, named maneuver-aided satellite tracking. The target satellite has potential non-cooperative maneuver, making tracking time window and tracking performances worsened. For tracking such a space target, we developed a general control-based tracking method which is called the spacecraft maneuver-aided tracking scheme (SMATS).

The SMATS has a generalized structure consisting of tracking algorithm, orbital and attitude control law, and some other coordinate and variables matching mechanisms. The objective of the SMATS is to keep chaser satellite staying with desired relative geometry autonomously, realizing long-term tracking with high precision.

The SMATS is a control-tracking-integrated closed-loop system. The system stability conditions are required to be clarified before control design. For this reason, the nonlinear error propagated in the system closed loop is analyzed, based on which an optimal dynamic inversion control (ODIC) law is developed with six DOFs. The control law, using a dimension-projected precise feedback linearization method, is an optimal solution for both orbital and attitude control and has the least impact to the tracking precision.

This chapter is organized as follows. Firstly, a framework of SMATS is presented, followed by the relative motion models and the tracking algorithm. The analysis for nonlinear error propagation is then provided to derive the system stability condition. After that, the six-DOF ODIC law is designed. Both the SMATS and ODIC law are verified by simulations finally. The presentation of this chapter is primarily based on [17, 18, 20].

Z. Jing et al., *Non-Cooperative Target Tracking, Fusion and Control*,
Information Fusion and Data Science, https://doi.org/10.1007/978-3-319-90716-1_17

17.2 Control-Based Spacecraft Tracking

17.2.1 MTT for Spacecraft

Maneuver target tracking (MTT) is a well-known topic for motion-body systems. There have been lots of MTT methods developed, mainly for two purposes. One is to improve the tracking precision under target maneuver, such as adaptive Gaussian sum (AGS) based EKF [15] and interacting multiple model (IMM) based UKF [3]. The other is to restrain the error propagation during the nonlinear dynamic process, such as cubature Kalman filter (CKF) [2], sparse Gauss–Hermite quadrature filter (SGHQF) [14], and other Bayesian filters.

Spacecraft tracking is a type of MTT working at a non-parallel gravitational field, where the relative motion follows orbital dynamics with unknown target maneuver and other external disturbances like natural perturbations. For the MTT problem for spacecraft, adaptive and robust tracking methods are necessary to use.

In previous works, Xiong provided an adaptive robust EKF [25] by introducing a self-switching attenuation factor to tune gains adaptively according to disturbances, and later presented a robust UKF [26] by a similar way, but Xiong's methods lead to failure of the switching mechanism under modeling errors. So, a redundant adaptive robust EKF (RAREKF) method was developed [18], where a flexible redundancy to modeling errors and disturbances is built by introducing a tunable redundant factor. The RAREKF can achieve tracking result with improved filtering performance.

Adaptive and robust tracking provides a feasible way to active satellite tracking, but the detective range and sight angles of the chaser satellite are generally limited. For a target satellite with potential uncooperative maneuver, the tracking continuity may probably become questionable.

Therefore, a control-based tracking method called follow-up tracking is proposed in [19], which has been presented in Chap. 16. In the method, chaser maneuver is allowed to compensate for the potential change of relative motion and to keep target in sight. The similar idea has already shown great potential in aerial visual tracking [6, 22–24]. With the detective performance of spacecraft onboard sensors improved, the follow-up tracking is a promising scheme for the MTT problem of spacecraft.

17.2.2 Maneuver-Aided Tracking

In this chapter, we discuss control-based tracking for spacecraft in detail and consider a generalized problem, called maneuver-aided tracking (MAT). The problem of MAT allows chaser maneuvering for active spacecraft tracking, so target tracking performance and chaser maneuver decision are supported by each other.

To give a structure of MAT applicable for spacecraft, we will present a spacecraft maneuver-aided tracking scheme (SMATS). It can be specified by four main parts: tracking algorithm, ORO coordinate matching, six-DOF maneuver control law, and attitude transformation, forming a closed control-tracking loop.

In SMATS, the aim of ORO coordinate matching is to update the reference frame during state estimation so that the quality of tracking results can be independent of the coupling between tracking and control. The objective of attitude transformation is to build the relationship between relative orbital states and chaser attitude, giving the desired attitude for control design of the chaser.

17.2.3 Problems of the Control Scheme in MAT

Through error propagation analysis, we found that SMATS is closed-loop robust stable when the chaser control law can keep each component of the propagated nonlinear errors bounded.

For the control design, besides the target maneuver, nonlinear modeling error is the most significant factor we need to consider to satisfy the robust stability condition. Typically, we use relative motion model to describe dynamics of spacecraft tracking where the exact model before linearization is nonlinear and non-affine. The chaser maneuver is an inherent variable of the reference orbit elements, varying with target maneuver and other system uncertainties. For such model, we have to approximate it to an affine nonlinear model and further to a linearized time-variant model. During the process, nonlinear modeling error is yielded unavoidably. The error will reduce the control accuracy and weaken the system stability, leading to increased fuel cost.

17.2.4 ODIC: An Optimal Control Method

Considering the problem above, a six-DOF control law based on optimal dynamic inversion control (ODIC) is built for applying in the SMATS. The control method is developed based on Keplerian models of the chaser and the target satellites. Given navigation information and relative measurements of the chaser, the Kepler models can help describe the relative orbital motion without any linearized approximation so that the nonlinear modeling error is avoided. On the other hand, to avoid the error caused by control design, dynamic inversion control is a more suitable method than others due to its precise analytical form and has been widely used in many fields, including ascent and reentry of aerospace flight [8, 9], disturbance attenuation of missiles [7], formation control [5], planetary entry [13], etc. By using the method based on dynamic inversion, the nonlinear model of relative motion can be projected into a decoupled linear space with reduced dimensions, providing an optimal control solution for the SMATS. Without any information loss of the dynamics, the ODIC has the best control precision and the least fuel cost and is easy for implementation.

17.3 Framework of SMATS

Consider a Chaser satellite and a Target satellite running on two proximity orbits, respectively. The Target is active with potential maneuver unknown to the Chaser. The aim of the Chaser is to achieve sustainable autonomous tracking to the Target with high precision. That necessitates the tracking scheme of the Chaser considering two aspects: robust tracking and maneuver control. Robust tracking aims to obtain desirable tracking performance under outer disturbances such as target maneuver, whereas maneuver control is to maintain the relative position and sensor sight range for tracking sustainability.

The overall structure of the SMATS is shown in Fig. 17.1. Because target maneuver and trajectory are unknown to the Chaser, dynamics of the Target is actually a black box, only bearings and range detectable for the Chaser onboard radar. The SMATS, as a functional module established on the Chaser, can be divided into four parts.

First is the block of robust tracking algorithm, aiming to produce fast and precise tracking results and give feedback of relative motion for the Chaser control. The ORO-based model is used to describe the dynamics of the relative motion. Due to the existence of both nonlinear modeling error and target maneuver, the RAREKF is adopted to provide a redundancy to those uncertainties.

Second is the block of ORO coordinate matching. During the Target maneuvering, the Chaser needs to adjust its current orbit, leading to change of relative motion. In order to guarantee the tracking continuity, reference orbital coordination should be adjusted in real time, so we use time-variant ORO of the Chaser as the reference frame.

Third is the block of six-DOF Chaser control law. The objective of orbit control is to keep relative motion at the desired formation geometry in the detectable range. To achieve that, the Chaser orbit should be controlled so that the real state of relative motion that can get from the tracking block may converge to the desired relative

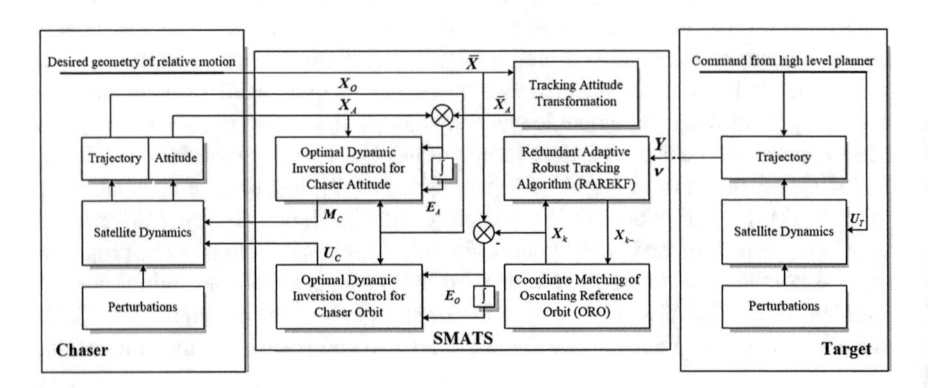

Fig. 17.1 The structure of the SMATS

trajectory. Also, the Chaser attitude should be controlled to maintain the Target at detectable eyesight. Here we assume the relative sensor fixed on the chaser satellite along in-track direction of the body-fixed frame. Then, the attitude control objective is equivalent to letting the sensor in-track axis follow the Target sightline.

Considering location varying of the assembled sensors, we need the fourth block for tracking attitude transformation. This block is to produce desired attitude curve according to the desired orbital geometry and the specified sensor position.

In the SMATS structure it is clear to see the control block maintains the expected relative motion to supply the Chaser a desirable tracking condition. That determines the feasibility of sustainable tracking to active spacecraft. For designing the control law, we focus on a six-DOF ODIC method based on dynamic inversion and integral optimization and the exact Keplerian and quaternion models to obtain the precise nonlinear optimal solution. By using ODIC, the chaser can keep target staying with desired status precisely if not considering the tracking error caused by noises from relative sensors and navigation devices.

17.4 System Model

In the SMATS, both satellite dynamics and relative orbital dynamics are required. The former defines the orbit and attitude motion, used for chaser control. The latter, describing the relative motion, is for target tracking.

17.4.1 Dynamics of Orbital Motion

Write the spacecraft Keplerian equation as

$$\ddot{\rho}_i = -\frac{\mu}{r_i^3}\rho_i + U_i + D_i; i = C, T \tag{17.1}$$

where ρ_i, U_i, and D_i denote spacecraft position, thrust acceleration, and disturbance in the Earth centered frame (ECF), respectively. Subscript C or T represents chaser or target. r is the geocentric distance, and μ the geocentric gravitational constant. Choose the osculating orbit of the chaser as the reference orbit. Denote the relative position vector as $\rho_R = [\rho_{Rx}, \rho_{Ry}, \rho_{Rz}]$, defined in radial, in-track, and cross-track (RIC) frame of the reference orbit (see Fig. 17.2). Then, the dynamic model of relative motion can be written as [10]

$$\ddot{\rho}_R = -2\omega_C \times \dot{\rho}_C - \dot{\omega}_T \times \rho_R - \omega_T \times (\omega_T \times \rho_R) + \frac{\mu}{r_i^3} \times \left[\rho_C - \frac{r_C^3}{r_T^3}\rho_T\right] - U_C + D \tag{17.2}$$

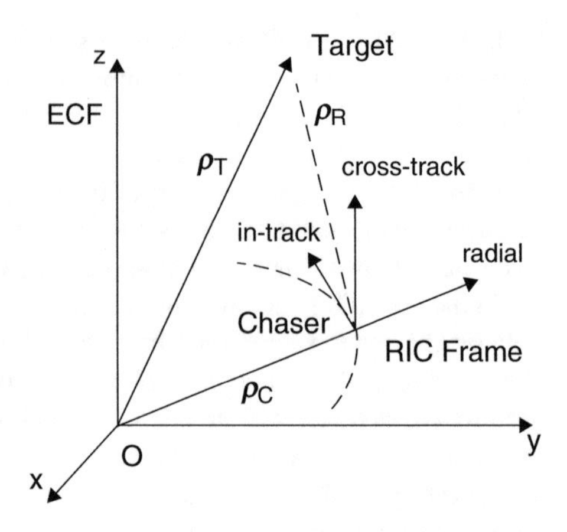

Fig. 17.2 Coordination system of orbital motion

in which ω is the orbital angular rate. External disturbances such as J_2 perturbation and other functional bounded uncertainties (e.g., target maneuver U_T) are included by D. Equation (17.2) describes the relative orbital motion precisely, but is inconvenient to use because of the nonlinearity and the unknown actual target status. A usual method is taking Taylor series to linearize r_T, assuming that the relative range compared with geocentric radial of the chaser is enough small, $r_R \ll r_T$. Denote relative state vector as $X_R = [\rho_R, \dot{\rho_R}]^T$. Then, Eq. (17.2) can be linearized as [4, 28]

$$\dot{X}_R = AX_R - BU_C + BD \tag{17.3}$$

Due to the weakest propagation of the nonlinear modeling error in orbital element space [1], the state transfer matrix in Eq. (17.3) is preferred to use differential orbital elements (DOE). To avoid the singularity problem of classical orbital elements and the calculation problem of small eccentricity, define the orbital element set as $\sigma = [a, \theta, u_1, u_2, \Omega]^T$. $\theta = u + f$, $u_1 = \cos u$ and $u_2 = e \sin u$. u and f represent the argument of periapsis and the true anomaly, respectively. The difference set is $\delta\sigma = \sigma_T - \sigma_C$. When $D = 0$, we may gain the transfer function of $\delta\sigma$ and the geometric transformation function between $\delta\sigma$ and X_R as

$$\delta\sigma(t) = \varphi(t, t - \delta)\delta\sigma(t - \delta) \tag{17.4}$$

$$X_R(t) = \phi(t)\delta\sigma(t) \tag{17.5}$$

where $\delta > 0$ is small enough. The detailed expressions are shown in [12]. Then, for arbitrary t_1 and $t_2 = t_1 + \delta$, the DOE-based model of relative orbital motion is

$$X_R(t_2) = \Phi(t_2, t_1)X_R(t_1) \tag{17.6}$$

with the transfer matrix $\Phi(t_2, t_1) = \phi(t_2)\varphi(t_2, t_1)\varphi^{-1}(t_1)$.

17.4.2 Dynamics of Attitude Rotation

The attitude dynamic equation with chaser control torque is governed by [27]

$$\dot{\omega}_A = -J^{-1}\tilde{\omega}_A J\omega_A + J^{-1}N_{GGT} + J^{-1}M_C \tag{17.7}$$

$$N_{GGT} = 3\mu(\tilde{\rho}_C^e J\rho_C^e)/r_C^3 \tag{17.8}$$

where $J \in \mathbb{R}^{3\times 3}$, ω_A, N_{GGT}, and M_C the rotational inertia, attitude angular rate, gravity gradient torque and control torque, respectively. Superscript e denotes the unit vector, and over tilde is for skew symmetric matrix. Clearly, rotation of the chaser is coupled with orbital motion through N_{GGT}.

The attitude kinematic equation using quaternion is well-known as

$$\dot{q}_A = \frac{1}{2}\begin{bmatrix} -\tilde{\omega}_A & \omega_A \\ -\omega_A^T & 0 \end{bmatrix} q_A \tag{17.9}$$

where quaternion $q_A = [q, q_0]^T$ and $q = [q_1, q_2, q_3]^T$. It can avoid the singularity problem of small Euler angles, and can be reformed as

$$\dot{q}_A = \frac{1}{2}\begin{bmatrix} -\tilde{q} & q \\ -q^T & q_0 \end{bmatrix}\begin{bmatrix} \omega_A \\ 0 \end{bmatrix} \tag{17.10}$$

which clarifies the relation between quaternion derivative and Euler angular rate.

17.5 Robust Tracking Algorithm

In the SMATS, robustness is necessary for tracking algorithm due to the potential target maneuver and chaser control. So, in this section, a robust tracking algorithm based on RAREKF [18] is presented. The algorithm is established on an ORO-based model, including two independent parts: filtering and matching.

17.5.1 The RAREKF

Denote the radar measurements (relative range, azimuth and elevation) in the kth time period as $Z_k(k)$ with noise $v_k(k)$. The subscript refers to the reference orbital frame. The unbiased converted measurement (UCM) technique [21] is used to keep measurements unbiased. Write the measurement model at the kth time period as

$$Y_k(k) = H_k(k)X_{Rk}(k) + v_k(k) \tag{17.11}$$

$$Y_k(k) = C_m Z_k(k) \tag{17.12}$$

where C_m is the conversion matrix. Y and υ are unbiased converted measurement and noise with covariance V. Detailed expressions of C_m and V can be found in [10]. Note $k\_$ and k_+ as the last and the next sampling time. For system (17.6), the standard steps of the RAREKF can be expressed as follows [18]:

Prediction: one-step prediction and covariance are

$$X_{Rk}(k|k\_) = \Phi_k(k, k\_)X_{Rk}(k\_|k\_) \tag{17.13}$$

$$P_k(k|k\_) = \Phi_k(k, k\_)P_k(k\_|k\_)\Phi_k^T(k, k\_) \tag{17.14}$$

Switch: replace prediction covariance with the adaptive switching function

$$\Sigma_k(k|k\_) = \begin{cases} (P_k^{-1}(k|k\_) - \gamma^{-2}L_k^T(k)L_k(k))^{-1} & \bar{P}_{Yk}(k) > \alpha P_{Yk}(k) \\ P_k(k|k\_), & \text{otherwise} \end{cases} \tag{17.15}$$

where γ and α are tunable attenuation factor and redundancy factor. P_Y and $\bar{P}_Y$ are the estimated and real measurement noise covariance. Take

$$L_k(k) = \gamma \left(P_k^{-1}(k|k\_) - \left[\left(\text{diag} \frac{\bar{P}_{Yk}(k)}{P_{Yk}(k)} \right) \otimes I_2 \right]^{-1} \alpha P_k^{-1}(k|k\_) \right)^{1/2} \tag{17.16}$$

$$\bar{P}_{Yk}(k) = \begin{cases} \hat{Y}_k(k)\hat{Y}_k^T(k), & k = 0 \\ \frac{1}{\eta+1}\left(\eta P_{Yk}(k\_) + \hat{Y}_k(k)\hat{Y}_k^T(k)\right), & k > 0 \end{cases} \tag{17.17}$$

in which $\otimes$ and I_2 represent the Kronecker product and two-dimensional identity matrix. η is the forgetting factor. Typically, let $\eta = 0.98$. $\hat{Y}$ denotes the measurement residue. Equation (17.15) indicates that the filtering mode switches from optimal to robust when $\bar{P}_Y$ exceeds α times of the predicted value P_Y.

Innovation: measurement residue and covariance are

$$\bar{Y}_k(k) = Y_k(k|k\_) - H_k(k)X_{Rk}(k|k\_) \tag{17.18}$$

$$P_{Yk}(k) = H_k(k)\Sigma_k(k|k\_)H_k^T(k) + V_k(k) \tag{17.19}$$

Update: filtering gain, state estimates and covariance matrix are

$$K_k(k) = \Sigma_k(k|k\_)H_k^T(k)P_{Yk}^{-1}(k) \tag{17.20}$$

$$X_{Rk}(k|k) = X_{Rk}(k|k\_) + K_k(k)\hat{Y}_k(k) \tag{17.21}$$

$$P_k(k|k) = (\Sigma_k^{-1}(k|k\_) + H_k^T(k)V_k(k)H_k(k))^{-1} \tag{17.22}$$

Update with $k = k_+$, and Eqs. (17.13)–(17.22) complete an iteration cycle of RAREKF.

17.5.2 ORO Coordination Matching

ORO refers to the osculating orbit of the chaser, which is taken as the real-time updated reference frame, so the state estimates in the current iterative cycle should be matching to the next reference frame at the end of each iterative time. Notice that the state estimates input to Eqs. (17.13) and (17.14) need to update their reference frame, so coordinate matching is necessary to connect each two neighboring cycles, by

$$X_{Rk+}(k|k) = I_R(k_+, k)X_{Rk}(k|k) \tag{17.23}$$

$$P_{k+}(k|k) = \phi_{k+}(k_+)\phi_k^{-1}(k)P_k(k|k)\phi_k^{-T}(k)\phi_{k+}^T(k_+) \tag{17.24}$$

where I_R is a conversion matrix of the state vector, formed as

$$I_R(k+1,k) = \begin{bmatrix} I^T(k+1)I(k) & 0 \\ \dot{I}^T(k+1)I(k) + I^T(k+1)\dot{I}(k) & I^T(k+1)I(k) \end{bmatrix} \tag{17.25}$$

in which

$$I = \begin{bmatrix} \cos\Omega_C & -\sin\Omega_C & 0 \\ \sin\Omega_C & \cos\Omega_C & 0 \\ 0 & 0 & 1 \end{bmatrix} \begin{bmatrix} 1 & 0 & 0 \\ 0 & \cos i_C & -\sin i_C \\ 0 & \sin i_C & \cos i_C \end{bmatrix} \begin{bmatrix} \cos\theta_C & -\sin\theta_C & 0 \\ \sin\theta_C & \cos\theta_C & 0 \\ 0 & 0 & 1 \end{bmatrix} \tag{17.26}$$

is the rotation matrix from RIC to ECF.

Combined with standard RAREKF, the robust filtering algorithm for the SMATS can be fully described by Eqs. (17.13)–(17.24). The RAREKF is convergent, guaranteeing the tracking stability of the SMATS.

17.6 Six-DOF Maneuver Control

Stable tracking is insufficient to achieve the closed-loop stability of the SMATS. A desirable maneuver control law of the chaser is also required. In this section, we will analyze the errors propagated in the closed control-tracking loop to derive a set of error bounded conditions for control design. If those conditions are satisfied, the SMATS can achieve closed-loop stability. Before that, we first present the attitude transformation to build up the relationship between the desired chaser attitude and the relative motion, making six-DOF control feasible.

17.6.1 Attitude Transformation

Assume that sensor sight axis is along in-track. Note its unit vector as X_{In}^{e}. Euler axis vector is calculated by $X_{Eu} = \rho_R \times X_{In}^{e}$. Then, the attitude quaternion can be derived from relative position by

$$q_A \triangleq g_p(\rho_R) = \begin{bmatrix} -\rho_{Rz} \sin\frac{\sigma}{2} / \sqrt{\rho_{Rx}^2 + \rho_{Rz}^2} \\ 0 \\ \rho_{Rx} \sin\frac{\sigma}{2} / \sqrt{\rho_{Rx}^2 + \rho_{Rz}^2} \\ \cos\frac{\sigma}{2} \end{bmatrix}. \tag{17.27}$$

where $\sigma = \tan^{-1}\left(\sqrt{\rho_{Rx}^2 + \rho_{Rz}^2} / \sigma_{Ry}\right)$. We may get the derivative attitude quaternion as

$$\dot{q}_A \triangleq g_d(\rho_R, \dot{\rho}_R) = \begin{bmatrix} \rho_{xz} q_3 - \rho_{Rz} \rho_\sigma q_0 \\ 0 \\ -\rho_{xz} q_1 + \rho_{Rx} \rho_\sigma q_0 \\ -\frac{\dot{\sigma}}{2} \cos\frac{\sigma}{2} \end{bmatrix} \tag{17.28}$$

where

$$\rho_{xz} = (\dot{\rho}_{Rx} \rho_{Rz} - \rho_{Rx}) \dot{\rho}_{Rz} / (\rho_{Rx}^2 + \rho_{Rz}^2);\ \rho_\sigma = \frac{\dot{\sigma}}{2} / \sqrt{\rho_{Rx}^2 + \rho_{Rz}^2}$$

$$\dot{\sigma} = -\left(\dot{\rho}_{Ry} \sqrt{\rho_{Rx}^2 + \rho_{Rz}^2} + \rho_{Ry} (\rho_{Rx} \dot{\rho}_{Rx} - \rho_{Rz} \dot{\rho}_{Rz}) (\rho_{Rx}^2 + \rho_{Rz}^2)^{-3/2}\right) / \|\rho_R\|^2$$

If using Eq. (17.10), the Euler angular rate ω_A can be calculated from relative state. The functions g_p and g_d are variant with sensor location, implying that the desired relative motion geometry governs the expected chaser attitude. That is to say, in the SMATS, the attitude control objective is unified with the orbital one. Thus, the six-DOF control law can be deemed as an integrated solution with two layers: inner for attitude and outer for position.

17.6.2 Error Analysis

For closed-loop stability of the SMATS, we analyze the propagation of tracking and control errors and derive a quantitative set of error bounded conditions for control design to guarantee bounded stability of the entire SMATS.

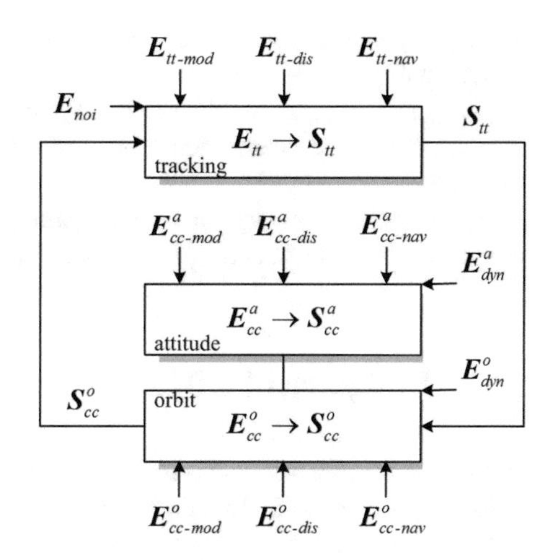

Fig. 17.3 Error propagation in the SMATS

Typically, the errors propagated in the SMATS are caused by model, disturbance, stochastic noise, and navigation, as shown in Fig. 17.3. The gross state errors during target tracking and chaser control can be written as

$$E_{tt} = E_{tt\text{-}mod}(\hat{M}) + E_{tt\text{-}dis}(D) + E_{tt\text{-}nav}(\hat{X}_C) + E_{noi} + S^o_{cc}(E^o_{cc}) \tag{17.29}$$

$$E^o_{cc} = E^o_{cc\text{-}mod}(\hat{M}) + E^o_{cc\text{-}dis}(D) + E^o_{cc\text{-}nav}(\hat{X}_C) + E^o_{dyn} + S_{tt}(E_{tt}) \tag{17.30}$$

$$E^a_{cc} = E^a_{cc\text{-}mod}(\hat{M}) + E^a_{cc\text{-}dis}(D) + E^a_{cc\text{-}nav}(\hat{X}_C) + E^a_{dyn} \tag{17.31}$$

where $\hat{M}$ and $\hat{X}_C$ denote the modeling and navigation errors with caused state errors $E_{tt\text{-}mod}$, $E_{tt\text{-}nav}$ of tracking and $E_{cc\text{-}mod}$, $E_{cc\text{-}nav}$ of control, and disturbance D as well. E_{noi} and E_{dyn} are the state errors from measurement noise and dynamic process of control. Superscript o or a represents orbit or attitude. Define S_{tt} and $S_{cc} = [S^o_{cc}, S^a_{cc}]^T$ as the propagated result of E_{tt} and E_{cc}, respectively.

From Fig. 17.3, we may know that tracking error and orbital control error are coupled through S_{tt} and S^o_{mc} and independent with attitude control. Clearly, if S_{tt} and S_{mc} are both bounded, the overall SMATS is stable. As a sufficient condition, it can be satisfied by the following three sub-conditions: (1) S_{tt} is bounded if S^o_{mc} bounded; (2) S^o_{mc} is bounded if S_{tt} bounded; and (3) S^a_{mc} is bounded.

The first sub-condition is already formed with using the robust tracking algorithm aforementioned. Because in the tracking algorithm, E_{noi} converges to the residue, i.e., $E_{noi} \to E_{res}$, and other terms at the right-hand side of Eq. (17.29) are external disturbance errors, restrained by the robust filtering mechanism in a small bounded E_{est} if S^o_{cc} bounded, that is, $E_{tt\text{-}mod}(\hat{M}) + E_{tt\text{-}dis}(D) + E_{tt\text{-}nav}(\hat{X}_C) + S^o_{cc}(E^o_{cc}) \to E_{est}$. Then, there is $E_{tt} \to E_{res} + E_{est} \triangleq S_{tt}$. Thus, S_{tt} is bounded.

Sub-conditions 2 and 3 express the requirements to orbit and attitude control for the system stability. When S_{tt} is bounded, Eqs. (17.30) and (17.31) indicate that S_{cc} will be stable only if the control law can let $E_{cc\text{-}mod}(\hat{M})$, $E_{mc\text{-}dis}(D)$, $E_{mc\text{-}nav}(\hat{X}_C)$

and E_{dyn} bounded with limited energy cost. Therefore, a sufficient condition set of system stability to design the control law is generated, as follows: (1) inherent errors $E_{mc\text{-}mod}$ and $E_{mc\text{-}dis}$ are bounded and better convergent asymptotically; (2) noises $E_{mc\text{-}nav}$ and S_{tt} are restrained robustly; and (3) the dynamic error E_{dyn} converges optimally.

Thus, it is necessary to find a control method for satisfying the sufficient condition set above to achieve the global stability of the entire SMATS.

17.7 Designing of ODIC

In this section, we provide a nonlinear optimal six-DOF control method based on ODIC for the SMATS. To gain the optimal control precision, the original two-body motion model and integral control are also used.

17.7.1 Orbital Control

For the SMATS, chaser maneuver is built on its own dynamics and hard to track the target status directly, but can follow its desired incremental curve, which can be obtained from the tracking and navigation results and the desired relative geometry. For simplified description, the navigation error of the chaser is not considered.

Considering Eq. (17.1) and noting $X_O = [\rho_C, \dot{\rho}_C]^T$, the Kepler model of chaser can be written as

$$\dot{X}_O = f_O(X_O) + g_O(X_O)U_C \tag{17.32}$$

Notice that the Lie bracket of the two vector fields in Eq. (17.32) has the form as

$$ad_{f_O}g_O(X_O) = \begin{bmatrix} 0 & -I_3 \\ \mu/r_C^3 - \mu\rho_C^2/r_C^3 & 0 \end{bmatrix}.$$

such that $N_O = [g_O, ad_{f_O}g_O]$ has full rank. Meanwhile, span$\{g_O\}$ is involutive, so Eq. (17.32) can be fed back linearized precisely.

Consider the three-dimensional vector space in Eq. (17.32) as a scalar space, that is, $y_O = \rho_C, y_O \in R$. Take

$$V_C = U_C - \mu\rho_C/r_C^3 \tag{17.33}$$

Then Eq. (17.32) can be reformed into a linearization form as

$$\dot{z}_O = Az_O + BV_C \tag{17.34}$$

$$z_O = \begin{bmatrix} y_O \\ \dot{y}_O \end{bmatrix}, A = \begin{bmatrix} 0 & 1 \\ 0 & 0 \end{bmatrix}, B = \begin{bmatrix} 0 \\ 1 \end{bmatrix}$$

Considering that the chaser will stop maneuvering if reaching the desired relative trajectory noted as $\bar{X}_O$ (variable with an over bar represents its desired value), let

$$\bar{V}_C = -\mu\bar{\rho}_C/r_C^3 \tag{17.35}$$

We have

$$\dot{\bar{z}}_O = A\bar{z}_O + B\bar{V}_C \tag{17.36}$$

and there is

$$\bar{X}_O = X_O + X_R - \bar{X}_R \tag{17.37}$$

where X_R and $\bar{X}_R$ are the target states relative to real and desired states of the chaser, i.e., $X_R = X_T - X_O$ and $\bar{X}_R = X_T - \bar{X}_O$. From Eqs. (17.34) and (17.36), it has

$$\dot{e}_O = Ae_O + BV_e \tag{17.38}$$

where $e_O = z_O - \bar{z}_O$ and $V_e = V_C - \bar{V}_C$. From Eq. (17.37), there is $e_O = \bar{X}_R - X_R$. To improve robustness, introduce the integral equation

$$\dot{e}_O^i = e_O \tag{17.39}$$

Note $E_O = [\int(y_O - \bar{y}_O)dt, e_O]^T$, then, the linearized error model of Eq. (17.1) can be derived from Eqs. (17.38) and (17.39), expressed as

$$\dot{E}_O = A_eE_O + B_eV_e \tag{17.40}$$

where A_e and B_e are constant matrices with controllable canonical form

$$A_e = \begin{bmatrix} 0\,1\,0 \\ 0\,0\,1 \\ 0\,0\,0 \end{bmatrix}, B_e = \begin{bmatrix} 0 \\ 0 \\ 1 \end{bmatrix}$$

Through minimizing the following cost function

$$\Lambda_O = \int (E_O^T Q_O E_O + V_e^T R_O V_e)dt \tag{17.41}$$

the integral optimal control law for Eq. (17.40) can be gained easily. Project it back to the vector space to obtain

$$V_e = -R_O^{-1}B_e^T(P_O \otimes I_3)E_O \tag{17.42}$$

in which P_O is symmetric definite positive, satisfying

$$P_OA_e + A_e^TP_O - P_OB_eR_O^{-1}B_e^TP_O + Q_O = 0 \tag{17.43}$$

where the weight matrix $Q_O \in R^{3\times 3}$, $R_O \in R$. Substitute Eqs. (17.33) and (17.35) into Eq. (17.42), we get the ODIC law for the chaser orbital motion as

$$U_C = \frac{\mu}{r_C^3}\rho_C - \frac{\mu}{\bar{r}_C^3}\bar{\rho}_C - R_O^{-1}B_e^T(P_O \otimes I_3)E_O \tag{17.44}$$

Applied with the control law above, the orbit motion in centric gravitational field can be linearized into a paralleled force field and optimized by an integral quadratic regulator without any error introduced from modeling or control process.

17.7.2 Attitude Control

Similarly, the attitude control law of the chaser can be derived. Equations (17.7) and (17.9) and noting $X_A = [q_A, \omega_A]^T$, the rotational dynamics has the form as

$$\dot{X}_A = f_A(X_A) + g_A(X_A)U_C \tag{17.45}$$

Also, we have the Lie bracket of f_A and g_A as

$$ad_{f_A}g_A(X_A) = \begin{bmatrix} \frac{1}{2}\begin{bmatrix} -\tilde{\omega}_A & \omega_A \\ -\omega_A^T & 0 \end{bmatrix} & \frac{1}{2}\begin{bmatrix} \tilde{q} & 0 \\ 0 & 0 \end{bmatrix} \\ 0 & -J^{-1}(\tilde{\omega}_A J + \dot{\tilde{\omega}}_A J\omega_A) \end{bmatrix}$$

Equation (17.45) can be fed back linearized precisely if the rank of $N_A = [g_A, ad_{f_A}g_A]$ equals to six for there is one redundant dimension in q_A, that is, only if $\omega_A \neq 0$.

If $\omega_A = 0$, Eq. (17.7) shows that $M_C = -N_{GGT}$. If $\omega_A \neq 0$, regard four-dimensional vector q_A as a scalar $y_A \in R$. From Eqs. (17.7) and (17.10), we may get $\ddot{y}_A = N_C$ where

$$N_C = \frac{1}{4}\begin{bmatrix} -\tilde{\omega}_A & \omega_A \\ -\omega_A^T & 0 \end{bmatrix} q_A + \frac{1}{2}\begin{bmatrix} \tilde{q} \\ -q^T \end{bmatrix} J^{-1}(-\tilde{\omega}_A J\omega_A + N_{GGT}) + \frac{1}{2}\begin{bmatrix} \tilde{q} \\ -q^T \end{bmatrix} J^{-1}M_C \tag{17.46}$$

Noting $z_A = [y_A, \dot{y}_A]^T$, Eq. (17.45) can be linearized as

$$\dot{z}_A = Az_A + BN_C \tag{17.47}$$

and the equation of the chaser having reached $\bar{X}_A$ is

$$\dot{\bar{z}}_A = A\bar{z}_A + B\bar{N}_C \tag{17.48}$$

where $\bar{X}_A$ can be acquired by Eqs. (17.27) and (17.28), and

$$\bar{N}_C = \frac{1}{4}\begin{bmatrix} -\tilde{\bar{\omega}}_A & \bar{\omega}_A \\ -\bar{\omega}_A^T & 0 \end{bmatrix}^2 \bar{q}_A + \frac{1}{2}\begin{bmatrix} \tilde{\bar{q}} \\ -\bar{q}^T \end{bmatrix} J^{-1}(-\tilde{\bar{\omega}}_A J\omega_A + N_{GGT}) \tag{17.49}$$

Define $E_A = [\int (y_A - \bar{y}_A)dt, z_A - \bar{z}_A]^T$. From Eqs. (17.47) and (17.48), we can get the error model as

$$\dot{E}_A = A_e E_A + B_e N_e \tag{17.50}$$

in which $N_e = N_C - \bar{N}_C$. Based on the projection back to the vector space, the chaser attitude can be controlled optimally by

$$N_e = -R_A^{-1} B_e^T (P_A \otimes I_4) E_A \tag{17.51}$$

by minimizing the cost function

$$\Lambda_A = \int (E_A^T Q_A E_A + N_e^T R_A N_e) dt \tag{17.52}$$

where $Q_A \in R^{3\times 3}$, $R_A \in R$ and P_A satisfies

$$P_A A_e + A_e^T P_A - P_A B_e R_A^{-1} B_e^T P_A + Q_A = 0 \tag{17.53}$$

Considering that the quaternion forms

$$\begin{bmatrix} \tilde{q}^T & -q \end{bmatrix} \begin{bmatrix} \tilde{q} \\ -q^T \end{bmatrix} = (1 - q_0^2) I_4$$

and substituting Eqs. (17.46) and (17.49) into Eq. (17.51), we can get the attitude control law of the chaser as

$$M_C = M_1 + M_2 - 2J(1 - q_0^2)^{-1} [\tilde{q}^T, -q] R_A^{-1} B_e^T (P_A \otimes I_4) E_A \tag{17.54}$$

where

$$M_1 = \frac{1}{2} J (1 - q_0^2)^{-1} [\tilde{q}^T, -q] \left[\begin{bmatrix} -\tilde{\bar{\omega}}_A & \bar{\omega}_A \\ -\bar{\omega}_A^T & 0 \end{bmatrix}^2 \bar{q}_A - \begin{bmatrix} -\tilde{\omega}_A & \omega_A \\ -\omega_A^T & 0 \end{bmatrix}^2 q_A \right]$$

$$M_2 = (\tilde{\omega}_A J \omega_A - N_{GGT}) - (1 - q_0^2)^{-1} (\tilde{q}^T \tilde{\bar{q}} + q \bar{q}^T)(\tilde{\bar{\omega}}_A J \bar{\omega}_A - N_{GGT})$$

Then, the six-DOF control law based on ODIC for the SMATS can be completely described by Eqs. (17.44) and (17.54). It is easy to verify that the control law satisfies the sufficient condition set to keep S_{cc} bounded. Cooperated with the robust tracking algorithm to guarantee a bounded S_{tt}, the SMATS can achieve stable.

17.8 Simulations

We use two simulation cases to verify the ODIC method and the entire SMATS. One is the in-track formation keeping [16] and the other is the spacecraft hovering. In both cases, assume that the target satellite has potential uncooperative maneuver and the chaser uses the SMATS to achieve sustainable tracking with high precision. In the SMATS, the six-DOF ODIC law is used for the chaser control and the ORO-based RAREKF is for the target tracking.

The first case is to compare the ODIC with the DOSMC presented in Chap. 16. The second case is to demonstrate the control–tracking interaction and the feasibility of the overall SMATS.

Both cases use the same initial classical orbital elements as below

$$a_T = a_C = 6678.317\,\text{km}, e_T = e_C = 0.05, i_T = i_C = 28.5^\circ$$

$$\Omega_T = \Omega_C = 28.5^o, f_T = f_C = 133.912^\circ, u_T = 0.5^\circ, u_C = 0$$

The initial orbital period of the target equals to 5432 s approximately. The simulations run 2 periods. The radar measurement variances $(1\,\text{m})^2$ for range and $(0.1\,\text{rad})^2$ for elevation and azimuth. The initial relative state variances are $(1\,\text{m})^2$ for position and $(0.1\,\text{m/s})^2$ for velocity. Take $\alpha = \text{diag}(3.5, 2.5, 10)$ for the RAREKF.

17.8.1 *Case A: Formation Keeping*

Two satellites are at in-track formation flying. The target satellite is supposed to maneuver 500 s from the time 4000 s, along radial and in-track with the acceleration $0.001\,\text{m/s}^2$. Take $Q_O = Q_A = 10^{-9} I_3$ and $R_O = R_A = 1$.

Figure 17.4 shows that the relative trajectory can hold successfully by the chaser control during and after target maneuvering, when the SMATS plays the essential role. The dash line shows the trajectory of the chaser uncontrolled, for which target maneuver breaks the geometry of relative motion, making the chaser lose the target. With the SMATS applied on the chaser, the in-track formation holds (thick solid line) so that the tracking condition becomes sustainable. The orbit control law and the error are given in Figs. 17.5 and 17.6, where ODIC is compared with DOSMC. Due to the chattering caused by the switching mechanism in DOSMC, we only consider the sliding surface, generated by optimal PD control (OPDC). From Fig. 17.5, we can see a large steady state error appears for the DOSMC due to linearization of the relative motion model, and cannot be reduced by OPDC and the switching law, but such large error does not occur for the ODIC. Moreover, the OPDC for DOSMC spends more fuel and will be magnified several times more than the ODIC if switching control is added. Figure 17.6 shows the result when J_2

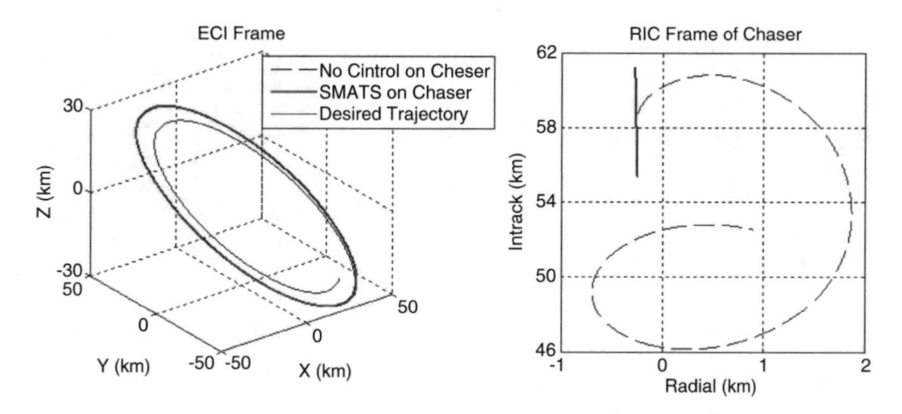

Fig. 17.4 Relative orbit of target to chaser in ECI and RIC frame of the chaser

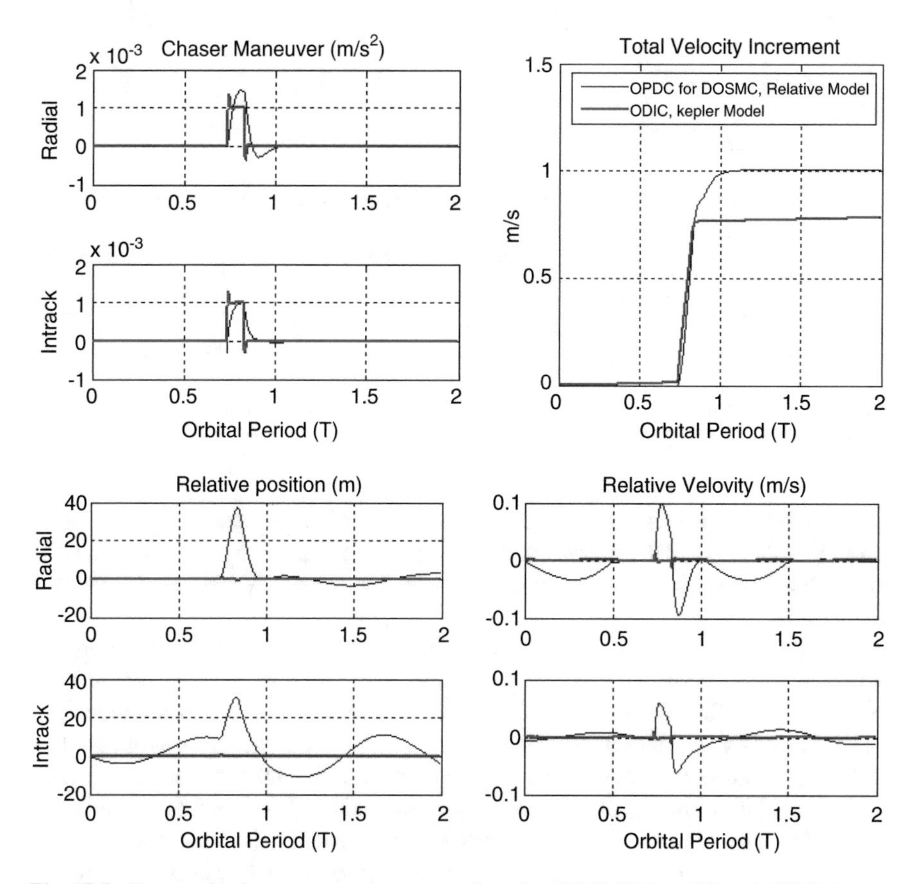

Fig. 17.5 Chaser orbital maneuver and state error by using ODIC (thick solid) and OPDC (thin)

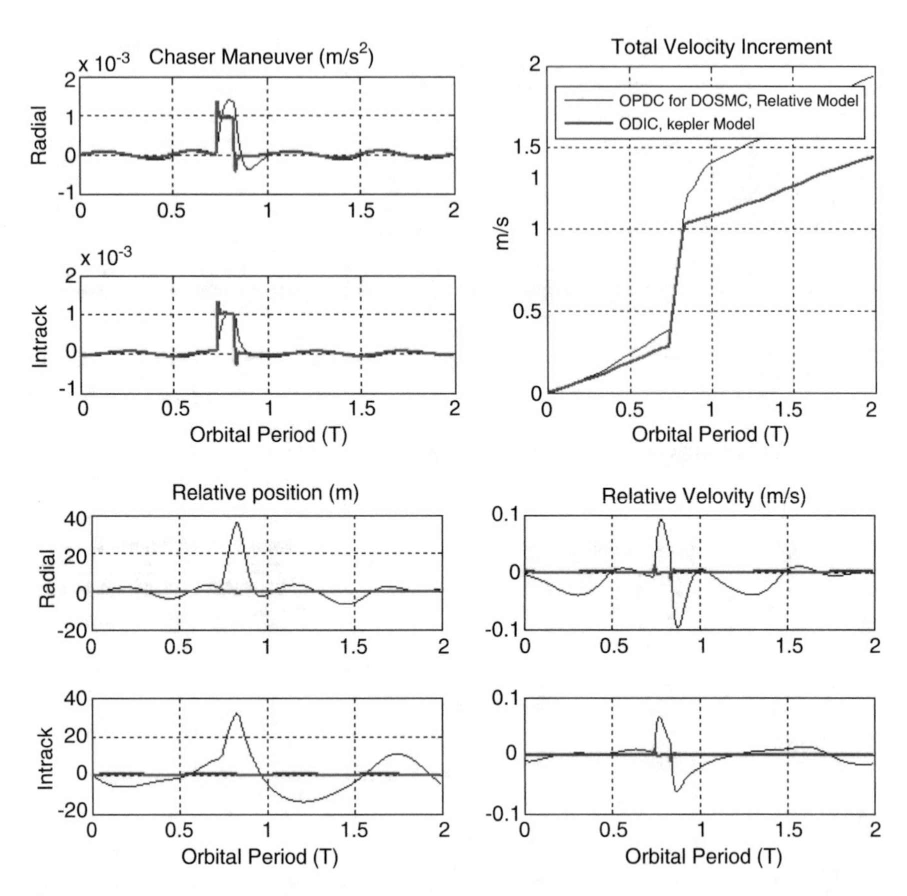

Fig. 17.6 Chaser orbital maneuver and state error by using ODIC (thick solid) and OPDC (thin) under J_2 perturbation

perturbation is considered besides target maneuver. It illustrates that the OPDC need more fuel than the ODIC to restrain the external disturbances. So, it is clear that the ODIC has better robustness, showing little effect to the steady state error under the J_2 perturbation.

Figure 17.7 demonstrates the attitude part of the chaser control law. The pointing error can be explicated by the angular error of Euler rotation axis, decreased from $\pm 30°$ for the DOSMC to $\pm 1°$ for the ODIC. The tracking error of using the ORO-based RAREKF is shown in Fig. 17.8. Because the tracking model is built based on linearized relative motion, the tracking result contains residue of the modeling error aside from the measurement noise. Nevertheless, the RAREKF regards the modeling error as a type of tolerable disturbances and can produce the state estimates fast following the real measurements. Thus, it has little effect to the filtering result. Tracking precision is kept at 10 m level for position and 1 m/s for velocity.

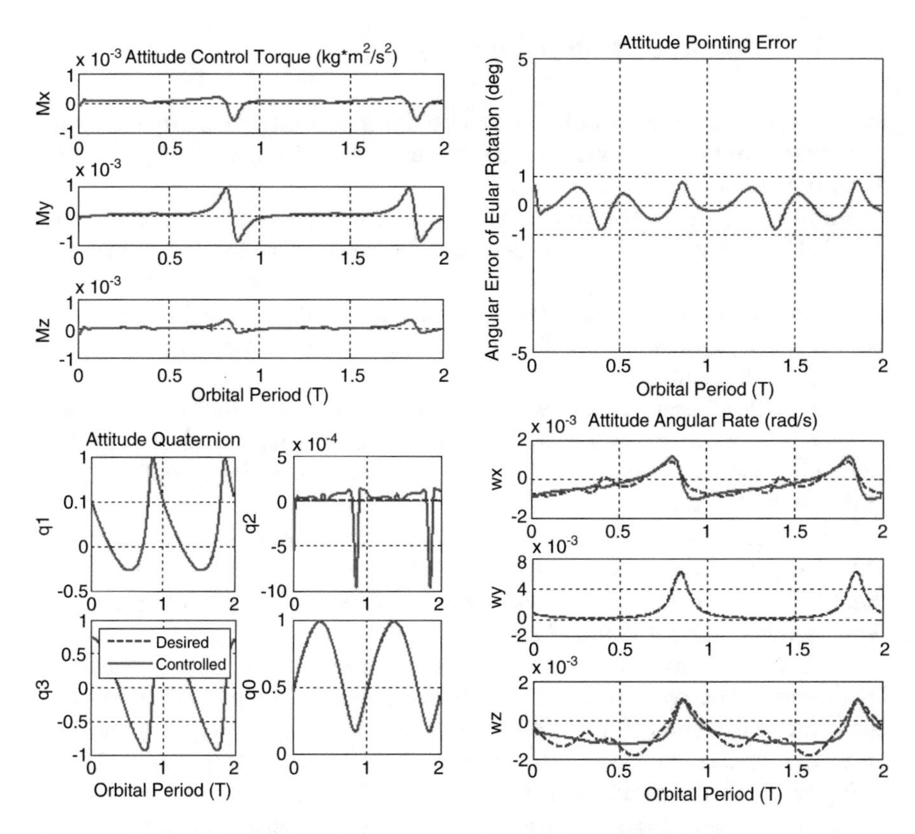

Fig. 17.7 Chaser attitude torque and angular state error generated by ODIC

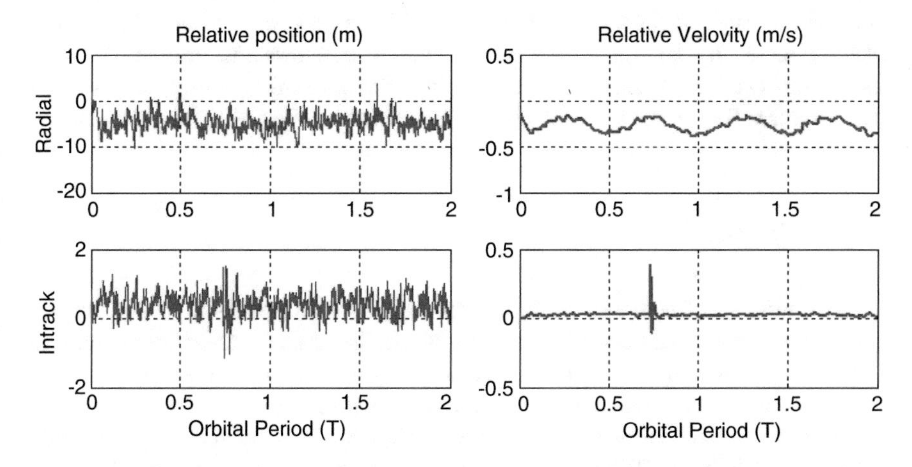

Fig. 17.8 Error in orbit estimate relative to the truth for SMATS with ORO-RAREKF

17.8.2 Case B: Spacecraft Hovering

In case A, tracking and control are considered independently to clarify the control performance of ODIC. In this case, control–tracking interaction will be illustrated to verify the entire SMATS.

For hovering around the target satellite, the chaser needs keep the relative range in its RIC frame and relative velocity of zero unchanged from the initial time.

In case A, in order to present the control performance of ODIC clearly, interaction of tracking and control is neglected. That is, for Eqs. (17.29) and (17.30), assume $S^o_{mc}(E^o_{mc}) = S_{tt}(E_{tt}) = 0$. In this case, the impact of these two terms to the entire SMATS is demonstrated.

Suppose that the target maneuver lasts 1000 s from the time 7000 s with the same magnitude as in case A. Applied by SMATS without changing the values of tracking and control parameters in case A, the chaser can achieve hovering around the target, as shown in Fig. 17.9.

In the figure we can learn, the position error to the hovering point is at 2 m. The orbit control and state error in radial and in-track directions are illustrated in Fig. 17.10. When the target maneuvers, an over offset is yielded to compensate for the deviation of relative motional states.

Figures 17.11 and 17.12 show the control and tracking results with their interaction. Compared with Fig. 17.10, the control error in Fig. 17.11 is increased due to the introduction of the tracking error. Nevertheless, the error propagation in the control-tracking loop is not enough to break up the closed-loop stability [11].

Figure 17.12 demonstrates that the tracking error has more immunity to the control error than control to tracking. That is to say, tracking algorithm plays a dominant role for the precision of the SMATS, whereas control law provides detectability support for sustainable tracking. The dotted line provides the tracking result effected by control error and the solid line shows the original. Clearly, interaction between tracking and control does not worsen the closed-loop system performance. Hence, the feasibility and effectiveness of the SMATS is verified.

17.9 Conclusion

For the problem of active satellite tracking, the SMATS presented in this chapter provides a generalized control-based solution. In the SMATS, the chaser control is introduced to guarantee the tracking sustainability with desirable tracking precision under the condition of potential uncooperative target maneuver.

In order to establish the SMATS, the ORO coordinate matching is built and combined with the RAREKF to develop a robust tracking algorithm. The algorithm can achieve fast and precise tracking under target maneuver and disturbances. Also, a six-DOF chaser control law based on ODIC is given to maintain the desired relative motion and guarantee the tracking sustainability. The ODIC law is derived

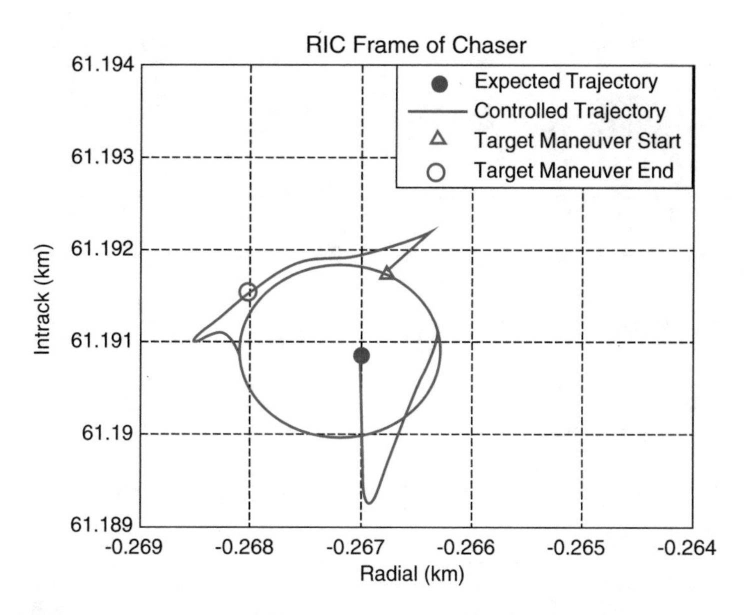

Fig. 17.9 Relative trajectory with the desired hovering point in RIC frame of the chaser

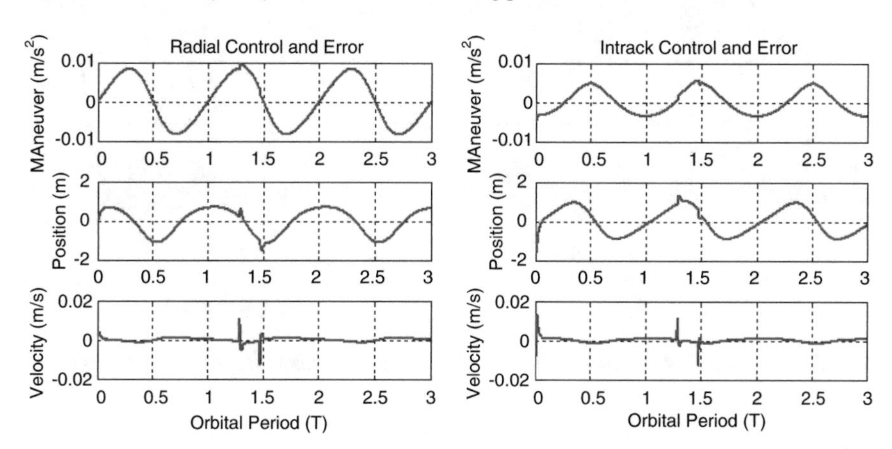

Fig. 17.10 Chaser orbit control and state error relative to the desired trajectory without impact from tracking error

with the Kepler models of both satellites, providing a nonlinear optimal solution for chaser control. The error propagation analysis has proven that the six-DOF ODIC law and the ORO-based RAREFK can guarantee the closed-loop stability of the SMATS.

Acknowledgements This work was supported in part by Natural Science Foundation of China (61175028, 60775022 and 60674107) and Aviation Science Foundation of China (2009ZC57003), and in part by the Canada Research Chair program.

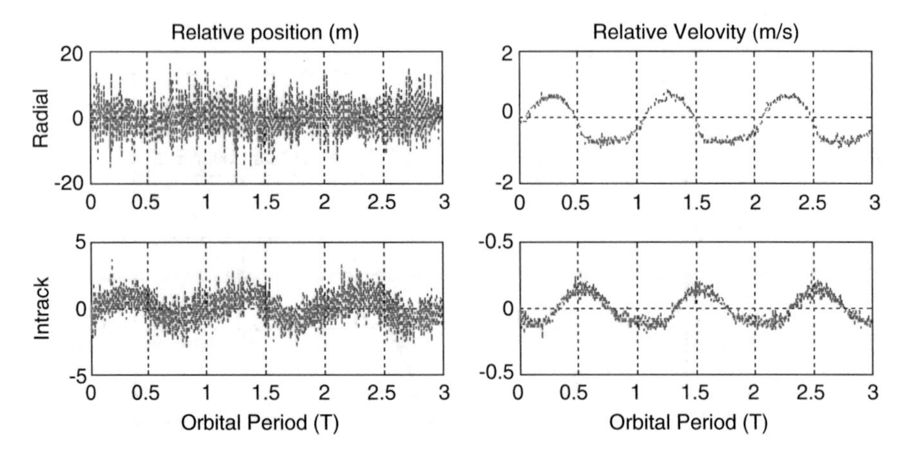

Fig. 17.11 Control error relative to the desired trajectory impacted by the tracking error

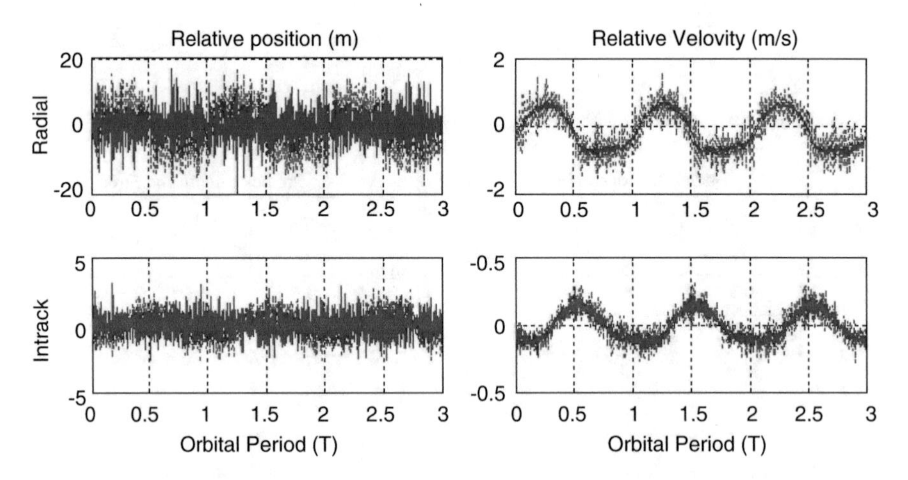

Fig. 17.12 Comparison of tracking results with (dotted) and without (solid) impact from control

References

1. Alfriend K, Yan H (2005) Evaluation and comparison of relative motion theories. J Guid Control Dyn 28(2):254–261
2. Arasaratnam I, Haykin S (2009) Cubature Kalman filters. IEEE Trans Autom Control 54(6):1254–1269
3. Babu VS, Shankar MR, Majumdar S, Rao SK (2011) Imm-unscented Kalman filter based tracking of maneuvering targets using active sonar measurements. In: International conference on communications and signal processing (ICCSP). IEEE, Piscataway, pp 126–130
4. Broucke RA (2003) Solution of the elliptic rendezvous problem with the time as independent variable. J Guid Control Dyn 26(4):615–621
5. Caicedo R, Valasek J, Junkins J (2004) Preliminary results of vehicle formation control using dynamic inversion. In: 42nd AIAA aerospace sciences meeting and exhibit, p 295

6. Cao C, Hovakimyan N, Evers J (2006) Active control of visual sensor for aerial tracking. In: AIAA guidance, navigation, and control conference and exhibit, p 6610
7. Chen WH (2003) Nonlinear disturbance observer-enhanced dynamic inversion control of missiles. J Guid Control Dyn 26(1):161–166
8. Da Costa R, Chu Q, Mulder J (2003) Reentry flight controller design using nonlinear dynamic inversion. J Spacecr Rocket 40(1):64–71
9. Doman DB, Ngo AD (2002) Dynamic inversion-based adaptive/reconfigurable control of the x-33 on ascent. J Guid Control Dyn 25(2):275–284
10. Duan Z, Han C, Li XR (2004) Comments on "unbiased converted measurements for tracking". IEEE Trans Aerosp Electron Syst 40(4):1374
11. Fujimoto K, Scheeres D, Alfriend K (2012) Analytical nonlinear propagation of uncertainty in the two-body problem. J Guid Control Dyn 35(2):497–509
12. Gim DW, Alfriend KT (2003) State transition matrix of relative motion for the perturbed noncircular reference orbit. J Guid Control Dyn 26(6):956–971
13. Hormigo T, Araújo J, Câmara F (2008) Nonlinear dynamic inversion-based guidance and control for a pinpoint mars entry. In: AIAA guidance, navigation and control conference and exhibit, p 6817
14. Jia B, Xin M, Cheng Y (2011) Sparse gauss-hermite quadrature filter with application to spacecraft attitude estimation. J Guid Control Dyn 34(2):367–379
15. Kramer KA, Stubberud SC (2005) An adaptive gaussian sum approach for maneuver tracking. In: IEEE aerospace conference. IEEE, Piscataway, pp 2083–2091
16. Lane CM, Axelrad P (2006) Formation design in eccentric orbits using linearized equations of relative motion. J Guid Control Dyn 29(1):146–160
17. Li Y (2010) Autonomous follow-up tracking and control of noncooperative space target. Ph.D. dissertation, Shanghai Jiao Tong University
18. Li Y, Jing Z, Hu S (2010) Redundant adaptive robust tracking of active satellite and error evaluation. IET Control Theory Appl 4(11):2539–2553
19. Li Y, Jing Z, Hu S (2011) Dynamic optimal sliding-mode control for six-DOF follow-up robust tracking of active satellite. Acta Astronaut 69(7):559–570
20. Li Y, Jing Z, Liu G (2014) Maneuver-aided active satellite tracking using six-DOF optimal dynamic inversion control. IEEE Trans Aerosp Electron Syst 50(1):704–719
21. Longbin M, Xiaoquan S, Yiyu Z, Kang SZ, Bar-Shalom Y (1998) Unbiased converted measurements for tracking. IEEE Trans Aerosp Electron Syst 34(3):1023–1027
22. Shirzi MA, Hairi-Yazdi M (2011) Active tracking using intelligent fuzzy controller and kernel-based algorithm. In: IEEE international conference on fuzzy systems (FUZZ). IEEE, Piscataway, pp 1157–1163
23. Stepanyan V, Hovakimyan N (2007) Adaptive disturbance rejection controller for visual tracking of a maneuvering target. J Guid Control Dyn 30(4):1090–1106
24. Stepanyan V, Hovakimyan N (2008) Visual tracking of a maneuvering target. J Guid Control Dyn 31(1):66–80
25. Xiong K, Zhang H, Liu L (2008) Adaptive robust extended Kalman filter for nonlinear stochastic systems. IET Control Theory Appl 2(3):239–250
26. Xiong K, Liu L, Zhang H (2009) Modified unscented Kalman filtering and its application in autonomous satellite navigation. Aerosp Sci Technol 13(4):238–246
27. Xu Y (2005) Sliding mode control and optimization for six DOF satellite formation flying considering saturation. J Astronaut Sci 53(4):433–443
28. Yoon H, Agrawal BN (2009) Novel expressions of equations of relative motion and control in Keplerian orbits. J Guid Control Dyn 32(2):664–669

Index

Z. Jing et al., *Non-Cooperative Target Tracking, Fusion and Control*,
Information Fusion and Data Science, https://doi.org/10.1007/978-3-319-90716-1

Printed in the United States
By Bookmasters